INHIBITION OF POLYAMINE METABOLISM

Biological Significance and Basis for New Therapies

INHIBITION OF POLYAMINE METABOLISM
Biological Significance and Basis for New Therapies

Edited by

Peter P. McCann
Merrell Dow Research Institute
Cincinnati, Ohio

Anthony E. Pegg
Department of Physiology and Cancer Research Center
Pennsylvania State University Medical School
Milton S. Hershey Medical Center
Hershey, Pennsylvania

Albert Sjoerdsma
Merrell Dow Research Institute
Cincinnati, Ohio

1987

ACADEMIC PRESS, INC.
Harcourt Brace Jovanovich, Publishers
Orlando San Diego New York Austin
Boston London Sydney Tokyo Toronto

ACADEMIC PRESS, INC.
Orlando, Florida 32887

United Kingdom Edition published by
ACADEMIC PRESS INC. (LONDON) LTD.
24–28 Oval Road, London NW1 7DX

Library of Congress Cataloging in Publication Data

Inhibition of polyamine metabolism.

Includes index.
1. Polyamines—Synthesis—Inhibitors—Physiological effect. 2. Polyamines—Synthesis—Inhibitors—Therapeutic use. 3. Polyamines—Metabolism—Regulation. I. McCann, Peter P. II. Pegg, Anthony E. III. Sjoerdsma, Albert, Date . [DNLM: 1. Polyamines—metabolism. QU 60 I55]
QP801.P638I55 1987 574.2'192 86-32147
ISBN 0-12-481835-8 (alk. paper)

PRINTED IN THE UNITED STATES OF AMERICA

87 88 89 90 9 8 7 6 5 4 3 2 1

Contents

4 Molecular and Cellular Functions of the Polyamines

Laurence J. Marton and David R. Morris

5 The Use of Inhibitors to Study the Biochemistry and Molecular Biology of Polyamine Biosynthesis and Uptake

Anthony E. Pegg

6 Inhibitors of Polyamine Biosynthesis: Cellular and *in Vivo* Effects on Tumor Proliferation

Prasad S. Sunkara, Stephen B. Baylin, and Gordon D. Luk

7 Polyamine Inhibition *in Vivo* and in Organ Growth and Repair

Charles Danzin and Pierre S. Mamont

8 Polyamine Synthesis Inhibitors Act as both Inducers and Suppressors of Cell Differentiation

Olle Heby, Gordon D. Luk, and Joel Schindler

9 The Contragestational Effects of Ornithine Decarboxylase Inhibition

J. R. Fozard

10 Modulation of Antineoplastic Drug Action by Inhibitors of Polyamine Biosynthesis

Carl W. Porter and Juhani Jänne

11 Inhibition of Carcinogenesis by Inhibitors of Putrescine Biosynthesis

Ajit K. Verma and R. K. Boutwell

12 Inhibition of Polyamine Biosynthesis in Microorganisms

Alan J. Bitonti and Peter P. McCann

13 Inhibition of Viral Polyamine Biosynthesis

A. Stanley Tyms and John D. Williamson

14 Inhibition of Polyamine Biosynthesis in Plants and Plant Pathogenic Fungi

Robert D. Slocum and Arthur W. Galston

15 Parasitic Protozoa and Polyamines

Cyrus J. Bacchi and Peter P. McCann

16 Clinical Aspects of Inhibition of Ornithine Decarboxylase with Emphasis on Therapeutic Trials of Eflornithine (DFMO) in Cancer and Protozoan Diseases

Paul J. Schechter, John L. R. Barlow, and Albert Sjoerdsma

Preface

Research into the biochemistry and cellular physiology of polyamines has been a steadily growing field over the last twenty years. Progress during this period and overviews of the area have been presented in a number of books and a much larger number of review articles. Although the function of polyamines is still not well understood at the molecular level, there is a great deal of information which attests to the importance of polyamines in cellular function. More recently, there has been a growing awareness that the polyamine biosynthetic pathway provides a useful target for the design of inhibitors which have value as pharmacological agents. The availability of such inhibitors has led to a rapid increase in knowledge of the importance of polyamines in a wide variety of living organisms, and there is now convincing evidence of clinical utility of these inhibitors in certain situations. In view of the ubiquitous distribution of polyamines and their key role in a variety of cellular processes, it seems likely that further uses for these inhibitors as chemotherapeutic agents and in the treatment of plant and animal pathology may be expected.

The large increase in information on the inhibition of polyamine metabolism has led to a need for a detailed summary of knowledge in this area. The purpose of the present volume is to provide an overview of the field concerning polyamine biosynthesis inhibitors at the present time. Chapters are included which describe the compounds which are currently in use; the rationale for the design of such inhibitors; their synthesis and action; their biological effects on mammalian cells and tissues, plants, and microorganisms including protozoal parasites; and a review of the present clinical experience with these inhibitors. These chapters also demonstrate clearly the major beneficial effect which the wide availability of these inhibitors has had on the field of polyamine research in general. It is hoped that the presentation of this material in a single volume will stimulate further research on both the basic and applied aspects of polyamines. The editors are most grateful to the authors of the individual chapters for the care with which their topic has been covered, the comprehensive nature of their contributions within the space available, and submitting the material in a timely manner.

Peter P. McCann
Anthony E. Pegg
Albert Sjoerdsma

Introduction: Polyamine Metabolism

All cells contain significant amounts of the polyamines, putrescine (a diamine), spermidine, and spermine. Although the physiological function of these amines is still not well understood at the molecular level, many studies summarized in the following chapters have shown that their concentration is highly regulated and that normal cellular growth and differentiation require polyamines. Many bacteria and protozoa contain putrescine and spermidine as their only polyamines while mammalian cells, yeast, other fungi, and other protozoa synthesize all three polyamines. This introduction outlines the general metabolic reactions responsible for the maintenance of polyamine content.

The general pathway for polyamine biosynthesis is shown in Fig. 1. Detailed description of the metabolic reactions responsible for polyamine synthesis are given in the various subsequent chapters. Many microorganisms and higher plants are able to produce putrescine from agmatine produced by decarboxylation of arginine, but all mammalian cells, lower eukaryotes including fungi, and most probably protozoa lack arginine decarboxylase. In these species, therefore, the only route to putrescine is via the enzyme ornithine decarboxylase. In contrast, plants and many bacteria are able to synthesize putrescine by at least two routes, either directly from ornithine, through the activity of ornithine decarboxylase, or indirectly, via arginine and arginine decarboxylase and one or two intermediates, i.e., agmatine and in some instances *N*-carbamoyl putrescine.

In mammalian cells, ornithine, the substrate for ornithine decarboxylase, is available from the plasma and can also be formed within the cell from arginine by the action of arginase. It is possible that arginase, which is much more widely distributed than other enzymes of the urea cycle, is present in extrahepatic tissues to ensure the availability of ornithine for polyamine production. Arginase can, therefore, be thought of as an initial step in the polyamine biosynthetic pathway.

Both ornithine decarboxylase and arginine decarboxylase are pyridoxal phosphate-dependent enzymes. Ornithine decarboxylase is present in very small amounts in quiescent cells, and its activity can be increased many fold within a few hours of exposure to trophic stimuli. Such stimuli include hormones, drugs, tissue regeneration, and growth factors. Even after such stimulation, ornithine decarboxylase is only a very small fraction of the total cellular protein. There is evidence for a macromolecular inhibitor of ornithine decarboxylase that may regulate the activity of the enzyme under some circumstances.

To convert putrescine into spermidine, an aminopropyl group must be added. This aminopropyl moiety is derived from methionine, which is first converted into *S*-adenosylmethionine and is then decarboxylated. The resulting decarboxy-*S*-adenosylmethionine is used as an aminopropyl donor in a manner analogous to the use of *S*-adenosylmethionine itself as a methyl donor. Once it has been decarboxylated, *S*-adenosylmethionine is committed to polyamine synthesis because the decarboxylated form is virtually inactive as a methyltransferase substrate. Therefore, the production of decarboxy-*S*-adenosylmethionine is kept low and constitutes the rate-limiting factor in spermidine formation. Mammalian *S*-adenosylmethionine decarboxylase is activated by putrescine and repressed by spermidine, linking the supply of decarboxy-*S*-adenosylmethionine to the need for spermidine and the availability of the other substrate (putrescine) for spermidine synthesis (Fig. 1). *S*-adenosylmethionine decarboxylase (which has an enzyme-bound pyruvate as a cofactor) resembles ornithine decarboxylase in that it is also present in mammalian tissues in very small amounts and that its activity is regulated by many hormones and other growth-promoting stimuli.

Nonmammalian *S*-adenosylmethionine decarboxylases also have a covalently attached pyruvate group but differ in other ways. *S*-Adenosylmethionine decarboxylase from bacteria and plants are activated by Mg^{2+}, unlike *S*-adenosylmethionine decarboxylases from yeast and other fungi which resemble the mammalian enzyme in being strongly activated by putrescine. Protozoal *S*-adenosylmethionine decarboxylases on the other hand appear to be both Mg^{2+} insensitive and relatively putrescine insensitive.

The transfer of the aminopropyl group from decarboxy-*S*-adenosylmethionine to putrescine is catalyzed by spermidine synthase. Another aminopropyl group is needed to convert spermidine into spermine, and this also comes from decarboxy-*S*-adenosylmethionine in a reaction catalyzed by a second aminopropyltransferase termed spermine synthase. Despite the similarity between these reactions, spermidine synthase and spermine synthase are discrete enzymes each specific for its own particular substrate. The aminopropyltransferases are present in many cells in amounts much greater than the decarboxylases and are thought to be

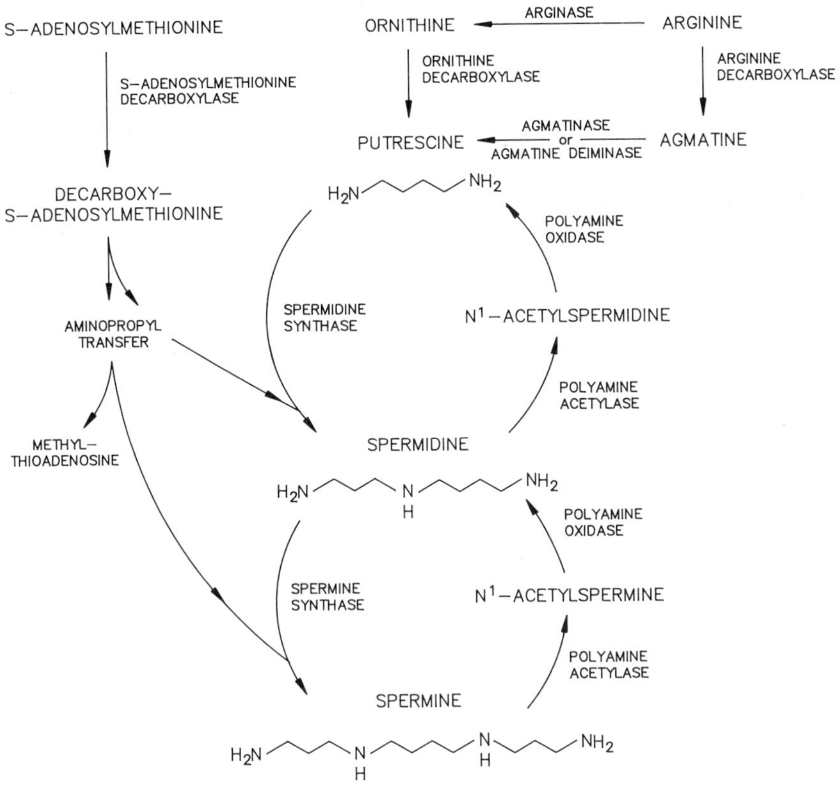

Fig. 1. Schematic overview of polyamine metabolism.

regulated by the availability of their substrates, particularly the aminopropyl donor, decarboxy-*S*-adenosylmethionine.

The other product of aminopropyltransferase reactions is 5′-methylthioadenosine. Although this nucleoside is produced in stoichiometric amounts with the polyamines, its concentration in the cell is very low due to rapid degradation by a phosphorylase producing adenine and 5′-methylthioribose-1-phosphate. The adenine is then converted to 5′-AMP by action of adenine phosphoribosyltransferase and the 5′-methylthioribose-1-phosphate is converted back to methionine in a reaction that conserves the methylthio group and all but the C-1 of the carbon atoms of this sugar. Therefore, all of the *S*-adenosylmethionine molecule not used for polyamine production is effectively salvaged. 5′-Methylthioadenosine phosphorylase is a widely distributed enzyme that is present in all normal tissues examined in amounts sufficient to maintain 5′-methylthioadenosine at very low levels. Certain mammalian tumor cell lines including some derived from humans have lost this enzyme, and these appear to excrete 5-methylthioadenosine to reduce the intracellular content of this nucleoside. It cannot be ruled out that 5′-methylthioadenosine (which has a number of effects on cellular physiology when applied exogenously) rather than the polyamines is the critical product of the pathway in Fig. 1, but the rapid degradation of more than 99% of this product argues against this possibility.

The aminopropyltransferase reactions which form spermidine and spermine are effectively irreversible but these polyamines can be converted back into putrescine by the combined actions of two enzymes, spermidine/spermine-N^1-acetyltransferase (polyamine acetylase) and polyamine oxidase (FAD dependent) (Fig. 1). Polyamine acetylase catalyzes the conversion of spermine into N^1-acetylspermine which is then degraded by polyamine oxidase to form spermidine and 3-acetamidopropanal. Similarly, spermidine is a substrate for polyamine acetylase which forms N^1-acetylspermidine and this is split by polyamine oxidase to form putrescine and 3-acetamidopropanal. The acetylase/oxidase system and either of the aminopropyltransferases can therefore be regarded as forming a cycle which leads to the production of 5′-methylthioadenosine from decarboxy-*S*-adenosylmethionine. The importance of these cycles and the extent to which they occur in unstimulated cells are unknown at present.

The regulation of the polyamine biosynthesis and interconversion pathway shown in Fig. 1 is accomplished by changes in the activity of three of these enzymes: ornithine decarboxylase, *S*-adenosylmethionine decarboxylase, and polyamine acetylase. The other enzymes (spermidine synthase, spermine synthase, and polyamine oxidase) appear to be regulated via the availability of their limiting substrates which are the decarboxylated and the acetylated polyamine derivatives. Polyamine concentration may also be affected by degradation of putrescine by diamine and monoamine oxidase. Ornithine decarboxylase, *S*-adenosylmethionine decarboxylase, and polyamine acetylase can undergo large changes in their activity which appear to be brought about by alterations in the amount of enzyme protein present. All three of these enzymes have an extremely short half-life, which enables a new level of enzyme protein to be reached very rapidly after the application of an appropriate stimulus.

Peter P. McCann
Anthony E. Pegg
Albert Sjoerdsma

1

Inhibition of Basic Amino Acid Decarboxylases Involved in Polyamine Biosynthesis

PHILIPPE BEY,* CHARLES DANZIN,† AND MICHEL JUNG†

Merrell Dow Research Institute
**2110 East Galbraith Road*
Cincinnati, Ohio 45215

†Merrell Dow Research Institute
Strasbourg Centre
16, rue d'Ankara
F 67084 Strasbourg Cedex, France

I. INTRODUCTION

After ornithine decarboxylase (ODC) was established to be one of the rate-limiting enzymes in the biosynthesis of polyamines in eucaryotes, and Russell and Synder (1968) had reported an extraordinary stimulation and decay of ODC activity in regenerating rat liver and in other situations of active cellular and tissue growth, the design and synthesis of specific inhibitors of ODC became a logical approach to unravel the physiological role of polyamines and to control rapid cell proliferation. The first publication dealing with inhibitors of ODC appeared in 1972 (Skinner and Johansson). This chapter is intended to discuss the extensive work since that date aimed at the inactivation of ODC, focusing on the mechanistic considerations which eventually led to the design and synthesis of potent ODC inhibitors acting directly at the enzyme active site. Extension of this work to the inhibition of arginine decarboxylase (ADC) and lysine decarboxylase (LDC), two other α-amino acid decarboxylases involved in the biosynthesis of polyamines in plants and bacteria (for a recent review, see Tabor and Tabor, 1984), will also be briefly reported. Comprehensive reviews on ODC and its inhibition have appeared regularly in the literature (Williams-Ashman *et al.*, 1976; Stevens and Stevens, 1980; Pegg and Williams-Ashman, 1981; Pegg and McCann, 1982; Pegg, 1986). Not covered in this chapter are agents which control *in vivo* ODC activity at the transcriptional, posttranscriptional, or posttranslational level. It is now firmly established that *in vivo* the polyamines can control ODC activity. The search for unnatural polyamine analogs capable of decreasing ODC activity *in vivo,* but devoid of the physiological properties of the natural polyamines, has been and continues to be a line of active research in many laboratories.

II. PROPERTIES OF ODC

The first ODC activity was detected in an extract of *Escherichia coli* (Gale, 1946). Later it was recognized that *E. coli* strains can contain two distinct enzymes capable of catalyzing the decarboxylation of ornithine; these were classified as the biodegradative and the biosynthetic enzymes (for a review see Morris and Fillingham, 1974). The biodegradative ODC is active only at low pH and appears to be involved in the regulation of the pH of the medium, whereas the biosynthetic ODC is responsible for the synthesis of putrescine and, ultimately, spermidine. In contrast to the biodegradative enzyme, which is found only in about 1 out of 10 *E. coli* strains, the presence of the biosynthetic ODC has been confirmed practically in all strains of *E. coli* and also in other unicellular organisms, including the lower ones such as *Physarum polycephalum* (Mitchell and Carter, 1977) and *Euglena gracilia* (Lafarge-Frayssinet *et al.*, 1978).

The biodegradative ODC has been purified to homogeneity (Applebaum *et al.*, 1977) and the biosynthetic enzyme has been obtained with a purity of about 85% as judged by polyacrylamide gel electrophoresis (Applebaum *et al.*, 1977). Both enzymes are dimers of 80,000 to 82,000 molecular weight and exhibit similar kinetic properties. The biosynthetic ODC has a K_m for ornithine of 9.6 mM and a specific activity of 6000 μmol of CO_2 liberated per milligram of protein per hour, compared with 3.6 mM and 7800 μmol of CO_2 per milligram of protein per hour for biodegradative ODC. The two enzymes, however, differ markedly in their pH profile and in their activation patterns by nucleotides. The optimal pH of biodegradative ODC is 6.9, whereas it is 8.1 for the biosynthetic enzyme. Both enzymes are activated by GTP, but CTP, UTP, and GMP have opposite effects on the activity of the two enzymes. The K_m and V_{max} are strongly influenced by the presence of nucleotides.

ODC activity has also been found in plants (Heimer *et al.,* 1979; Birecka *et al.,* 1985), protozoa (Chang and Cheng, 1976), and a variety of mammalian tissues. In eukaryotes ODC is essentially a cytosolic enzyme; experiments with enucleated cells have shown that more than 90% of ODC activity resides in the cytoplasts (McCormick, 1977). The basal level of ODC activity in resting cells is very low. It has been estimated that there may be as few as 100 to 200 molecules of ODC per quiescent cell (Pegg *et al.,* 1982). Rapid induction of enzyme activity can be achieved with a variety of different stimuli such as stress, cell poison, hormones, or growth factors (for a review see McCann, 1980). However, even though the basal activity can be enhanced by many orders of magnitude, the amount of ODC protein present still represents a minute fraction of the total cytosolic proteins; it ranges from 0.01% in androgen-stimulated mouse kidney to 0.00012% in thioacetamide-stimulated rat liver. Because of its low amount, the purification of ODC has proven a formidable task. The first attempt was described in 1972. From 7 kg of liver from thioacetamide-treated rats, Ono and co-workers (1972) obtained 1.02 mg of a 5400-fold enriched protein showing a single band on polyacrylamide gel electrophoresis. A specific activity of 12 μmol of CO_2 per milligram of protein per hour and a molecular weight of approximately 100,000 were determined for this protein. Since then, ODCs enriched 6000-, 8000-, and 10,350-fold from the liver of thioacetamide-treated rats (Pegg and McGill, 1979), Simian-virus-40-transformed 3T3 mouse fibroblast (Boucek and Lembach, 1977), and kidney from androgen-treated mice (Seely *et al.,* 1982) were obtained. Their specific activities were 14, 100, and 3000 μmol of CO_2 per milligram of protein per hour, respectively, and their molecular weights were in the range of 100,000, thus confirming the results of Ono *et al.* (1972). More recently, ODCs isolated from Morris hepatoma 7777, thioacetamide-treated rat liver, and androgen-stimulated mouse kidney were purified in parallel (Seely *et al.,* 1985).

Mammalian ODCs are highly unstable proteins. Their activity and stability are strongly dependent upon the presence of low-molecular-weight thiols (Jänne and Williams-Ashman, 1971). In general, dithiols are much more effective than monothiols to activate ODC. It has been suggested that in the absence of thiols purified ODC polymerizes through thiol disulfide interchanges and that the polymerized forms have much less or no catalytic activity at all. Addition of thiols to polymerized ODC can under certain conditions partially restore activity. In contrast with the bacterial ODCs, no other small-molecular-weight molecules have been found to regulate the *in vitro* activity of mammalian ODCs. A range of optimal pH from 7.0 to 8.1 has been reported for rat liver ODC depending upon the state of purity of the enzyme (Ono *et al.*, 1972; Friedman *et al.,* 1972;

Obenrader and Prouty, 1977) or the presence of thiols in the medium (Raina and Jänne, 1968).

All known ODC requires pyridoxal phosphate (PLP) for activity. The cofactor usually dissociates from the apoenzyme. For example, the K_m of PLP for the biosynthetic ODC of *E. coli* is 1 μM (Applebaum *et al.*, 1977) and 0.25 μM for thioacetamide-induced rat liver ODC (Obenrader and Prouty, 1977). There is evidence that ODC may exist in distinct forms which differ in their affinity for the cofactor and their overall activity. The existence of distinct forms of enzyme has been convincingly established only for ODC of *P. polycephalum* (Mitchell *et al.*, 1977, 1978, 1979). One form has a low affinity for PLP with an apparent K_m of 33 μM and the other one a high affinity with an apparent K_m of 0.13 μM. The two forms have been separated and have different isoelectric points. More than one form of protein may also exist for ODC of mouse 3T3 cells (Clark and Fuller, 1976), rat liver (Obenrader and Prouty, 1977), and mouse kidney (Seely *et al.*, 1985).

ODC shows high degree of selectivity for L-ornithine as substrate. Lysine, the higher homologue of ornithine, is also decarboxylated by ODC from rat liver (Pegg and McGill, 1979) and mouse kidney (Persson, 1977), but much less efficiently. The K_m and V_{max} for lysine are respectively 100 times greater and about 4 times lower than those for ornithine. The K_m values of ornithine and lysine vary with sources of ODC (Pegg and McGill, 1979). The lower homologue of ornithine, 2,4-diaminobutanoic acid, is apparently not a substrate for rat liver ODC. It is interesting to note that D-ornithine has some affinity for rat liver ODC (Ono *et al.*, 1972). Putrescine, the product of the reaction, and the higher polyamines spermidine and spermine, are usually weak competitive inhibitors for ODC (Pegg and Williams-Ashman, 1968).

Direct mechanistic studies of ODC and its inhibition have been hampered and limited by the low amount of enzyme present in tissues and the considerable technical difficulties in purification of the enzyme. With only rare exceptions, the kinetic studies of ODC inhibition have been carried out with crude preparations of enzyme, the specific activity of which was usually in the range of 0.1 to 50 nmol CO_2 per milligram protein per hour. ODC activity was determined most of the time by measuring the release of $^{14}CO_2$ from L-[1-^{14}C]ornithine in buffer solution containing PLP and dithiothreitol. This experimental protocol is not without drawbacks since it is documented that nonenzymatic release of CO_2 can occur in the presence of PLP (Kalyankar and Snell, 1962) or can arise from the reaction of ornithine with hydrogen peroxide generated by other enzymes, such as diamine oxidase, which might be present in the crude preparation (Suresh and Adiga, 1977). Moreover, no adducts formed with irreversible inhibitors have yet been fully characterized and identified. However, cell lines grown under constant pressure of ODC inhibition have been found to overproduce ODC as a result of gene amplification (Kahana and Nathans, 1984; Pohjanpelto *et al.,* 1985; Alhonen-Hongisto *et al.,* 1985). In addition, molecular cloning of mouse ODC cDNA has been recently reported (McConlogue *et al.,* 1984; Kontula *et al.,* 1984; Berger *et al.,* 1984; Kahana and Nathans, 1984). It is therefore reasonable to expect that in the near future quantities of purified ODC will become available for mechanistic studies.

III. MECHANISM OF ACTION OF ODC

As already mentioned, ODCs require PLP for activity. PLP-dependent enzymes are known to catalyze an array of clearly distinguishable reactions on amine and amino acid substrates such as transamination, racemization, or β and γ replacement reaction, and decarboxylation. The mechanism of PLP-catalyzed enzymatic reactions are well established (for reviews see Davis and Metzler, 1972; Snell and DiMari, 1970) and can be explained essentially in terms of the chemical reactivity of the cofactor. Detailed studies of the enzymatic decarboxylation of glutamic acid, ly-

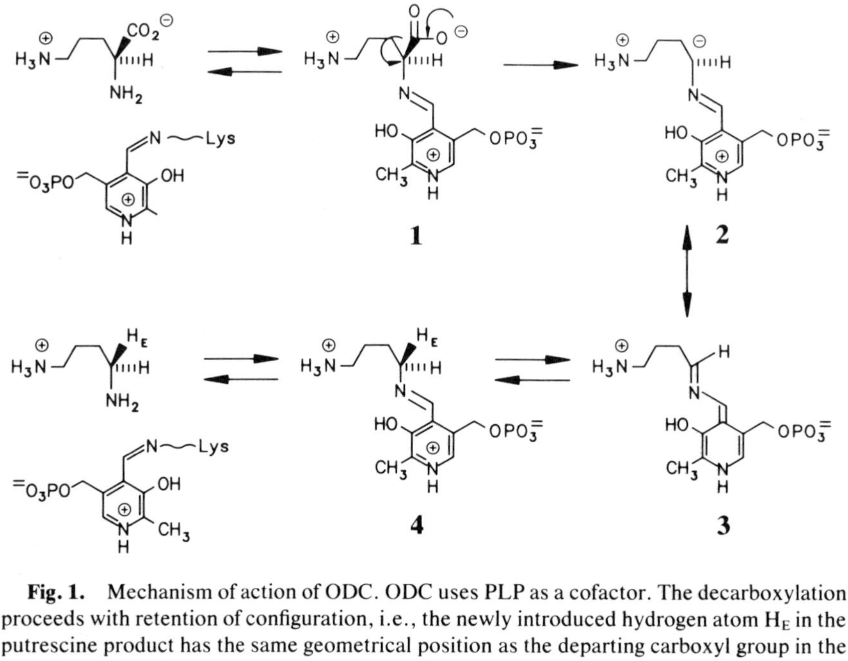

Fig. 1. Mechanism of action of ODC. ODC uses PLP as a cofactor. The decarboxylation proceeds with retention of configuration, i.e., the newly introduced hydrogen atom H_E in the putrescine product has the same geometrical position as the departing carboxyl group in the ornithine substrate.

sine, arginine, and L-aromatic amino acids (for a review see Boeker and Snell, 1972) pointed to an identical mechanism of action depicted in Fig. 1 which has been postulated to apply also to all other PLP-dependent α-amino acid decarboxylases, including ODC. The first intermediate is a Schiff's base adduct (**1**) formed between the α-amino group of ornithine and the aldehyde function of PLP. Such Schiff base intermediates are in fact common to all PLP-dependent enzymes. They are formed most probably by a transaldimination reaction since in absence of substrate the PLP usually is covalently linked through an internal Schiff's base to the terminal amino group of a lysine residue of the apoenzyme. The three bonds to the α-carbon atom of the substrate become weaker due to a displacement of their electrons toward the cationic nitrogen of the aldimine and the pyridoxal ring system. The $C\alpha$–CO_2H bond is then selectively cleaved. The selectivity of the cleavage is supposedly controlled by the apoenzyme which, through noncovalent interactions, orients the $C\alpha$–CO_2H bond perpendicularly to the plane formed by the aldimine and the aromatic ring of the cofactor so that delocalization of its electrons in the π system of the conjugated aldimine can be facilitated (Dunathan, 1971). The resulting carbanionic intermediate (**2**), which has a quinoid structure (**3**) in one of its reasonance forms, is then protonated on the α-carbon of the substrate to give the aldimine (**4**), which next undergoes a transaldimination reaction to yield the product putrescine and regenerate the original PLP–Schiff base with the terminal amino function of the lysine residue of the enzyme.

The PLP-utilizing α-amino acid decarboxylases whose stereochemistry has been studied thus far catalyze the decarboxylation with retention of configuration, i.e., the newly introduced hydrogen in the amine product has the same geometrical position [pro (*R*)] as that of the departing carboxyl group in the α-amino acid (for a review see Floss and Vederas, 1982). The only exception is the decarboxylation of *meso*-diaminopimelic acid to L-lysine by diaminopimelic acid decarboxylase which occurs with inversion of configuration (Asada *et al.*, 1981). The stereochemistry of ODC is in line with the general rule. L-Ornithine decarboxylation by ODC from *E. coli* (Orr and Gould, 1982) and kidney of androgen-treated mice (Orr *et al.*, 1984) has been demonstrated to proceed with retention of configuration in both cases.

IV. INHIBITION OF ODC

Three major approaches have been used to inhibit ODC: (1) synthesis of analogs of the substrate ornithine and the product putrescine as potential competitive inhibitors; (2) synthesis of molecules capable of interacting or combining with the cofactor; and (3) design of enzyme-activated inhibitors. Inhibitors of the first two groups are reversible whereas those of the last group are irreversible.

A. Reversible Inhibitors of ODC

1. Analogs of Ornithine and Putrescine

The analogs of ornithine and putrescine synthesized and tested as potential competitive inhibitors of ODC are listed in Table I along with their apparent dissociation constants. With the exception of the (*E*)-dehydro analogs of ornithine and putrescine, none of them turned out to be more potent than α-methylornithine, one of the first inhibitors to have been reported for ODC (Russell, 1973). These results provided, however, important information regarding the topography of the active site of ODC and the structural requirements for a ligand to bind ODC, which can be summarized as follows:

1. An L configuration is preferred for the α-amino acid. The D enantiomers are not totally devoid of affinity. The binding of D-α-methylornithine to ODC is about 1/70th that of the L isomer (Bey *et al.*, 1978).

2. The carboxyl group of the substrate is not essential for affinity. It can be replaced with a tetrazoyl group (Bey *et al.*, 1978) or even a hydrogen atom as some putrescine analogs are excellent competitive inhibitors; phosphonic or phosphinic groups (Sosnovsky *et al.*, 1985), however, are not tolerated.

3. A distance of about 6 Å between the two nitrogen atoms of the ligand is of critical importance for binding (Bey *et al.*, 1978). Most probably ornithine has a fully stretched conformation in the enzyme's active site, with the two nitrogen atoms as far away as possible. Structural features such as unsaturation which favor the stretched conformation of the ligand increase the affinity for the enzyme (Relyea and Rando, 1975).

4. Substitution of the α-hydrogen atom of ornithine for alkyl groups is allowed. The affinity of these analogs decreases rapidly as the size of the alkyl group augments from methyl to *n*-butyl. Further increase in the chain length of the alkyl group surprisingly produces a small enhancement of the activity (Abdel-Monem *et al.*, 1975b).

5. Introduction of hydrophilic functions on the β-carbon atom of ornithine or on an α-methyl substituent results in a total loss of affinity for the enzyme (Bey *et al.*, 1978; Inoue *et al.*, 1975). Lipophilic substituents such as halogen or methyl on the β-carbon atom are, however, tolerated (Skinner and Johansson, 1972; Bey, 1978).

6. The area of the enzyme which binds the terminal amino group is very sensitive to steric hindrance. Substantial loss of affinity results from substitution of the terminal amino group. Substituents on the carbon atom bearing the terminal amino function can, however, be accommodated as

TABLE I

Competitive Inhibitors of ODC

Analog	Enzyme source	$K_i(\mu M)$	Reference
Ornithine			
α-Methylornithine	Rat liver	40	Bey *et al.* (1978)
α-Methylornithine	Rat prostate	40	Abdel-Monem *et al.* (1975)
(*R*)-α-Methylornithine	Rat liver	1300	Bey *et al.* (1978)
(*S*)-α-Methylornithine	Rat liver	19	Bey *et al.* (1978)
α-Ethylornithine	Rat prostate	5300	Abdel-Monem *et al.* (1975b)
α-*n*-Propylornithine	Rat prostate	7800	Abdel-Monem *et al.* (1975b)
α-*n*-Butylornithine	Rat prostate	11,000	Abdel-Monem *et al.* (1975b)
α-*n*-Hexylornithine	Rat prostate	4700	Abdel-Monem *et al.* (1975b)
α-*n*-Octylornithine	Rat prostate	2300	Abdel-Monem *et al.* (1975b)
α-Benzylornithine	Rat prostate	5300	Abdel-Monem *et al.* (1975b)
α-Hydroxymethylornithine	Rat liver	No inhibition at 1 m*M*	Bey *et al.* (1978)
α-Methoxymethylornithine	Rat liver	No inhibition at 1 m*M*	Metcalf *et al.* (1978)
β-Chloroornithine	Rat liver	150	Bey (1978)
erythro-β-Hydroxyornithine	Mouse parotid gland	—	Inoue *et al.* (1975)
threo-β-Hydroxyornithine	Mouse parotid gland	—	Inoue *et al.* (1975)
β-Methylornithine	Rat liver	No inhibition at 0.1 m*M*	Skinner and Johansson (1972)
δ-*N*-Phenylornithine	Rat liver	No inhibition at 0.1 m*M*	Skinner and Johansson (1972)
δ-*N*-Aminopropylornithine	Bull prostate	4400	Bey *et al.* (1978)
δ-*N*-Aminopropyl-α-methylornithine	Bull prostate	5500	Bey *et al.* (1978)
δ-*N*-Methyl-α-methylornithine	Rat liver	500	Bey *et al.* (1978)
δ-*N*,*N*-Dimethyl-α-methylornithine	Rat liver	No inhibition at 1 m*M*	Bey *et al.* (1978)
α-Tetrazoylputrescine	Rat liver	120	Bey *et al.* (1978)
trans-1,4-Diaminocyclohexane-1-carboxylic acid	Rat liver	70	Bey *et al.* (1978)
cis-1,4-Diaminocyclohexane-1-carboxylic acid	Rat liver	No inhibition at 0.25 m*M*	Bey *et al.* (1978)
(*E*)-Dehydroornithine	Rat liver	4.4	Relyea and Rando (1975)
α-Methyllysine	Rat liver	No inhibition at 4 m*M*	Bey *et al.* (1978)
α-Methyl-α,δ-diaminobutyric acid	Rat liver	No inhibition at 4 m*M*	Bey *et al.* (1978)
Putrescine			
(*E*)-Dehydroputrescine	Rat liver	2	Relyea and Rando (1975)
1,4-Diaminobutanone	Rat liver	46	Stevens *et al.* (1978)

demonstrated by the good affinity of *trans*-1,4-diaminocyclohexane-1-carboxylic acid (Bey *et al.*, 1978).

Interestingly, α-methylornithine was demonstrated to be a substrate for ODC of *Lactobacillus* 30a. The decarboxylation is, however, 6000 times slower than that of ornithine (O'Leary and Herreid, 1978). In addition, α-methylornithine undergoes a decarboxylation-dependent transamination which transforms PLP into pyridoxamine phosphate and produces a gradual inactivation of the enzyme. Abdel-Monem *et al.* (1975b) had previously shown that rat prostate ODC is similarly inactivated by α-methylornithine but had failed to observe any decarboxylation, probably because their assay for detection of CO_2 was not sensitive enough. However, as the inactivation can be reversed by addition of PLP, the efficiency of this inhibitory mechanism *in vivo* may be questioned.

2. *Inhibitors Interacting with the Cofactor*

Reversible formation of a Schiff base between the α-amino function of the α-amino acid substrates and the aldehyde group of PLP is an obligatory step in the reaction catalyzed by the α-amino acid decarboxylases. Ketone reagents such as hydrazine or hydroxylamine, which form more stable adducts than do α-amino acids with PLP, have been used successfully to design inhibitors of PLP-dependent enzymes. For example, alkoxyamine brocresine (Reid and Shepherd, 1963), *N*-hydroxyglutamate (Cooper and Griffith, 1979), and α-hydrazino-3,4-dihydroxyphenylalanine, α-methyl-α-hydrazino-3,4-dihydroxyphenylalanine, and benserazide (Bartholini and Pletscher, 1969) are potent and fairly selective inhibitors of histidine decarboxylase, glutamate decarboxylase, and L-aromatic amino acid decarboxylase, respectively. By analogy, the analogs of ornithine and putrescine, listed in Table II, in which one of the amino groups is substituted for a hydrazino or an oxyamino group, were synthesized and evaluated as ODC inhibitors.

L-Canaline, an ornithine analog in which an oxygen atom replaces the γ-carbon atom, is only a weak inhibitor of ODC (IC_{50} = 0.1 m*M*) (Rahiala *et al.*, 1971). This result is not surprising in view of the fact that it is the α-amino and not the γ-amino group which forms a Schiff base with the cofactor. On the other hand, 1-aminoxy-3-aminopropane, the corresponding putrescine analog, is an extremely potent inhibitor of mouse kidney ODC (K_i = 3.2 n*M*) (Khomutov *et al.*, 1985) and of rat liver ODC (K_i = 1.4 n*M*) (M. Kolb and C. Danzin, personal communication). Interestingly, introduction of a methyl group on the carbon atom bearing the amino function, or extension of the chain by one carbon, only slightly reduces the inhibitory activity (C. Danzin and M. Kolb, personal communication).

The α-hydrazino analog of ornithine, 5-amino-2-hydrazinopentanoic acid, synthesized first by Johansson *et al.* (1973), is a potent inhibitor of *E. coli* and rat prostate ODC (Harik and Snyder, 1973). Sawayama *et al.* (1976), who developed a new method for the preparation of (DL)-5-amino-2-hydrazinopentanoic acid, demonstrated that the synthesis of Johansson *et al.* yielded in fact preferentially the D enantiomer. Accordingly, the K_i for the true racemic mixture was found to be four times lower than that for the Johansson *et al.* compound (Inoue *et al.*, 1975). The α-methyl analog of α-hydrazinoornithine, (DL)-5-amino-2-hydrazino-2-methylpentanoic acid, proved to be equipotent to α-hydrazinoornithine (Abdel-Monem *et al.*, 1975a).

All the compounds described in Table II were found to be competitive inhibitors with ornithine. The inhibition, however, is relieved by increasing the concentration of PLP in the assay medium. The reversal effect by PLP is an indication that these inhibitors do interact with the cofactor and is in agreement with the ready dissociation of the cofactor from the holoenzyme. In addition, these inhibitors often lack specificity. For example,

TABLE II

ODC Inhibitors Interacting with the Pyridoxal Phosphate Cofactor

Inhibitor	Enzyme source	$K_i (\mu M)$	[PLP] (*M*)	Reference
L-Canaline	Rat liver	100	2×10^{-4}	Rahiala *et al.* (1971)
α-Hydrazinoornithine	Rat liver	0.5	10^{-5}	Inoue *et al.* (1975)
α-Hydrazinoornithine	Rat ventral prostate	0.5	10^{-5}	Inoue *et al.* (1975)
α-Hydrazinoornithine	Rat ventral prostate	2.0	10^{-5}	Harik and Snyder (1973)
α-Hydrazinoornithine	*E. coli*	0.5	10^{-5}	Harik and Snyder (1973)
α-Methyl-α-hydrazino-ornithine	Rat ventral prostate	3	10^{-5}	Abdel-Monem *et al.* (1975b)
N-(5′-Phosphopyridoxyl)-ornithine	Rat liver	1.2	5×10^{-6}	Heller *et al.* (1975)
N-(5′-Phosphopyridoxyl)-ornithine	Rat liver	66	75×10^{-6}	Heller *et al.* (1975)
N-(5′-Pyridoxyl)ornithine	Rat liver	>1000	75×10^{-6}	Heller *et al.* (1975)
N-(5′-Pyridoxyl)lysine	Rat liver	>1000	75×10^{-6}	Heller *et al.* (1975)
1-Aminooxy-3-aminopropane	Mouse kidney	0.0032	—	Khomutov *et al.* (1985)
1-Aminooxy-3-aminopropane	Rat liver	0.0014	—	C. Danzin and M. Kolb (unpublished)
1-Aminooxy-4-aminobutane	Rat liver	0.0037	—	C. Danzin and M. Kolb (unpublished)
1-Aminooxy-3-aminobutane	Rat liver	0.0028	—	C. Danzin and M. Kolb (unpublished)

1-aminooxy-3-aminopropane, the most potent ODC inhibitor in this series of compounds, inactivates also *S*-adenosylmethionine decarboxylase and spermidine synthase (Khomutov *et al.*, 1985), and α-hydrazinoornithine inhibits many other PLP-dependent enzymes (Harik *et al.*, 1973).

Stable PLP–substrate adducts have been demonstrated to be inhibitors of the corresponding PLP-dependent enzymes (for a review see Rudd and Thonassi, 1981), possibly because they mimic some transition states along the reaction pathway. Heller *et al.* (1975) explored the possibility of using this interesting approach to inhibit ODC by synthesizing *N*-(5′-phosphopyridoxyl)ornithine, which unfortunately proved to be only a weak, noncompetitive inhibitor with respect to both ornithine and PLP.

3. Shortcomings of Competitive Inhibitors

The two reversible inhibitors α-methylornithine and α-hydrazinoornithine proved to be useful to block the biosynthesis of polyamines in cells in culture. α-Methylornithine was used to establish the first correlation between cell proliferation inhibition and polyamine depletion in HTC cells (Mamont *et al.*, 1976). *In vivo*, however, they were either not potent enough to cause a significant decrease of spermidine concentration or sufficiently selective to be devoid of untoward biological effects (Hölttä *et al.*, 1981). In addition, they share the major disadvantage of increasing the apparent half-life of ODC (Harik *et al.*, 1974; McCann *et al.*, 1977). Because of this effect, their administration leads to an accumulation of ODC. When the concentration of the inhibitor in the body decreases due to metabolic clearance, reactivation of the enzyme results in an "overshoot" of putrescine production. These different shortcomings of reversible inhibitors prompted the search for irreversible inhibitors.

B. Irreversible Inhibition of ODC

1. Concept of Enzyme-Activated Inhibition

The design of useful irreversible inhibitors of ODC has been closely associated with the development of the new concept of suicide inhibition of enzymes (Abeles and Maycock, 1976). Suicide inhibitors are also referred to as k_{cat} inhibitors (Rando, 1974b), Trojan horse inhibitors (Miesowicz and Bloch, 1979), enzyme-activated inhibitors (Seiler *et al.*, 1978), or mechanism-based inhibitors (Kalman, 1979). As the name enzyme-activated inhibitors was coined by Merrell scientists, we shall use this terminology in this review. Enzyme-activated inhibitors are chemically inert pseudosubstrates of the target enzymes which incorporate in their structures a latent group that is transformed, as a result of catalytic turnover, into a species that eventually inactivates the enzyme. This concept in enzyme inhibition evolved from a serendipitous discovery made in

Bloch's laboratory (Helmkamp *et al.*, 1968). It was found that β-hydroxydecanoyl thioester dehydrase is irreversibly inactivated by the *N*-acetylcysteamine thioester of 3-decanoic acid, the acetylenic analog of its substrate 3-decynoic acid *N*-acetylcysteamine thioester. Bloch and coworkers postulated a rearrangement, catalyzed by the dehydrase, of the chemically inert acetylenic thioester into an electrophilic conjugated allene which eventually alkylates an imidazole group of histidine residue in the active site. This mechanism was recently confirmed by carbon-13 NMR spectroscopy (Schwab *et al.*, 1984). The potential of Bloch's discovery for drug design did not remain unheeded. By virtue of the very mechanism of their action, which demands that both the affinity and the catalytic requirements of the target enzyme be satisfied, enzyme-activated inhibitors can be expected to be very selective *in vivo*. Application of this novel approach to inactivation of enzymes of therapeutic interest was immediately envisaged and the first enzymes to receive attention were those which use cofactors such as pyridoxal phosphate or flavin.

2. *Experimental Criteria for Characterization of Enzyme-Activated Inhibitors*

Demonstration that a compound is an enzyme-activated inhibitor relies on a combination of kinetic and chemical criteria. No single criterion suffices. Definitive evidence requires elaborate studies with purified enzyme to determine the stoichiometry of the inhibition, characterization of the structure of the enzyme inhibitor adduct after inactivation, and demonstration that the inhibitor is transformed by the target enzyme before the inactivation takes place. However, reasonable evidence can be obtained from relatively simple kinetic experiments performed with a crude preparation of enzyme.

In the first approximation, the kinetics of enzyme inactivation by enzyme-activated inhibitors can be accounted for by the minimum scheme represented in Eq. (1) where K_i is the apparent dissociation constant of the complex EI [Eq. (3)] and k the rate constant for the inactivation. This scheme is similar to that proposed by Kitz and Wilson (1962) for affinity labeling agents where a rapid formation of a complex between E and I is followed by a slow chemical modification, leading to inactivation. Under steady-state conditions, i.e., [E] >> [I], the rate for enzyme inactivation is expressed by Eq. (4), which leads to the solution given in Eq. (5).

$$\mathrm{E} + \mathrm{I} \rightleftharpoons \mathrm{EI} \rightarrow \mathrm{E}_{\mathrm{inact}} \tag{1}$$

$$\mathrm{E}_0 = [\mathrm{E}] + [\mathrm{EI}] + [\mathrm{E}_{\mathrm{inact}}] \tag{2}$$

$$[\mathrm{I}][\mathrm{E}]/[\mathrm{EI}] = K_i \tag{3}$$

$$d[\mathrm{E_{inact}}]/dt = k[\mathrm{EI}] \qquad (4)$$

$$\mathrm{Ln}\{[\mathrm{E_0}] - [\mathrm{E_{inact}}]\}/[\mathrm{E_0}] = -kt/(1 + K_i)/[\mathrm{I}] \qquad (5)$$

Equation (5) provides a family of first-order rate constants (k_{app}). The time of half-inactivation of enzyme at infinite concentration of I, τ_{50}, and the time of half-inactivation of enzyme, $t_{1/2}$, at a given concentration of inhibitor I can be expressed as a function of k_{app} and K [Eq. (6)] from which Eq. (7) can be derived.

$$t_{1/2} = \ln 2/k_{app}; \qquad \tau_{50} = \ln 2/k \qquad (6)$$

$$t_{1/2} = \tau_{50}(1 + K_i/[\mathrm{I}]) \qquad (7)$$

Experimentally the values of τ_{50} and K_i are determined from the intercepts with the y and x axis of the line obtained by plotting $t_{1/2}$ as a function of the reciprocal of the concentration of I. This graphical representation is similar to a Lineweaver–Burk representation.

More rigorous analyses of the kinetics of enzyme-activated inhibitors based on a minimum scheme represented by Eq. (8) were proposed successively by Malcom and Radda (1970), Jung and Metcalf (1975), Cornish-Bowden (1979), Brocklehurst (1979), Waley (1980), Tatsunami *et al.* (1981), and Carlson (1984); see also the review of Walsh (1982). None of the more elaborate mathematical solutions derived from these studies negates the general validity of Eq. (7) under most experimental conditions.

$$\begin{array}{l} \mathrm{E + I \rightleftharpoons EI \rightleftharpoons EI' \rightarrow E_{inact}} \\ \qquad\qquad\qquad\quad \downarrow \\ \qquad\qquad\qquad \mathrm{E + P} \end{array} \qquad (8)$$

In these analyses, however, the constants K_i and k appear usually as more complex functions than those defined by Kitz and Wilson; in particular, K_i is more akin to a Michaelis–Menten constant than to a dissociation constant.

The minimum kinetic criteria for enzyme-activated inhibitors can be summarized as follows (Abeles, 1983):

1. Inhibition is time dependent; usually, but not necessarily, pseudo first-order kinetics are observed.
2. The rate of enzyme inactivation decreases in the presence of a competitive inhibitor or substrate.
3. Saturation kinetics are usually observed.
4. The rate of enzyme inactivation is altered by the presence of nucleophiles in the incubation medium.

Most of the enzyme-activated inhibitors mentioned in this review met those experimental criteria and reported K_i and τ_{50} values were usually determined as described in Tables I and II, i.e., according to the analysis of Kitz and Wilson. In addition, dialysis or passage through a Sephadex column of the inactivated ODC usually did not lead to reactivation of the enzyme, which is compatible with formation of a covalent linkage between the enzyme and the inhibitor. It must be pointed out that, contrary to what is usually believed, enzyme-activated inhibitors do not uniformly lead to covalent bond formation (for a review see Silverman and Hoffman, 1984).

3. *Design of Enzyme-Activated Inhibitors for ODC*

PLP-dependent enzymes are almost ideal enzymes for the design of enzyme-activated inhibitors (Bey, 1981). Early successes in the design of enzyme-activated inhibitors for PLP-dependent enzymes are provided by the inactivation of aspartate aminotransferase, γ-cystathionase, and alanine racemase by vinylglycine (Rando, 1974a; Gehring *et al.*, 1977), propargylglycine (Abeles and Walsh, 1973), and β-chloroalanine (Manning *et al.*, 1974), respectively. In each case the assumption was that the fate of the obligatory quinoid intermediate **(3)** in Fig. 1 was altered by the presence of the unsaturation or the halogen atom to generate a conjugated imine, which eventually alkylates a residue within bonding distance to inactivate irreversibly the target enzyme. This proposed mechanism of inhibition, however, has been unambiguously demonstrated only for the inactivation of aspartate aminotransferase by vinylglycine through the structural identification of the apoprotein inhibitor adduct (Gehring *et al.*, 1977).

a. Ornithine Analogs. Similar mechanistic reasoning led Merrell scientists (Metcalf *et al.*, 1978; Bey, 1978; Danzin *et al.*, 1981) to propose that analogs of α-amino acids incorporating either a leaving group on an α-methyl substituent, or an acetylene or an ethylene function in place of the α-hydrogen atom, could be potential irreversible inhibitors of the corresponding α-amino acid decarboxylase. As some α-methyl-α-amino acids are known to serve as alternative substrates for the corresponding α-amino acid decarboxylases (Lovenberg *et al.*, 1963), it was hoped that these α-amino acid derivatives would be decarboxylated by the corresponding α-amino acid decarboxylase and generate the conjugated imines **(5)**, as depicted in path b of Fig. 2, which then could alkylate the apoprotein. As already discussed, α-methylornithine was subsequently demonstrated to be decarboxylated by ODC from *Lactobacillus* (O'Leary and Herreid, 1978) as shown in path a of Fig. 2. Interestingly, Merck scientists (Kollonitsch *et al.*, 1978) independently came up with the same general approach to inhibit α-amino acid decarboxylases. Later this approach was

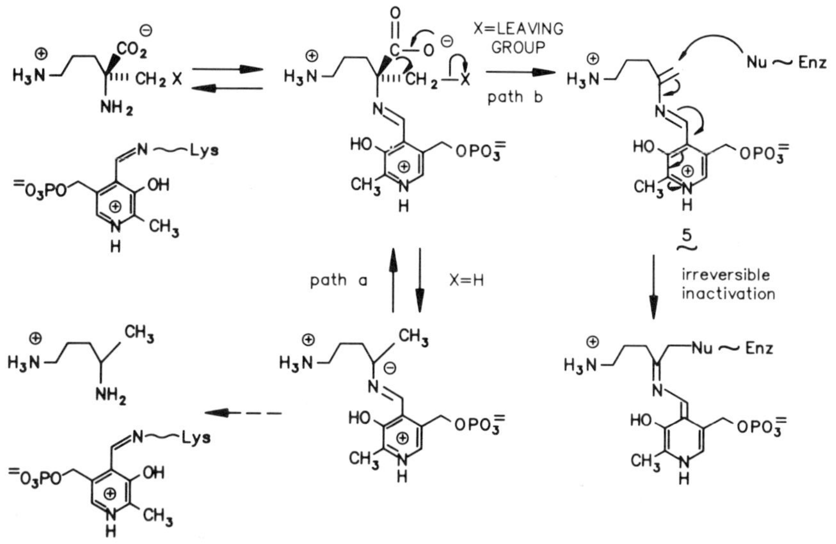

Fig. 2. Postulated mechanism for the irreversible inhibition of ODC by α-functionalized methylornithine analogs. Path a represents the decarboxylation of α-methylornithine to α-methylputrescine. When X is a leaving group (path b), the normal electron flow toward the PLP ring can be redirected to catalyze the elimination of X^- and generate an electrophilic imine which eventually could alkylate a residue (Nu) from the enzyme.

extended to include allene as a latent reactive group (Castelhano *et al.*, 1984; Casara *et al.*, 1984). Noteworthy is the fact that since these α-amino acid derivatives have no hydrogen atom left on the α-carbon atom, they cannot be enzyme-activated inhibitors for PLP-dependent enzymes such as racemases or transaminases, which proceed initially by abstraction of the α-hydrogen atom. These α-substituted amino acids represented new classes of α-amino acid derivatives whose preparation required the development of new synthetic methodologies (Bey *et al.*, 1979; Kollonitsch *et al.*, 1978; Metcalf and Jund, 1977; Taub and Patchett, 1977).

In accordance with expectation, most of these amino acid analogs in the ornithine series proved to be irreversible enzyme-activated inhibitors of ODC. As indicated by their kinetic constants listed in Table III and determined according to the method of Kitz and Wilson, the potency of these inhibitors depends upon the nature of the latent functionality. Increase in the bulkiness and/or decrease in the nucleofugality of the leaving group on the α-methyl substituent results in a lowering of the inhibitory activity: for example, α-cyanomethylornithine displays only weak inhibitory activity. Not unexpected in view of the physical and chemical properties of the fluorine atom [i.e., small size, high energy of the C–F bond which renders

TABLE III

Enzyme-Activated Irreversible Inhibitors of ODC: Substrate Analogs

Ornithine analog	Enzyme source	K_i (μM)	τ_{50} (min)	Reference
α-Chloromethyl	Rat liver	No saturation kinetics	$t_{1/2}$ = 22 min at 100 μM	Metcalf *et al.* (1978)
α-Fluoromethyl	Rat liver	75	1.6	Bey (1978)
α-Fluoromethyl	Rat liver	18	300	Kollonitsch *et al.* (1978)
α-Fluoromethyl	*E. coli*	360	12	Bitonti *et al.* (1982)
α-Fluoromethyl	*Trypanosoma brucei brucei*	104	1.1	Bitonti *et al.* (1985)
α-Cyanomethyl	Rat liver	8700	29	Metcalf *et al.* (1978)
α-Ethylenic	Rat liver	810	27	Danzin *et al.* (1981)
α-Ethynyl	Rat liver	10	8.5	Danzin *et al.* (1981)
α-Allenyl	—	No data available	—	Castelhano *et al.* (1984)
α-Chlorofluoromethyl	Rat liver	64	2.9	Bey *et al.* (1984)
α-Difluoromethyl	Rat liver	39	3.1	Metcalf *et al.* (1978)
α-Difluoromethyl	*E. coli*	No time-dependent inhibition at 5 mM	—	Kallio and McCann (1981)
α-Difluoromethyl	*Trypanosoma brucei brucei*	130	1	Bitonti *et al.* (1985)
α-Methyl dehydro	Rat liver	2.7	No time-dependent inhibition at 100 μM	Bey *et al.* (1983)
	E. coli	2300	11	C. Danzin (unpublished)
α-Fluoromethyl dehydro	Rat liver	2.7	2.6	Bey *et al.* (1983)
α-Fluoromethyl dehydro	*E. coli*	1700	3.7	C. Danzin (unpublished)
α-Fluoromethyl dehydro	*Trypanosoma brucei brucei*	14	1.1	Bitonti *et al.* (1985)
α-Difluoromethyl dehydro	Rat liver	30	2.6	Bey *et al.* (1983)
(+)-Difluoromethyl	Rat liver	130	3.0	C. Danzin (unpublished)
(−)-Difluoromethyl	Rat liver	18	3.9	Metcalf *et al.* (1978)
(+)-α-Fluoromethyl dehydro	Rat liver	20	0.6	C. Danzin (unpublished)
(−)-α-Fluoromethyl dehydro	Rat liver	2	2.9	C. Danzin (unpublished)

fluorine difficult to displace directly in S_N2 reactions, and good leaving-group ability as fluoride ion in ElcB elimination reactions (reviews see Walsh, 1983; Bey, 1984)], the α-fluoromethylornithines proved to be the most potent inhibitors of mammalian ODC among these ornithine analogs. Interestingly, although α-chloromethylornithine appears to be much less potent than α-fluoromethylornithine (in fact, it is even questionable that α-chloromethylornithine is an enzyme-activated inhibitor of ODC, as no apparent saturation kinetics were observed), the diastereomeric mixture of α-chlorofluoromethylornithine has kinetic inhibitory constants very similar to those of α-difluoromethylornithine (Bey *et al.,* 1984; C. Danzin *et al.,* unpublished). The inhibitory activity of α-ethynylornithine is similar to that of the fluoromethylornithines, whereas the α-ethylenic analog is practically devoid of activity (Danzin *et al.,* 1981).

Ornithine analogs incorporating the same latent reactive groups on the β-carbon atom could also be considered as potential enzyme-activated inhibitors of ODC. Indeed, upon enzymatic turnover, a conjugated imine similar to that depicted in Fig. 2 could be generated and alkylate the apoprotein. Along this line, (*E*)-dehydroornithine (Relyea and Rando, 1975), (*E*)-α-methyldehydroornithine (Bey *et al.,* 1983), and β-chloroornithine (Bey, 1978) were synthesized but none of these ornithine analogs inhibited mammalian ODC in a time-dependent manner. (*E*)-α-Methyldehydroornithine, however, was found to inhibit in a time-dependent manner ODC from *E. coli* (C. Danzin *et al.,* unpublished). (*E*)-Dehydroornithine, as already discussed, proved to be a very potent competitive inhibitor of mammalian ODC; its K_i is about 60 times lower than the K_m of L-ornithine for ODC. Although (*E*)-dehydroornithine was reported not to be a substrate for ODC (Relyea and Rando, 1975), its increased affinity compared to L-ornithine for the ODC active site suggested that introduction of a trans double bond in the side chain might improve the inhibitory activity of the α-functionalized methylornithine enzyme-activated inhibitors. Synthetic access to the (*E*)-dehydro analogs of fluoromethylornithines was consequently developed (Bey *et al.,* 1983). As indicated in Table III, the unsaturation dramatically increases the affinity of the monofluoromethyl analog for ODC, although the K_i of the difluoromethyl analog remains practically unchanged. In both cases the unsaturation had little influence on the rate of inactivation. (*E*)-α-Fluoromethyldehydroornithine is the most potent enzyme-activated irreversible inhibitor of mammalian ODC reported thus far in the ornithine series.

b. Putrescine Analogs. α-Ethynyl GABA, designed and synthesized in our laboratories as an enzyme-activated inhibitor of GABA transaminase (Jung and Metcalf, 1975), quite unexpectedly was also found to be an

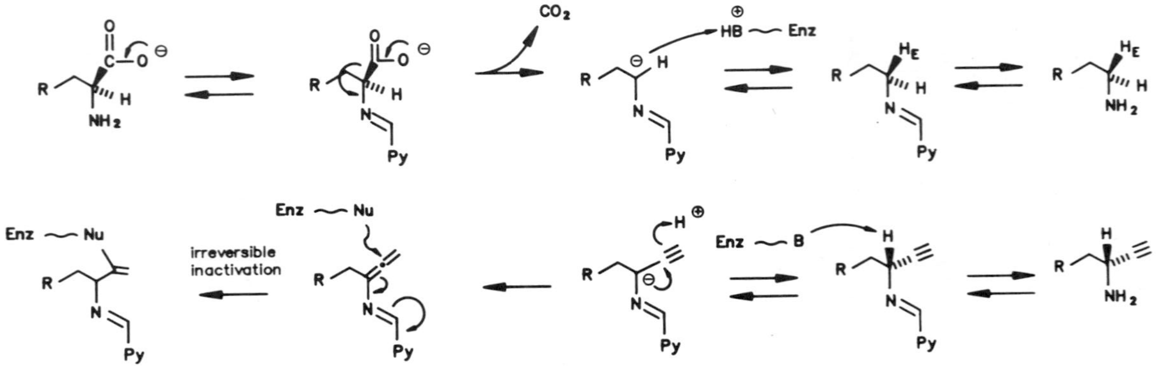

Fig. 3. Postulated mechanism for the irreversible inhibition of α-amino acid decarboxylases by product analogs. As a close analog of the amine product of the decarboxylase, the amine inhibitor (in this case the α-ethynylamine) is accepted by the decarboxylase active site and forms a Schiff base with the PLP cofactor. The same catalytic residue responsible for the protonation of the carbanionic intermediate in the normal decarboxylation reaction (top line) can abstract the α-hydrogen atom and generates a propargylic carbanion that can rearrange to an electrophilic allenyl imine, which eventually alkylates a residue of the enzyme.

enzyme-activated inhibitor of glutamic acid decarboxylase, the enzyme which catalyzes the decarboxylation of glutamic acid into GABA (Jung *et al.*, 1977). A mechanism relying on the microreversibility of the protonation step in the enzymatic decarboxylation of glutamic acid to GABA was proposed to rationalize this first example of an inhibition of an α-amino acid decarboxylase by an analog of its product (Fig. 3). It was realized immediately that such an approach could be generalized to the inactivation of other PLP-dependent enzymes. Consequently, the putrescine analogs incorporating in the α position latent reactive groups similar to the ones used to design the enzyme-activated inhibitors in the ornithine series were synthesized systematically as potential inhibitors of ODC (Metcalf *et al.*, 1978; Danzin *et al.*, 1982; Bey *et al.*, 1983; Danzin and Casara, 1984; Casara *et al.*, 1985). As evidenced from their kinetic constants listed in Table IV, all these putrescine analogs, with the exception of the trifluoromethyl analog (Sjoerdsma and Schechter, 1984), were found to be time-dependent inhibitors of ODC. In general, the relative potency of the putrescine inhibitors parallels that of the corresponding ornithine derivatives. In particular, introduction of a trans double bond in the chain consistently improved the inhibitory activity.

The mechanism based on the principle of microscopic reversibility of enzyme catalysis proposed for the inactivation of ODC by putrescine analogs implies that the α-hydrogen atom of the inhibitor is abstracted and that the stereochemistry of the abstracted hydrogen atom is the same as that of the newly added hydrogen atom in the product amine of the decarboxylation. Evidence that abstraction of the hydrgen atom does occur was provided by the inactivation of rat liver ODC with the α-deuterio analogs of α-ethynylputrescine (Metcalf *et al.*, 1978) and α-difluoromethylputrescine (Danzin *et al.*, 1982). In both cases a primary isotope effect was measured on the apparent dissociation constant K_i whereas the inactivation rate constant τ_{50} was not affected. These results also suggest that hydrogen abstraction is not the rate-limiting step in the inactivation of the enzyme. Using the (*R*) and (*S*) enantiomers of α-ethynylputrescine, prepared with a purity superior to 99.5% from the corresponding (*S*) and (*R*) isomers of α-ethynyl GABA, Casara *et al.* (1982) found that only (*R*)-ethynylputrescine irreversibly inhibits mammalian ODC, the (*S*) isomer being only a weak competitive inhibitor (K_i = 100 μM). As ODC has recently been demonstrated to introduce in the reaction product a pro (*R*) hydrogen with retention of configuration (Orr *et al.*, 1984), this stereospecific irreversible inactivation of ODC by (*R*)-ethynylputrescine is in full agreement with the mechanism proposed in Fig. 3 [R = $(CH_2)_2NH_2$].

In vivo, α-ethynylputrescine and the α-fluoromethylputrescines very efficiently inhibit not only ODC (Danzin *et al.*, 1979b, 1982) but also 4-

TABLE IV

Enzyme-Activated Irreversible Inhibitors of ODC: Product Analogs

Putrescine analog	Enzyme source	K_i (μM)	τ_{50} (min)	Reference
α-Methyl	Rat liver	520	No time-dependent inhibition	Danzin *et al.* (1982)
α-Monofluoromethyl	Rat liver	56	4.4	Danzin *et al.* (1982)
α-Monofluoromethyl	*E. coli*	110	2.1	Kallio *et al.* (1982)
α-Monofluoromethyl	*Trypanosoma brucei brucei*	25	2.8	Bitonti *et al.* (1985)
α-Difluoromethyl	Rat liver	30	7.4	Danzin *et al.* (1982)
α-Difluoromethyl	*E. coli*	No time-dependent inhibition at 2 mM		P. Bey and C. Danzin (unpublished)
α-Difluoromethyl	*Trypanosoma brucei brucei*	140	2.4	Bitonti *et al.* (1985)
α-Deuterio-α-difluoromethyl	Rat liver	90	7.3	Danzin *et al.* (1982)
α-Trifluoromethyl	Rat liver	No time-dependent inhibition at 2 mM		Sjoerdsma and Schechter (1984)
α-Trifluoromethyl	*E. coli*	No time-dependent inhibition at 2 mM		P. Bey and C. Danzin (unpublished)
(E)-α-Fluoromethyl dehydro	Rat liver	42	0.2	Bey *et al.* (1983)
(E)-α-Fluoromethyl dehydro	*E. coli*	At 200 μM, $t_{1/2}$ = 11 min		C. Danzin (unpublished)
(E)-α-Fluoromethyl dehydro	*Trypanosoma brucei brucei*	11	1.5	Bitonti *et al.* (1985)

(*E*)-α-Difluoromethyl dehydro	Rat liver	60	0.7	
α-Ethynyl	Rat liver	2.3	9.7	Metcalf *et al.* (1978)
α-Ethynyl	*E. coli*	Inhibitory at 1.25 m*M*		Kallio *et al.* (1982)
α-Deuterio-α-ethynyl	Rat liver	—	—	Metcalf *et al.* (1978)
(*R*)-α-Ethynyl	Rat liver	4.3	9.6	Casara *et al.* (1982)
(*R*)-α-Ethynyl	*E. coli*	At 1 m*M*, $t_{1/2}$ = 13 min		Danzin and Jung (1984)
(*S*)-α-Ethynyl	Rat liver	No time-dependent inhibition		Casara *et al.* (1982)
(*S*)-α-Ethynyl	*E. coli*	At 1 m*M*, $t_{1/2}$ = 51 min		
(*E*)-α-Ethynyl dehydro	Rat liver	1	5	Metcalf *et al.* (1978)
(*E*)-α-Ethynyl dehydro	*E. coli*	Inhibitory at 1.25 m*M*		Kallio *et al.* (1982)
α-Ethylenic	Rat liver	540	10	P. Casara and C. Danzin (unpublished)
α-Ethylenic	*E. coli*	No time-dependent inhibition at 2 m*M*		P. Casara and C. Danzin (unpublished)
α-Allenyl	Rat liver	160	8	Danzin and Casara (1984)
α-Allenyl	*E. coli*	300	1.4	Danzin and Casara (1984)
(*R*)-α-Allenyl	Rat liver	70	10	Danzin and Casara (1984)
(*R*)-α-Allenyl	*E. coli*	140	2	Danzin and Casara (1984)
(*S*)-α-Allenyl	Rat liver	3400	25	Danzin and Casara (1984)
(*S*)-α-Allenyl	*E. coli*	No time-dependent inhibition at 2 m*M*		Danzin and Casara (1984)
α-Ethynyl-δ-methyl	Rat liver	13.5	1.8	Danzin *et al.* (1983)
(*R*)-α-Ethynyl-(*R*)-δ-methyl	Rat liver	3	1.7	Danzin *et al.* (1983)
(*R*)-α-Ethynyl-(*R*)-δ-methyl	*Trypanosoma brucei brucei*	5	1.0	Bitonti *et al.* (1985)
(*S*)-α-Ethynyl-(*S*)-δ-methyl	Rat liver	2100	2.2	Casara *et al.* (1985)
(*S*)-α-Ethynyl-(*R*)-δ-methyl	Rat liver	1200	2.2	Casara *et al.* (1985)
(*R*)-α-Ethynyl-(*S*)-δ-methyl	Rat liver	300	1.2	Casara *et al.* (1985)
α-Ethynyl-δ,δ-dimethyl	Rat liver	1300	5.4	Casara *et al.* (1985)

aminobutyrate transaminase (GABA-transaminase) (EC 2.6.1.19). These putrescine analogs are good substrates for monoamine oxidase (EC 1.4.3.4) and are rapidly transformed to the corresponding GABA analogs, which are potent inhibitors of GABA-transaminase. This metabolism severely restricted the overall usefulness of putrescine analogs as ODC inhibitors since GABA-transaminase inhibitors are known to have a profound CNS effect. As it was known that α-methyl-substituted amines are poor substrates for monoamine oxidase (Blaschko, 1952) and that the ODC active site can accommodate substituents on the δ-carbon atom, δ-methyl-α-ethynylputrescine was synthesized in the hope of preventing the undesired metabolism of α-ethynylputrescine (Danzin *et al.*, 1983). As anticipated, δ-methyl-α-ethynylputrescine proved to be resistant to oxidation by monoamine oxidase and to retain an excellent affinity for ODC. The introduction of a methyl group on the carbon atom, however, adds a second asymmetric center to the molecule and δ-methyl-α-ethynylputrescine exists as four isomers. In an attempt to avoid this stereochemical complication, the δ-*gem*-dimethyl analog was synthesized, but was found to be only weakly active, indicating that the stereochemistry of the methyl substituent is obviously important for activity. Stereospecific syntheses of the four isomers of δ-methyl-α-ethynylputrescine were therefore developed and the inhibitory activity was found to reside essentially with the (2*R*,5*R*) isomer (Casara *et al.*, 1985).

c. Alternative Mechanisms for the Enzyme-Activated Inhibition. As already mentioned, due to the difficulties of obtaining purified ODC, with the exception of difluoromethylornithine, no mechanistic studies other than the simple kinetic experiments used to establish the enzyme-activated nature were performed with these inhibitors. Experiments with difluoromethylornithine labeled with ^{14}C in the 1-position and 5-position demonstrated that the inhibitor is decarboxylated by ODC with a partition ratio of decarboxylation to inactivation close to 2.5 (Pegg, Chapter 5, this volume) and that 1 mol of inhibitor binds to one subunit of ODC through covalent linkage with the lysine residue at position 298 (Pegg, 1986). These results confirm the first step of the mechanism proposed in Fig. 2. However, no structure of enzyme-inhibitor adduct has yet been determined, and the last steps depicted in Figs. 2 and 3 still remain highly speculative. In particular, the irreversible inhibition of ODC *in vitro* by the difluoromethyl analogs of putrescine and ornithine is not accounted for satisfactorily by the proposed mechanism depicted in Figs. 2 and 3. Indeed, in these cases the inhibition would be expected to be reversible since the enzyme–inhibitor adduct (**6**) as postulated in path a of Fig. 5 would have the degree of oxidation of an aldehyde; these adducts should

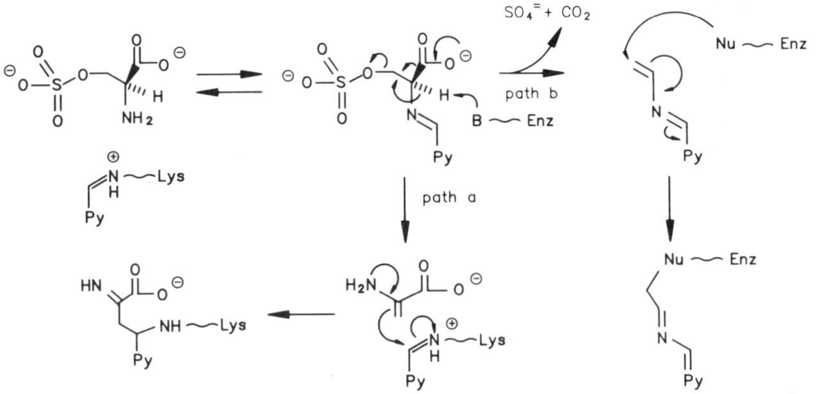

Fig. 4. Mechanism for the irreversible inhibition of glutamic acid decarboxylase by L-serine-*O*-sulfate (Likos *et al.*, 1982). Instead of catalyzing the decarboxylation of L-serine-*O*-sulfate (path b), glutamic acid decarboxylase removes the proton from the α-carbon (path a). The sulfate (SO_4^{2-}) group is subsequently eliminated to generate, after a transamination reaction with the PLP cofactor, an amino acrylate moiety, which then attacks the electrophilic internal Schiff base of PLP with a Lys residue to form a covalent tertiary adduct with PLP and the enzyme.

be fairly labile and readily hydrolyze during dialysis to regenerate the native enzyme and the corresponding aldehyde. Interestingly, Likos *et al.* (1982) demonstrated that the enzyme-induced inactivation of glutamic acid decarboxylase by L-serine-*O*-sulfate proceeds by the mechanism depicted in path a of Fig. 4 and not through path b as previously proposed (Sukhareva and Braunstein, 1971). They suggested that this alternative mechanism should also be considered for the inhibition of other PLP-dependent enzymes by enzyme-activated inhibitors which generate α,β-unsaturated amine or amino acid intermediates. This hypothesis was recently confirmed by Badet *et al.* (1984), who reported that β-fluoroalanine inhibits the PLP-dependent alanine racemase by this same mechanism. It is noteworthy that if Likos' mechanism were to be operative in the inactivation of ODC by the difluoromethyl analogs of putrescine and ornithine (path b, Fig. 5), the enzyme–inhibitor adduct would be expected to be stable to the dialysis conditions.

As enzymes are usually stereospecific in their mode of action and enzyme-activated inhibitors have to be substrates of the target enzymes, it has been taken for granted that the inhibitory activity of an enzyme-activated inhibitor should reside with only one stereoisomer. Interestingly, in contrast with mammalian ODC, bacterial ODC is inactivated by both enantiomers of α-ethynylputrescine. Similarly, both enantiomers of

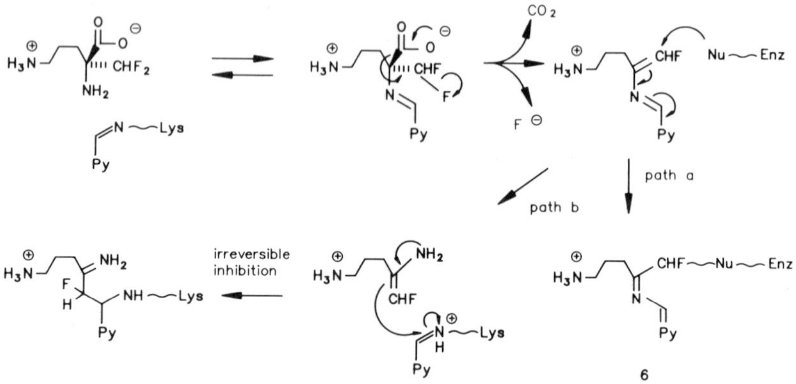

Fig. 5. Alternate mechanism for the irreversible inhibition of ODC by difluoromethylornithine. Path a represents the mechanism which was originally postulated for the inactivation of ODC. An alternate mechanism suggested by the work of Likos *et al.* (1982) is depicted in path b.

difluoromethylornithine and (*E*)-monofluoromethyldehydroornithine, which were recently obtained in our laboratories with a purity superior to 99.5%, were found to inhibit mammalian ODC. (−)-Difluoromethylornithine is markedly more potent than the corresponding (+) enantiomer. In the case of (*E*)-monofluoromethyldehydroornithine, however, the two enantiomers have similar inhibitory activity at a concentration of about 5 μM (C. Danzin *et al.*, unpublished). An explanation for this apparent lack of stereospecificity could be the poor substrate property of enzyme-activated inhibitors for their target enzymes. The substrate and catalytic specificity of enzymes is indeed never absolute. [For example, in addition to the decarboxylation of L-dopa and L-5-hydroxytryptophan, L-aromatic amino acid decarboxylase can catalyze the transamination of D-tryptophan and D-5-hydroxytryptophan, (Borri-Voltattorni *et al.*, 1983).] It is therefore quite conceivable that enzyme-activated inhibitors might be activated through chemical transformations which do not correspond to the main or any known reaction pathways catalyzed by the target enzyme. In fact, enzyme inactivation could be telltale for the existence of these secondary catalytic pathways, simply because, experimentally, enzyme inhibition can be followed much more easily than enzyme turnover. Such a possibility is particularly attractive for PLP-dependent enzymes. As already discussed, these enzymes are known to catalyze a variety of different reactions. They are also believed to be derived from a common ancestor protein (Dunathan and Voet, 1974). So it can be expected that they have all retained some capability of catalyzing these different reactions.

V. INHIBITION OF ARGININE DECARBOXYLASE AND LYSINE DECARBOXYLASE

In plants and bacteria, putrescine can be synthesized by an alternate pathway which involves as a first step the decarboxylation of arginine to agmatine, catalyzed by arginine decarboxylase (ADC). In addition, in bacteria, cadaverine can substitute for putrescine. Cadaverine is the product of lysine decarboxylase (LDC). As it could be expected on the basis of the similarity of the mechanism of action of PLP-dependent α-amino acid decarboxylases, approaches which were successful to design enzyme-activated inhibitors for ODC proved to apply also to the inactivation of ADC and LDC. Thus the α-fluoromethyl analogs of arginine (Kallio *et al.*, 1981) and lysine (Pösö *et al.*, 1984) as well as of agmatine and cadaverine Bitonti *et al.*, 1987) are potent inhibitors of ADC and LDC, respectively (see Table V). Interestingly, cadaverine and lysine analogs are also enzyme-activated inhibitors of rat liver ODC (C. Danzin *et al.*, unpublished), in agreement with the fact that lysine is a weak substrate for rat liver ODC.

VI. CONCLUSION

ODC is the enzyme in the polyamine biosynthetic pathway which has attracted the most attention as a target for inhibition. Of the various approaches which have been explored to inhibit ODC, the design of enzyme-activated irreversible inhibitors has been particularly successful. More than 30 enzyme-activated inhibitors for ODC have now been reported, both in the ornithine and the putrescine series. The exact mechanism of action of these inhibitors has yet to be established. The chemical structure of an enzyme–inhibitor adduct after inactivation of the enzyme remains to be determined. *In vitro,* these inhibitors display a great selectivity for ODC, as could be expected from their mode of action. *In vivo,* however, the inhibitors in the putrescine series are of limited value due to their oxidation to the corresponding GABA analogs, which are potent enzyme-activated irreversible inhibitors of GABA-transaminase and/or glutamic acid decarboxylase. It is also noteworthy that the relative potency of these inhibitors depends upon the source of the enzyme. Difluoromethylornithine, one of the first enzyme-activated inhibitors to have

TABLE V

Inhibitors of ADC and LDC

Inhibitor	Enzyme source	K_i (μM)	τ_{50} (min)	Reference
		ADC		
Arginine analogs				
α-Difluoro	*E. coli*	800	1	Kallio *et al.* (1982)
(*E*)-α-Monofluoro-methyl-dehydro	*E. coli*	34	1.1	Bitonti *et al.* (1987)
α-Monofluoromethyl	*E. coli*	65	1.0	
Agmatine				
α-Monofluoromethyl	*E. coli*	9	0.6	Bitonti *et al.* (1987)
α-Ethynyl	*E. coli*	11	12.6	Bitonti *et al.* (1987)
α-Allenyl	*E. coli*	7	3.0	
		LDC		
Lysine analogs				
α-Monofluoromethyl	*E. coli*	600	0.7	C. Danzin *et al.* (unpublished)
α-Monochloromethyl	*E. coli*	300	2.7	C. Danzin *et al.* (unpublished)
α-Difluoromethyl	*E. coli*	7000	No time-dependent inhibition	C. Danzin *et al.* (unpublished)
	Mycoplasma dispar	With 100 μM, 50% inhibition after 30 min		Pösö *et al.* (1984)
Cadaverine analogs				
α-Monofluoromethyl	*E. coli*	180	2.4	C. Danzin *et al.* (unpublished)
α-Difluoromethyl	*E. coli*	25	No time-dependent inhibition	C. Danzin *et al.* (unpublished)
α-Trifluoromethyl	*E. coli*	11	No time-dependent inhibition	C. Danzin *et al.* (unpublished)
α-Ethynyl	*E. coli*	2100	5.2	C. Danzin *et al.* (unpublished)

been synthesized, has been for almost a decade the compound of choice to block polyamine biosynthesis *in vivo*. Recently, two other enzyme-activated irreversible inhibitors of ODC which also show potential for *in vivo* applications have appeared. The first one is (*E*)-dehydrofluoromethyl-ornithine. Because of its poor ability to penetrate the cell membrane, this ornithine derivative has to be used as its methyl ester for maximum efficacy *in vivo* (Mamont *et al.*, 1986). The second one, the (2*R*,5*R*) isomer of δ-methyl-α-ethynylputrescine, is the first analog of the product putrescine to be of any interest as an ODC inhibitor. The methyl group on the δ-carbon atom prevents the oxidation of the terminal amino function and consequently the undesirable transformation *in vivo* to a GABA-transaminase inhibitor.

Extension of the enzyme-activated inhibitor approach to the inactivation of other α-amino acid decarboxylases involved in the polyamine biosynthetic pathway has also led to the preparation of fluoromethyl analogs of arginine and agmatine, and of lysine and cadaverine, which are potent irreversible inhibitors of arginine decarboxylase and lysine decarboxylase, respectively.

The use of these inhibitors to study the biochemical and pharmacological consequences of polyamine biosynthesis blockade in mammals, plants, bacteria, and protozoa is discussed at length in other chapters of this book.

REFERENCES

Abeles, R.H. (1983). *Chem. Eng. News* **61,** 48–56.

Abeles, R.H., and Maycock, A.L. (1976). *Acc. Chem. Res.* **9,** 313–319.

Abeles, R.H., and Walsh, C.T. (1973). *J. Am. Chem. Soc.* **95,** 6124–6125.

Abdel-Monem, M.M., Newton, N.E., and Weeks, C.E. (1975a). *J. Med. Chem.* **18,** 945–948.

Abdel-Monem, M.M., Newton, N.E., Ho, B.C., and Weeks, C.E. (1975b). *J. Med. Chem.* **18,** 600–604.

Alhonen-Hongisto, L., Kallio, A., Sinervirta, R., Seppänen, P., Kontula, K.K., Jänne, O.A., and Jänne, J. (1985). *Biochem. Biophys. Res. Commun.* **126,** 734–740.

Applebaum, D.M., Dunlap, J.C., and Morris, D.R. (1977). *Biochemistry* **16,** 1580–1584.

Asada, Y., Tanizawa, K., Sawada, S., Suzuki, T., Misono, H., and Soda, K. (1981). *Biochemistry* **20,** 6881–6886.

Badet, B., Roise, D., and Walsh, C.T. (1984). *Biochemistry* **23,** 5188–5194.

Bartholini, G., and Pletscher, A. (1969). *J. Pharm. Pharmacol.* **21,** 323–324.

Berger, F.G., Szymanski, P., Read, E., and Watson, G. (1984). *J. Biol. Chem.* **259,** 7941–7946.

Bey, P. (1978). *In* "Enzyme-Activated Irreversible Inhibitors" (N. Seiler, M.J. Jung, and J. Koch-Weser, eds.), pp. 27–41. Elsevier, New York.

Bey, P. (1981). *Chem. Ind.* 139–143.

Bey, P. (1984). *In* "Le Fluor et les Materiaux Fluores" (P.V. Huong, ed.), pp. 67–74. Masson, Paris.

Bey, P., Danzin, C., Van Dorsselaer, V., Mamont, P., Jung, M., and Tardif, C. (1978). *J. Med. Chem.* **21,** 50–55.

Bey, P., Vevert, J.P., Van Dorsselaer, V., and Kolb, M. (1979). *J. Org. Chem.* **44,** 2732–2742.

Bey, P., Gerhart, F., Van Dorsselaer, V., and Danzin, C. (1983). *J. Med. Chem.* **26,** 1551–1556.
Bey, P., Ducep, J.B., and Schirlin, D. (1984). *Tetrahedron Lett.* **25,** 5657–5660.
Birecka, H., Bitonti, A.J., and McCann, P.P. (1985). *Plant Physiol.* **79,** 515–519.
Bitonti, A.J., McCann, P.P., and Sjoerdsma, A. (1982). *Biochem. J.* **208,** 435–441.
Bitonti, A.J., Bacchi, C.J., McCann, P.P., and Sjoerdsma, A. (1985). *Biochem. Pharmacol.* **34,** 1773–1777.
Bitonti, A.J., Casara, P.J., McCann, P.P., and Bey, P. (1987). *Biochem. J.* **242,** 69–74.
Blaschko, H. (1952). *Pharmacol. Rev.* **4,** 415–158.
Boeker, E.A., and Snell, E.E. (1972). *In* "The Enzymes" (P. Boyer, ed.), Vol 6, pp. 217–253. Academic Press, New York.
Borri-Voltattorni, C., Minelli, A., and Dominici, P. (1983). *Biochemistry* **22,** 2249–2254.
Boucek, R.J., and Lembach, K.J. (1977). *Arch. Biochem. Biophys.* **184,** 408–415.
Brocklehurst, K. (1979). *Biochem. J.* **181,** 775–778.
Carlson, G.M. (1984). *Biochem. Biophys. Acta* **789,** 347–350.
Casara, P., Danzin, C., Metcalf, B.W., and Jung, M.J. (1982). *J. Chem. Soc. Chem. Commun.* 1190–1192.
Casara, P., Jund, K., and Bey, P. (1984). *Tetrahedron Lett.* **25,** 1891–1894.
Casara, P., Danzin, C., Metcalf, B., and Jung, M. (1985). *J. Chem. Soc. Perkin Trans. I,* 2201–2207.
Castelhano, A.L., Pliura, D.H., Taylor, G.J., Hsieh, K.C., and Krantz, A. (1984). *J. Am. Chem. Soc.* **106,** 2734–2735.
Chang, K.P., and Cheng, Y.C. (1976). *J. Parasitol.* **62,** (Suppl.), 33.
Clark, J.L., and Fuller, J.L. (1976). *Eur. J. Biochem.* **67,** 303–314.
Cooper, A.J.L., and Griffith, O.W. (1979). *J. Biol. Chem.* **254,** 2748–2753.
Cornish-Bowden, A. (1979). *Eur. J. Biochem.* **93,** 383–385.
Danzin, C., and Casara, P. (1984). *FEBS Lett.* **174,** 275–278.
Danzin, C., and Jung, M. (1984). *In* "Chemical and Biological Aspects of Vitamin B_6 Catalysis. Part A" (A.E. Evangelapoulos, ed.), pp. 377–385. Liss, New York.
Danzin, C., Jung, M.J., Metcalf, B.W., Grove, J., and Casara, P. (1979a). *Biochem. Pharmacol.* **28,** 627–631.
Danzin, C., Jung, M.J., Seiler, N., and Metcalf, B.W. (1979b). *Biochem. Pharmacol* **28,** 633–639.
Danzin, C., Casara, P., Claverie, N., and Metcalf, B.W. (1981). *J. Med. Chem.* **24,** 16–20.
Danzin, C., Bey, P., Schirlin, D., and Claverie, N. (1982). *Biochem. Pharmacol.* **31,** 3871–3878.
Danzin, C., Casara, P., Claverie, N., Metcalf, B.W., and Jung, M.J. (1983). *Biochem. Biophys. Res. Commun.* **116,** 237–243.
Davis, L., and Metzler, E.D. (1972). *In* "The Enzymes" (P. Boyer, ed.), Vol. 7, pp. 33–74. Academic Press, New York.
Dunathan, H.C. (1971). *Adv. Enzymol.* **35,** 79–134.
Dunathan, H.C., and Voet, J.G. (1974). *Proc. Natl. Acad. Sci. U.S.A.* **71,** 3888–3891.
Floss, H.G., and Vederas, J.C. (1982). *In* "Stereochemistry" (C. Tamm, ed.), pp. 161–199. Biomedical Press, New York.
Friedman, S.J., Halpern, X.V., and Canellakis, E.S. (1972). *Biochim. Biophys, Acta* **261,** 181–187.
Gale, E.F. (1946). *Adv. Enzymol.* **6,** 1–32.
Gehring, H., Rando, R.R., and Christen, P. (1977). *Biochemistry* **16,** 4832–4836.
Harik, S.I., and Snyder, S.H. (1973). *Biochim. Biophys. Acta* **327,** 501–509.
Harik, S.I., Pasternak, G.W., and Snyder, S.H. (1973). *In* "Polyamines in Normal and Neoplastic Growth" (D.H. Russell, ed.), pp. 307–321. Raven, New York.

Harik, S.I., Hollenberg, M.D., and Snyder, S.H. (1974). *Mol. Pharmacol.* **10,** 41–47.
Heimer, Y.M., Mizrahi, Y., and Bachrach, V. (1979). *FEBS Lett.* **104,** 146–148.
Heller, J.S., Canellakis, E.S., Bussolotti, D.L., and Coward, J.K. (1975). *Biochim. Biophys. Acta* **403,** 197–207.
Helmkamp, G.M., Rando, R.R., Brock, D.J.R., and Bloch, K. (1968). *J. Biol. Chem.* **243,** 3229–3231.
Hölttä, E., Korpela, H., and Hovi, T. (1981). *Biochim. Biophys. Acta* **677,** 90–102.
Inoue, H., Kato, Y., Takigawa, M., Adachi, K., and Takeda, Y. (1975). *J. Biochem.* (*Tokyo*) **77,** 879–893.
Jänne, J., and Williams-Ashman, H.G. (1971). *J. Biol. Chem.* **246,** 1725–1732.
Johansson, J.G., Alto, P., and Skinner, W.A. (1973). U.S. Patent 3754027; *Chem. Abstr.* **79,** 137504t.
Jung, M.J., and Metcalf, B.W. (1975). *Biochem. Biophys. Res. Commun.* **67,** 301–306.
Jung, M.J., Lippert, B., Metcalf, B.W., Schechter, P.J., Bohlen, P., and Sjoerdsma, A. (1977). *J. Neurochem.* **28,** 717–723.
Kahana, C., and Nathans, D. (1984). *Proc. Natl. Acad. Sci. U.S.A.* **81,** 3645–3649.
Kallio, A., and McCann, P.P. (1981). *Biochem. J.* **200,** 69–75.
Kallio, A., McCann, P.P., and Bey, P. (1981). *Biochemistry* **20,** 3163–3166.
Kallio, A., McCann, P.P., and Bey, P. (1982). *Biochem. J.* **204,** 771–775.
Kalman, T.I. (1979). *In* "Drug Action and Design: Mechanism-Based Enzyme Inhibitors" (T.I. Kalman, ed.), pp. 75–91. Elsevier, New York.
Kalyankar, G.D., and Snell, E.E. (1962). *Biochemistry* **1,** 594–600.
Khomutov, R.M., Hyvönen, T., Karvonen, E., Kauppinen, L., Paalanen, T., Paulin, L., Kloranta, T., Pajula, R.L., Andersson, L.C., and Pösö, H. (1985). *Biochem. Biophys. Res. Commun.* **130,** 596–602.
Kitz, R., and Wilson, I.B. (1962). *J. Biol. Chem.* **237,** 3245–3249.
Kollonitsch, J., Patchett, A.A., Marburg, S., Maycock, A.L., Perkins, L.M., Dolduras, G.A., Duggan, D.E., and Aster, S.D. (1978). *Nature* (*London*) **274,** 906–908.
Kontula, K.K., Torkheli, T.K., Bardin, C.W., and Jänne, O.A. (1984). *Proc. Natl. Acad. Sci. U.S.A.* **81,** 731–735.
Lafarge-Frayssinet, C., Bertaux, O., Valencia, R., and Frayssinet, C. (1978). *Biochim. Biophys. Acta* **539,** 435–444.
Likos, J.J., Veno, H., Feldhaus, R.W., and Metzler, D.E. (1982). *Biochemistry* **21,** 4377–4386.
Lovenberg, W., Barchas, J., Weissbock, H., and Udenfriend, S. (1963). *Arch. Biochem. Biophys.* **103,** 9–14.
McCann, P.P. (1980). *In* "Polyamines in Biomedical Research" (J.M. Gaugas, ed.), pp. 109–123. Wiley, New York.
McCann, P.P., Tardif, C., Duchesne, M.C., and Mamont, P.S. (1977). *Biochem. Biophys. Res. Commun.* **76,** 893–899.
McConlogue, L.C., Gupta, M., Wu, L., and Coffino, P. (1984). *Proc. Natl. Acad. Sci. U.S.A.* **81,** 540–544.
McCormick, F. (1977). *J. Cell. Physiol.* **93,** 285–292.
Malcom, A.D.B., and Radda, G.K. (1970). *Eur. J. Biochem.* **15,** 555–561.
Mamont, P., Bohlen, P., McCann, P.P., Bey, P., Schuber, F., and Tardif, C. (1976). *Proc. Natl. Acad. Sci. U.S.A.* **73,** 1626–1630.
Mamont, P.S., Danzin, C., Kolb, M., Gerhart, F., Bey, P., and Sjoerdsma, A. (1986). *Biochem. Pharmacol.* **35,** 159–165.
Manning, J.M., Merrifield, N.E., Jones, W.M., and Gotschlich, E.C. (1974). *Proc. Natl. Acad. Sci. U.S.A.* **71,** 417–421.
Metcalf, B.W., and Jund, K. (1977). *Tetrahedron Lett.* **41,** 3689–3692.

Metcalf, B.W., Bey, P., Danzin, C., Jung, M., Casara, P., and Vevert, J.P. (1978). *J. Am. Chem. Soc.* **100,** 2551–2553.
Miesowicz, F.M., and Bloch, K. (1979). *J. Biol. Chem.* **254,** 5868–5874.
Mitchell, J.L.A., and Carter, D.D. (1977). *Biochim. Biophys. Acta* **483,** 425–434.
Mitchell, J.L.A., and Kottas, G.E. (1979). *FEBS Lett.* **102,** 265–268.
Mitchell, J.L.A., Carter, D.D., and Rybski, J.A. (1978). *Eur. J. Biochem.* **92,** 325–331.
Morris, D.R., and Fillingham, H. (1974). *Annu. Rev. Biochem.* **43,** 303–326.
Obenrader, M.F., and Prouty, W.F. (1977). *J. Biol. Chem.* **252,** 2866–2872.
O'Leary, M.H., and Herreid, R.M. (1978). *Biochemistry* **17,** 1010–1014.
Ono, M., Inoue, H., Suzuki, F., and Takeda, Y. (1972). *Biochim. Biophys. Acta* **284,** 285–297.
Orr, G.R., and Gould, S.J. (1982). *Tetrahedron Lett.* **23,** 3139–3142.
Orr, G.R., Gould, S.J., Pegg, E.A., Seely, I.E., and Coward, J.K. (1984). *Bioorg. Chem.* **12,** 252–258.
Pegg, A.E. (1986). *Biochem. J.* **234,** 249–262.
Pegg, A.E., and McCann, P.P. (1982). *Am. Physiol. Soc.* **243,** (*Cell. Physiol.* **12,**), C212–C221.
Pegg, A.E., and McGill, S. (1979). *Biochim. Biophys. Acta* **568,** 416–427.
Pegg, A.E., and Williams-Ashman, H.G. (1968). *Biochem. J.* **108,** 533–539.
Pegg, A.E., and Williams-Ashman, H.G. (1981). *In* "Polyamines in Biology and Medicine" (D.R. Morris and L.J. Marton, eds.), pp. 3–42. Dekker, New York.
Pegg, A.E., Seely, J.E., Pösö, H., Ragione, F.D., and Zagon, I.S. (1982). *Fed. Proc., Fed. Am. Soc. Exp. Biol.* **41,** 3065–3072.
Persson, L. (1977). *Acta Physiol. Scand.* **100,** 424–429.
Pohjanpelto, P., Hölttä, E., Jänne, O.A., Knuutila, S., and Alitalo, X. (1985). *J. Biol. Chem.* **260,** 8532–8537.
Pösö, H., McCann, P.P., Tanskanen, R., Bey, P., and Sjoerdsma, A. (1984). *Biochem. Biophys. Res. Commun.* **125,** 205–210.
Rahiala, E.L., Kekomaki, M., Jänne, J., Raina, A., and Räihä, N.C.R. (1971). *Biochim. Biophys. Acta* **227,** 337–343.
Raina, A., and Jänne, J. (1968). *Acta Chem. Scand.* **22,** 2375–2378.
Rando, R.R. (1974a). *Biochemistry* **13,** 3859–3863.
Rando, R.R. (1974b). *Science* **185,** 320–324.
Reid, J.D., and Shepherd, D.M. (1963). *Life Sci.* **1,** 5–8.
Relyea, N., and Rando, R.R. (1975). *Biochem. Biophys. Res. Commun.* **67,** 392–402.
Rudd, E.A., and Thonassi, J.W. (1981). *Biochemistry* **20,** 7469–7475.
Russell, D.H. (1973). *In* "Polyamines in Normal and Neoplastic Growth" (D.H. Russell, ed.), pp. 1–13. Raven, New York.
Russell, D.H., and Snyder, S.H. (1968). *Proc. Natl. Acad. Sci. U.S.A.* **60,** 1420–1427.
Sawayama, T., Kinugasa, H., and Nishimora, H. (1976). *Chem. Pharm. Bull.* (*Tokyo*) **24,** 326–329.
Schwab, J.M., Li, W., and Ho, C. (1984). *J. Am. Chem. Soc.* **106,** 7293–7294.
Seely, J.E., Pösö, H., and Pegg, A.E. (1982). *Biochemistry* **21,** 3394–3399.
Seely, J.E., Persson, L., Sertich, G.J., and Pegg, A.E. (1985). *Biochem. J.* **226,** 577–586.
Seiler, N., Jung, M.J., and Koch-Weser, J. (1978). *In* "Enzyme-Activated Irreversible Inhibitors" (N. Seiler, M.J. Jung, and J. Koch-Weser, eds.), Elsevier, New York.
Silverman, R.B., and Hoffman, S.J. (1984). *Med. Chem. Rev.* 415–447.
Sjoerdsma, A., and Schechter, P. (1984). *Clin. Pharmacol. Ther.* **35,** 287–299.
Skinner, W.A., and Johansson, J.G. (1972). *J. Med. Chem.* **15,** 427.
Snell, E.E., and DiMari, S.J. (1970). *In* "The Enzymes" (P. Boyer, ed.), Vol. 2, pp. 335–370. Academic Press, New York.

Sosnovsky, G., Lukszo, J., Gravela, E., and Zuretti, M.F. (1985). *J. Med. Chem.* **28,** 1350–1354.
Stevens, L., and Stevens, E. (1980). *In* "Polyamines in Biomedical Research" (J.M. Gaugas, ed.), pp. 167–183. Wiley, New York.
Stevens, L., McKinnon, I.M., Turner, R., and North, M.J. (1978). *Trans. Biochem. Soc.* **6,** 407–409.
Sukhareva, B.S., and Braunstein, A.E. (1971). *Mol. Biol.* (*Moscow*) **5,** 302.
Suresh, M.R., and Adiga, P.R. (1977). *Eur. J. Biochem.* **79,** 511–518.
Tabor, C.W., and Tabor, H. (1976). *Annu. Rev. Biochem.* **45,** 285–306.
Tabor, C.W., and Tabor, H. (1984). *Annu. Rev. Biochem.* **53,** 749–790.
Tatsunami, S., Yago, N., and Hosoe, M. (1981). *Biochim. Biophys. Acta* **662,** 226–235.
Taub, D., and Patchett, A.A. (1977). *Tetrahedron Lett.* **32,** 2745–2748.
Waley, S.G. (1980). *Biochem. J.* **185,** 771–773.
Walsh, C. (1982). *Tetrahedron* **38,** 871–909.
Walsh, C. (1983). *Adv. Enzymol.* 197–285.
Williams-Ashman, H.G., Corti, A., and Taldoni, B. (1976). *Ital. J. Biochem.* **25,** 5–32.

2

Pharmacologic Interference with Enzymes of Polyamine Biosynthesis and of 5′-Methylthioadenosine Metabolism

ANTHONY E. PEGG*
AND H. GUY WILLIAMS-ASHMAN†

* *Department of Physiology and Cancer Research Center*
Milton S. Hershey Medical Center
Pennsylvania State University College of Medicine
Hershey, Pennsylvania 17033

† *Ben May Institute and*
Department of Pharmacological and Physiological Sciences and
Department of Biochemistry and Molecular Biology
University of Chicago
Chicago, Illinois 60637

I. INTRODUCTION

Much of the recent progress in the use of specific inhibitors to interfere with polyamine metabolism has been brought about by the availability of inhibitors of ODC,[1] which are covered extensively in other chapters of

[1] Abbreviations: ODC, ornithine decarboxylase; DFMO, α-difluoromethylornithine; AdoMet, *S*-adenosylmethionine; AdoMetDC, *S*-adenosylmethionine decarboxylase; MGBG, methylglyoxal bis(guanylhydrazone); EGBG, ethylglyoxal bis(guanylhydrazone); AdoHcy, *S*-adenosylhomocysteine; AdoDap, 9-[6(*R*,*S*,),8-diamino-5,6,7,8-tetradeoxy-*β*-D-*ribo*-octofuranosyl]-9*H*-purin-6-amine; AdoS$^+$(CH$_3$)$_2$, *S*-methyl-5′-methylthioadenosine;

this book. These ODC inhibitors are clearly the compounds of choice for depleting cellular polyamines at present since they are highly specific and because the prevention of the production of putrescine should theoretically lead to the depletion of all of the cellular polyamines. However, in practice this is not the case and spermine levels are actually affected only slightly by inhibitors such as DFMO. Furthermore, the very large increase in the content of decarboxylated AdoMet which is brought about by DFMO and other inhibitors of ODC complicates the interpretation of results obtained with these compounds. Also, ODC as a target for the design of inhibitors has some disadvantages in that the very rapid turnover of the enzyme allows cells to recover quickly from inhibition, and ODC gene amplification, which confers resistance to DFMO, appears to occur rather readily (see Chapter 5).

It is clear that the design of experiments to provide a detailed understanding of the physiological function of polyamines would be enhanced by the availability of compounds to provide selective blocks at all stages of the polyamine metabolic pathway. Although many of the available compounds are still in the developmental stage and much more work must be done, it appears that it will soon be possible to interfere with almost every step in the polyamine biosynthetic pathway and ancillary reactions. The combination of some of these inhibitors with ODC inhibitors such as DFMO may provide useful therapeutic regimes. This chapter summarizes the present knowledge of inhibitors of the enzymes in the polyamine biosynthetic pathway apart from ODC and is divided into sections according to the enzyme under consideration.

II. *S*-ADENOSYLMETHIONINE DECARBOXYLASE (AdoMetDC)

The structure of AdoMetDC and the regulation of its activity have been covered in a number of recent reviews (Williams-Ashman and Pegg, 1981; Pegg, 1984; Tabor and Tabor, 1984). AdoMetDC has been purified to homogeneity from *Escherichia coli*, yeast, and mammalian tissues and to a lesser extent from a variety of sources including plants and the slime mold *Physarum polycephalum* (Williams-Ashman and Pegg, 1981; Pegg, 1984; Tabor and Tabor, 1984; Yamanoha and Cohen, 1985). These en-

---

AdoDato, *S*-adenosyl-1,8-diamino-3-thiooctane; MTA, 5′-deoxy-5′methylthioadenosine; SIBA, 5′-deoxy-5′-isobutylthioadenosine; MTR, 5-methylthioribose; MTR-1-P, 5-methylthioribose-1-phosphate; APRT, adenine phosphoribosyltransferase; KMTB, 2-keto-4-*S*-methylthiobutyrate.

zymes differ in their subunit structure and cation requirement but all of the AdoMetDC preparations so far examined contain pyruvate as a covalently bound prosthetic group. Mammalian AdoMetDC, which is activated by putrescine, has two subunits of 32,000 M_r. The *E. coli* enzyme was reported to consist of six subunits of 17,000 M_r. In both cases it appeared that each subunit contains a pyruvate prosthetic group (reviewed by Pegg, 1984; Tabor and Tabor, 1984).

In mammalian cells the amount of AdoMetDC protein is increased in response to polyamine depletion and is greatly reduced when spermidine or spermine levels are elevated. These changes are brought about both by alterations in the half-life of the protein and by changes in its rate of synthesis which are mediated by changes in the mRNA (Pegg, 1984; Shirahata and Pegg, 1985, 1986).

Early studies of inhibitors of AdoMetDC have been reviewed by Porter *et al.* (1981) and by Pegg (1983a). Much interest was stirred by the discovery by Williams-Ashman and Schenone (1972) that the antileukemic agent MGBG is a potent inhibitor of this enzyme. MGBG is a powerful competitive inhibitor of the putrescine-activated mammalian AdoMetDC with a K_i of less than 1 μM (Williams-Ashman and Pegg, 1981; Williams-Ashman and Seidenfeld, 1986). MGBG is less active toward the Mg^{2+}-activated AdoMetDC from *E. coli* and the AdoMetDC from Chinese cabbage (*Brassica pekinensis*) or from slime mold (*P. polycephalum*), which do not require putrescine or metal cations, but these enzymes are still sensitive to the drug (Pegg, 1984; Yamanoha and Cohen, 1985; Pegg and Jacobs, 1983). A number of analogs of MGBG have been tested as potential AdoMetDC inhibitors. Some of these, such as dimethylglyoxal bis(guanylhydrazone) and ethylglyoxal bis(guanylhydrazone) (EGBG), are even more potent than MGBG (Corti *et al.*, 1972; Pegg and Jacobs, 1983), whereas others, such as glyoxal bis(guanylhydrazone) and di-*N‴*-methylglyoxal bis(guanylhydrazone), are less active (Corti *et al.*, 1972; Seppänen *et al.* 1984). All of these analogs and MGBG are reversible inhibitors of the enzyme, but 1,1′-[(methylethanediylidene)dinitrilo]bis(3-aminoguanidine) (MBAG), a derivative which has additional amino groups at each end, is an irreversible inactivator of AdoMetDC (Pegg, 1978). This inactivation is prevented by either MGBG or the substrate AdoMet and is probably due to the reaction of the hydrazine moiety of MBAG with the pyruvate prosthetic group of the enzyme. Attempts to produce inhibitors of AdoMetDC based on the synthesis of various pyruvate derivatives were not successful (Pankaskie and Abdul-Monem, 1980a), but this is not surprising since the pyruvate prosthetic group is clearly covalently attached to the enzyme protein and is not subject to replacement with such derivatives.

Detailed reviews of the biochemistry and pharmacology of MGBG have been published (Mihich, 1975; Porter *et al.*, 1981; Williams-Ashman and Seidenfield, 1986). Many studies have shown that MGBG can bring about changes in cellular polyamine content when administered to intact animals or to cells in culture. Also, it is known that the administration of spermidine can prevent the inhibition of the growth of malignant cells by MGBG. However, it appears unlikely that the antitumor effects of MGBG are due to its action on polyamine biosynthesis. As described below, MGBG has many other actions and is not specifically an inhibitor of AdoMetDC. Its transport into the cell uses the same carrier system as polyamines and the prevention of its antitumor action by spermidine may be due to competition for this transport. EGBG is more potent than MGBG as an inhibitor of AdoMetDC but is much less active an antileukemic agent.

Furthermore, the inhibition of AdoMetDC activity brought about by MGBG *in vivo* may be quite transient because the drug brings about a large increase in the amount of the enzyme protein. This increase appears to be produced by means of the stabilization of the enzyme against breakdown and the half-life of AdoMetDC is increased by more than 10-fold (Williams-Ashman and Pegg, 1981; Pegg, 1984; Tabor and Tabor, 1984; Shirahata and Pegg, 1985, 1986). Although an additional effect on the synthesis of the protein cannot be entirely ruled out (and it is known that such an increase occurs in response to spermidine depletion), there was no increase of the mRNA level for AdoMetDC in rat prostate even though the enzyme protein increased by more than 20-fold (Shirahata and Pegg, 1986).

MGBG is actively transported into mammalian cells (Porter *et al.*, 1981; Williams-Ashman and Seidenfeld, 1986; Pegg, 1986) and can reach very high intracellular concentrations. In addition to its action on AdoMet decarboxylase, the drug has a variety of other effects. Some of these, such as the ability to inhibit diamine oxidase, to interfere with polyamine transport, and to induce spermidine/spermine-N^1-acetyltransferase, may have direct relevance to changes in intracellular polyamine concentrations brought about by the drug (Porter *et al.*, 1981; Williams-Ashman and Seidenfeld, 1986; Pegg, 1986). In particular, the inhibition of diamine oxidase may contribute to the rise in putrescine in some conditions (Kallio and Jänne, 1983) and evidence has been published indicating that in rats treated with high doses of MGBG the decrease in spermidine and spermine and the increase in putrescine may be mediated principally by the acetylase/oxidase pathway via induction of spermidine/spermine-N^1-acetyltransferase (Pegg, *et al.*, 1985). Other changes, for example the production of severe mitochondrial damage, may not be directly related to the depletion of polyamine levels (Porter *et al.*, 1981; Williams-Ashman and Seidenfeld, 1986; Pegg, 1986).

It is clear that MGBG is not suitable for investigation of the function of AdoMetDC and its importance in cellular physiology. Attempts to obtain more specific inhibitors have been made and are continuing. It is well known that AdoMetDC is quite strongly inhibited by its product, decarboxylated AdoMet. Tests of a number of nucleosides related to AdoMet or decarboxylated AdoMet were reported by Pankaskie and Abdul-Monem (1980b), Yamanoha and Samejima (1980), Kolb *et al.* (1982), Pegg and Jacobs (1983), and Pegg *et al.* (1986). Although some substances with inhibitory activity were described and some requirements for binding to the enzyme were indicated by these studies, none of the compounds described was nearly as potent as MGBG. It was found that the terminal amino and carboxyl groups were not required for inhibition and in fact the best competitive inhibitor found among the nucleosides tested was *S*-methyl-5′-methylthioadenosine [$AdoS^{+}(CH_3)_2$], which had a K_i of 2 (Kolb *et al.*, 1982) or 6 μM (Pegg and Jacobs, 1983), but this nucleoside is a more potent inhibitor of spermine synthase (see below). Inhibitory activity clearly required the presence of adenine or a close derivative as the base (Pegg and Jacobs, 1983), but it was noted by Pankaskie and Abdul-Monem (1980b) and by Kolb *et al.* (1982) that the sulfonium center in AdoMet could be replaced with a nitrogen. The most promising current attempts to make nucleoside inhibitors of AdoMetDC involve the synthesis of analogs which might act as irreversible inhibitors. The synthesis of 5′-[(3-aminooxypropyl)amino]-5′-deoxyadenosine (Kolb and Barth, 1985) and of *S*-(5′-deoxy-5′adenosyl)-methylthioethylhydroxylamine (Khomutov *et al.*, 1983) has been reported. These nucleosides would be expected to be irreversible inactivators of the enzyme by forming an oxime with its pyruvoyl group and preliminary evidence that the latter compound does inactivate AdoMetDC has been published (Khomutov *et al.*, 1983; Artamonova *et al.*, 1986).

The synthesis of the α-difluoromethyl derivative of AdoMet has been described (Bey *et al.*, 1978), but, unfortunately, in view of the great value of DFMO as an ODC inhibitor, it appears that this compound is inactive as an inhibitor of AdoMetDC (Kolb *et al.*, 1982). Whether this relates to the absence of an appropriate site for reaction of an activated species at the active site of AdoMetDC or to the fact that this enzyme uses pyruvate rather than pyridoxal phosphate as cofactor is as yet unknown.

Other substances which have been reported as inhibitors of AdoMetDC include 1-aminooxy-3-aminopropane (Khomutov *et al.*, 1985), guanethidine (Johnson and Taylor, 1980), phenothiazines including chlorpromazine and imipramine (Hietala *et al.*, 1983, 1984), and Berenil and pentamidine (Karvonen *et al.*, 1985; Bitoni *et al.*, 1986). The specificity of these inhibitors toward AdoMetDC has not been established. It is not yet clear whether the pharmacological actions of these drugs are in any way related to their interference with the action of AdoMetDC. The inhibition of AdoMetDC by the antiprotozoal agents Berenil and pentamidine is probably due to their similarity to MGBG.

III. SPERMIDINE SYNTHASE

This aminopropyltransferase has been isolated in homogeneous form from *E. coli* (Bowman *et al.*, 1973), bovine brain, rat prostate (Raina *et al.*, 1984), rat liver (Samejima and Yamanoha, 1982), and pig liver (Yamanoha *et al.*, 1984). It consists of two subunits of M_r 35,000 to 37,000 and has no known cofactors. Partially purified preparations of spermidine synthase have been obtained from *Trypanosoma brucei brucei* (Bitonti *et al.*, 1984) and from the cabbage *B. pekinensis* (Yamanoha and Cohen, 1985). Aminopropyltransferase activities have been measured in crude extracts from a considerable number of microorganisms (e.g., Paulin *et al.*, 1983, 1985; Pösö *et al.*, 1983; Pegg *et al.*, 1983); some of these may differ in properties from the *E. coli* enzyme but none has been fully characterized. It is worth emphasizing that although the mammalian spermidine synthase is in many respects similar to the aminopropyltransferase of *E. coli,* there are some important differences. In particular, the mammalian enzyme does not form spermine at all whereas the *E. coli* aminopropyltransferase will use spermidine as a substrate under some conditions, although putrescine is its favored substrate. It is therefore not unexpected that there are significant differences between the mammalian and the *E. coli* enzyme in the sensitivity toward certain inhibitors and care should be taken in the extrapolation of results where studies have not been made directly with the enzyme in question.

Early work on the development of aminopropyltransferase inhibitors has been reviewed and summarized (Samejima and Nakazawa, 1980; Hibasami *et al.*, 1980a; Pegg and Coward, 1981; Coward *et al.*, 1983; Pegg, 1983b). A variety of nucleosides having some inhibitory action against spermidine synthase were identified in early surveys and, of these, *S*-adenosyl-L-homocysteine sulfone (Hibasami *et al.*, 1980a) and *S*-5′-deoxyadenosyl-(5′)-2-methylthioethylamine (Samejima and Nakazawa, 1980) were the most active. More recently, 5′-[3-aminopropyl)amino]-5′-deoxyadenosine was found to be the best spermidine synthase inhibitor of a series of nucleosides tested by Pegg *et al.* (1986), but none of the aforementioned compounds is nearly as potent as the mechanism-based inhibitor, AdoDato (Tang *et al.*, 1980; Coward *et al.*, 1983), which is described below.

A number of amine analogs have been tested as potential substrates and inhibitors of spermidine synthase (reviewed in Pegg, 1983b), but no useful inhibitors have emerged from these studies. Recently, 1-aminooxy-3-aminopropane was reported to be a potent competitive inhibitor of spermidine synthase (Khomutov *et al.*, 1985). This compound is not specific in that it also interferes with the action of ODC and AdoMetDC but it may have value as an agent for depletion of polyamines.

Hibasami and colleagues (Hibasami *et al.*, 1980b) discovered that cyclohexylamine was a strong inhibitor of mammalian spermidine synthase. Subsequent work indicated that spermidine synthases from a variety of sources including trypanosomes, plants, and various bacteria were also sensitive to inhibition by this compound (Pegg *et al.*, 1983; Pösö *et al.*, 1983; Paulin *et al.*, 1983; Mattila *et al.*, 1984; Sindhu and Cohen, 1984), although spermidine synthases from other bacteria were not affected (Pösö *et al.*, 1983; Mattila *et al.*, 1984). Some confusion in the literature has been generated by the fact that cyclohexylamine is sold by the Sigma Chemical Company as dicyclohexylammonium sulfate, and this substance has been described incorrectly by workers publishing results with it as dicyclohexylamine. The recent paper by Batchelor *et al.* (1986) clarifies this situation and shows clearly that only cyclohexylamine is effective. Cyclohexylamine is readily taken up by cells and leads to a depletion of spermidine in mammalian tissues and cells (Hibasami and Ito, 1981; Ito *et al.*, 1982; Porta *et al.*, 1984; Feurstein *et al.*, 1985; Mitchell *et al.*, 1985), plants (Balint and Cohen, 1985; Greenberg and Cohen, 1985), and in susceptible bacteria (Bitonti *et al.*, 1982, 1984; Mattila *et al.*, 1984). Detailed studies of the specificity and metabolic stability of cyclohexylamine are lacking and it is by no means ruled out that this compound might affect other aspects of cellular physiology. However, it is worth noting that the changes produced in the polyamine content of HTC cells by cyclohexylamine (Mitchell *et al.*, 1985) are virtually identical to the changes produced by AdoDato (Pegg *et al.*, 1982).

It has been reported that *N*-chlorosulfonyldicyclohexylamine is also an inhibitor of spermidine synthase and can be used to deplete spermidine and reduce the growth rate of Ehrlich ascites tumor cells (Hibasami *et al.*, 1985), but a detailed chemical characterization of this compound is clearly needed in view of the evidence described above that the purported inhibitory effects of dicyclohexylamine are due to cyclohexylamine.

Because of the close similarity in the reactions catalyzed by spermidine synthase and spermine synthase, it is not at all surprising that most of the inhibitors described for one enzyme also affect the other. However, the pioneering work of J.K. Coward on the synthesis of mechanism-based inhibitors of these reactions has demonstrated that excellent discrimination can be achieved by the rational synthesis of appropriate compounds (Tang *et al.*, 1980, 1981; Coward, 1982; Pegg *et al.*, 1982; Coward *et al.*, 1983). This work has resulted in the production of *S*-adenosyl-1,8-diamino-3-thiooctane (AdoDato), a mechanism-based inhibitor which has an IC_{50} of less than 20 n*M* for mammalian spermidine synthase under assay conditions approximately comparable to those *in vivo* (Pegg and Coward, 1981; Pegg *et al.*, 1983). The spermidine synthases from various mammalian sources, bacteria, trypanosomes, and plants are all sensitive to inhibition by AdoDato although the sensitivity differs slightly (Pegg *et al.*, 1983; Bitonti *et al.*, 1984; Sindhu and Cohen, 1984; Yamanoha *et al.*, 1984). AdoDato does not significantly inhibit spermine synthase.

The administration of AdoDato to cultured cells leads to a striking depletion of spermidine but to an increase in putrescine, spermine, and decarboxylated AdoMet (Pegg *et al.*, 1982). There is little total change in the cellular polyamine concentration. These results indicate that (1) there is a compensatory increase in the production of decarboxylated AdoMet (brought about by an increase in the amount of AdoMet decarboxylase) and that (2) spermidine synthase and spermine synthase compete for the available decarboxylated AdoMet and, in the presence of AdoDato, the spermine synthase is favored in this competition.

The spermidine synthase from mycobacteria has been reported to be highly sensitive to inhibition by the dextro isomer of ethambutol and it was suggested that this inhibition may represent the mode of action of this drug against tuberculosis (Pösö *et al.*, 1983; Paulin *et al.*, 1985). There was a striking difference between the spermidine synthases from *E. coli* and *Pseudomonas aeruginosa,* which were sensitive to cyclohexylamine and insensitive to ethambutol, and the spermidine synthase from *Myocobacteria bovis,* which was unaffected by 1 m*M* cyclohexylamine and strongly inhibited by 100 μM ethambutol (Pösö *et al.*, 1983).

Spermidine synthase from *E. coli* is inhibited by 5′-methylthioformycin (IC_{50} of 7 μM), whereas the mammalian spermidine synthase was much less affected (IC_{50} of 300 μM) (Carteni-Farina *et al.*, 1984).

IV. SPERMINE SYNTHASE

Spermine synthase has been purified to homogeneity from bovine brain (Pajula *et al.*, 1979; Pajula, 1983). The enzyme has two apparently identical subunits of M_r 44,000 and has no known cofactors or prosthetic groups. Unlike spermidine synthase it is not inhibited by excess of the substrate, decarboxylated AdoMet (Pajula, 1983), but it is very strongly inhibited by the product, MTA (Pajula and Raina, 1979; Pajula *et al.*, 1979). This inhibition is sufficiently strong that it was necessary to add MTA phosphorylase in order to prevent MTA accumulation during kinetic studies. Under these conditions the enzyme had apparent K_m values of 60 μM for spermidine and 0.1 μM for decarboxylated AdoMet (Pajula, 1983).

A variety of synthetic analogs of MTA have been found to inhibit spermine synthase, including 5′-ethylthioadenosine, 5′-isobutylthioadenosine (SIBA), 5′-methylthiotubercidin (5′-deoxy-5′-methylthio-7-deazaadenosine), and 5′-methylthioformycin (Hibasami *et al.*, 1980b; Pegg *et al.*, 1981; Hibasami *et al.*, 1982; Raina *et al.*, 1982; Della Ragione and Pegg, 1983; Carteni-Farina *et al.*, 1984). 5′-Methylthiotubercidin is somewhat less active than MTA *in vitro* but it is much more stable in the cell since it is not a substrate for MTA phosphorylase (Coward, 1982). Although spermine synthase is more susceptible to MTA than spermidine synthase, these nucleosides are not very selective inhibitors of spermine synthesis (Hibasami *et al.*, 1980b). In fact, when SV-3T3 or BHK21 cells were treated with 5′-methylthiotubercidin, spermidine was depleted more than spermine (Pegg *et al.*, 1981; Raina *et al.*, 1982).

Pegg *et al.* (1986) have identified three nucleosides having strong inhibitory actions toward spermine synthase. Thse are 5′-[(3-aminopropyl)amino]-5′-deoxyadenosine, 9-[6(*R,S*),8-diamino-5,6,7,8-tetradeoxy-β-D-*ribo*-octofuranosyl]-9*H*-purin-6-amine (AdoDap), and dimethyl(5′-adenosyl)sulfonium perchlorate [$AdoS^+(CH_3)_2$], which had IC_{50} values of 17, 12, and 8 μM, respectively, when tested in assays containing 5 μM decarboxylated AdoMet—a concentration above the normal physiological substrate concentration (Pegg and Coward, 1985; Pegg *et al.*, 1986).

As mentioned above, 5′-[3-aminopropyl)amino]-5′-deoxyadenosine was also a strong inhibitor of spermidine synthase but $AdoS^+(CH_3)_2$ and AdoDap were quite specific toward spermine synthase. Both compounds were able to block the synthesis of spermine in SV-3T3 cells but there was a compensatory increase in the production of decarboxylated AdoMet and in the concentration of spermine and there was no effect on cell growth (Pegg and Coward, 1985; Pegg *et al.*, 1986). At high concentrations of $AdoS^+(CH_3)_2$ (above 400 μM) there was a decrease in both spermidine and spermine, but at these levels it is likely that the inhibition of AdoMetDC becomes a factor (Pegg and Coward, 1985). Neither $AdoS^+(CH_3)_2$ nor AdoDap was able to reduce the residual spermine content of cells treated with DFMO (Pegg *et al.*, 1986), but it is likely that this is due to their ineffectiveness as inhibitors in the presence of very high levels of decarboxylated AdoMet. The combination of these spermine synthase inhibitors with DFMO and an effective AdoMet DC inhibitor would therefore be of great interest.

At present $AdoS^{+}(CH_3)_2$ appears to be the most useful compound for blocking spermine synthesis. However, the development of mechanism-based inhibitors is currently underway and preliminary evidence has been obtained that *S*-adenosyl-1,12-diamino-3-thio-9-azadodecane is a very effective inhibitor (J. K. Coward and A. E. Pegg, unpublished observations). The rationale for the synthesis and testing of this compound is the same as that which underlies AdoDato for spermidine synthase (Tang *et al.*, 1980; Coward, 1982), and preliminary results indicate that it should be equally valuable as a specific inhibitor.

V. ENZYMATIC DEGRADATION OF 5′-METHYLTHIOADENOSINE (MTA)

MTA and polyamines are produced in equimolar concentrations in the spermidine and spermine synthase reactions. In all higher animal nonmalignant cells that have been scrutinized, MTA does not normally accumulate in more than trace amounts (<10 nmol/g) in comparison with the cellular levels of spermidine and spermine. This is because MTA in a wide variety of noncancerous cells is very rapidly metabolized via a series of reactions, as reviewed by Williams-Ashman *et al.* (1982), Schlenk (1983), and Williams-Ashman (1985). The disposal of MTA is of value to the cellular metabolic economy for two reasons. First, degradation of MTA may effectively prevent this nucleoside serving as a cytostatic agent (Pegg *et al.*, 1981; Kubota *et al.*, 1983) or as an inhibitor of certain methyltransferases (for references see Williams-Ashman *et al.*, 1982), or of spermidine and especially spermine synthases (Pegg *et al.*, 1981; Raina *et al.*, 1982), or of AdoHcy hydrolase (Della Ragione and Pegg, 1983). Second, the reactions of MTA metabolism permit metabolic salvage of the ATP and L-methionine consumed in the biosynthesis of the cellular AdoMet molecules that are utilized for the production of decarboxylated AdoMet and hence of spermidine and spermine.

Although MTA is converted to 5′-methylthioinosine by the nonspecific adenosine deaminase of *Aspergillus oryzae* (Zappia *et al.*, 1974), there is no evidence indicating that MTA is directly deaminated in multicellular eucaryotes. The first step in MTA metabolism involves enzymatic rupture of the *N*-ribose bond. In some bacteria, including Enterobacteriaceae and Bacillaceae, MTA is cleaved hydrolytically to 5-methylthioribose and adenine by an enzyme that also utilizes AdoHcy as substrate (Shimizu *et al.*,

1986). The irreversible AdoHcy/MTA nucleosidase of *E. coli* has been purified extensively and characterized by Della Ragione *et al.* (1985). The reaction product adenine is only a feeble inhibitor.

Hydrolytic MTA nucleosidases that do not degrade AdoHcy have been described in several higher plants (Schlenk, 1983). Guranowksi *et al.* (1981) purified an enzyme of the latter class 9000-fold to homogeneity from yellow lupine seedlings; this nucleosidase is quite strongly inhibited by its reaction product adenine, and also splits 5′-methylthioinosine at about 2% of the rate with MTA as substrate.

By contrast, other prokaryotic microorganisms including certain thermophilic archebacteria, lower eukaryotes such as *Penicillium chrysogenum,* higher animal cells, and *Drosophila melanogaster* initially degrade MTA by a reversible phosphorolytic reaction catalyzed by enzymes designated as MTA phosphorylases (Pegg and Williams-Ashman, 1969; Williams-Ashman *et al.,* 1982; Schlenk, 1983; Shimizu *et al.,* 1986). MTA phosphorylases are entirely distinct from the classical purine nucleoside phosphorylases in their substrate specificity and various other properties. The products of MTA breakdown by MTA phosphorylases are (1) adenine (which is a strong inhibitor) and (2) the α-anomer of 5-methylthioribose-1-phosphate (MTR-1-P). It is noteworthy that some bacteria and higher plant tissues that contain hydrolytic MTA nucleosidases have been shown to possess a specific, ATP-dependent MTR kinase which converts MTR to MTR-1-P (Ferro *et al.,* 1978; Kushad *et al.,* 1982). Thus, MTA phosporylases catalyze the production of MTR-1-P in a single, reversible reaction, whereas in cells containing hydrolytic MTA nucleosidases the formation of MTR-1-P occurs by the sequential actions of the latter enzymes and MTR kinase. The coexistence of MTA phosphorylase and a hydrolytic MTA nucleosidase in the same type of cell has never been reported. As considered shortly, MTR-1-P is effectively recycled to L-methionine in a variety of bacterial, plant, and animal cells and free MTR can be converted to L-methionine only if it is first phosphorylated on C-1.

The adenine moiety of MTA that is produced by the actions of MTA phosphorylases or hydrolytic MTA nucleosidases is readily converted to 5′-AMP by reaction with phosphoribosyl pyrophosphate in reactions catalyzed by adenine phosphoribosyltransferase (APRT), an enzyme of almost ubiquitous biological occurrence. The 5′-AMP formed by the aforementioned pathways can then be phosphorylated to yield ADP and ATP by obvious enzyme reactions. Thus ATP can be effectively salvaged from the adenine that is liberated when MTA is sundered enzymatically. In accord with these notions is the demonstration by Kamatani and Carson (1981) that proliferating human lymphoblast WIL2 cells deficient in APRT but replete with MTA phosphorylase readily generated adenine at rates equivalent to the formation of MTA by an MTA phosphorylase-deficient human lymphoblastoid CCRF-CEM cell line. Kamatani and Carson (1981) estimated that more than 85% of the adenine derived from MTA in the WIL2 cell line originated from the MTA formed during the biosynthesis of spermidine and spermine. Of related interest are the findings of Savarese *et al.* (1983, 1985) that various 5′-halogenated derivatives of 5′-deoxyadenosine, as well as certain congeners of MTA in which the adenine moiety is suitably substituted (e.g., 5′-deoxy-5′ethylthio-2-fluoroadenosine), are (1) highly effective substrates for MTA phosphorylase, (2) competitive inhibitors of the cleavage of MTA by its specific phosphorylase, and (3) very cytotoxic to cell lines that contain MTA phosphorylase but not to cell lines lacking MTA phosphorylase, probably by virtue of the conversion of the substituted adenine liberated by enzymatic cleavage to the corresponding ATP analogs and subsequent incorporation into nucleic acids.

MTR-1-P formed from MTA can be isomerized enzymatically to 5-methylthioribulose-1-phosphate (Trackman and Abeles, 1983). This represents the first of the series of reactions by which MTR-1-P can be converted to 2-keto-4-*S*-methylthiobutyrate (KMTB). The KMTB so produced can undergo enzymatic transamination with either glutamine or asparagine to yield L-methionine. Although various mechanisms have been proposed (Trackman and Abeles, 1981; Schlenk, 1983), the precise nature and enzymology of the reactions involved in the transformation of 5-methylthioribulose-1-phosphate into KMTB remain uncertain. Nonetheless, there is no doubt that many different normal and malignant cells that contain MTA phosphorylase can convert MTA to methionine efficiently, whereas certain malignant cells that are essentially devoid of MTA phosphorylase are unable to synthesize L-methionine from MTA (Backlund and Smith, 1982; Carson *et al.*, 1983).

The only reactions of MTA metabolism for which specific inhibitors are known are those catalyzed by the enzymes that split MTA at the N–ribose linkage. These inhibitors will now be discussed with reference to various types of enzymes that promote the first step in MTA degradation.

The action of the homogeneous AdoHcy/MTA hydrolase of *E. coli* on MTA as substrate was found by Della Ragione *et al.* (1985) to be very strongly inhibited by 5′-methylthioformycin (K_i of 27 n*M*) and *S*-formycinylhomocysteine (K_i of 9.7 n*M*). 5′-Methylthioformycin is a structural analog of MTA in which a C–C bond replaces the glycosidic N–C bond. The same enzyme acting on MTA was inhibited less powerfully by 5′-deoxy-5′-chloroformycin (K_i of 0.32 μM), 5′-methylthiotubercidin (K_i of 4.1 μM), *S*-tubercidinylhomocysteine (K_i of 1.9 μM), and *S*-8-azaadenosylhomocysteine (K_i of 0.93 μM). None of these inhibitors is a substrate for the AdoHcy/MTA hydrolase of *E. coli*. Della Ragione *et al.* (1985)

raise the possibility that 5′-methylthioformycin might be of value in bacterial chemotherapy; however, if this is proved to be the case, it could conceivably relate to effects of the latter inhibitor on the degradation of AdoHcy (a powerful inhibitor of the majority of methyltransferases) rather than on the metabolism of MTA, even though the apparent K_m values for MTA and AdoHcy are 0.4 and 4.3 μM, respectively.

Considerable attention has been given to the development of inhibitors of mammalian MTA phosphorylases. Coward *et al.* (1977) reported that 5′-methylthiotubercidin is not a substrate for rat ventral prostate MTA phosphorylase but did act as a competitive inhibitor of the enzyme, though not an extremely effective one (K_m MTA/K_i of 0.25). Subsequently, Savarese *et al.* (1979) found that 5′-deoxy-5′-chloroformycin is a more powerful inhibitor, but also is not a substrate for the MTA phosphorylase of mouse sarcoma 180 (K_m MTA/K_i of 30). Della Ragione and Pegg (1983) also found that 5′-methylthiotubercidin and more pronouncedly 5′-deoxy-5′-chloroformycin are nonsubstrate inhibitors of rat liver MTA phosphorylase. However, both of these compounds have lesser yet definite inhibitory actions toward rat liver spermine synthase but do not inactivate rat liver AdoHcy hydrolase, for which MTA is not a substrate but does serve as an irreversible inactivator (Della Ragione and Pegg, 1983). SIBA is a good substrate for mammalian MTA phosphorylases, and, as might be expected, competitively inhibits the degradation of MTA by these enzymes (Carteni-Farina *et al.*, 1979; Pegg *et al.*, 1981; Parks *et al.*, 1981). Unfortunately, as described in Section IV, SIBA is also an effective inhibitor of mammalian AdoHcy hydrolase and spermine synthase, though less active than MTA in these respects (Pegg *et al.*, 1981; Raina *et al.*, 1982; Hibasami *et al.*, 1982; Della Ragione and Pegg, 1983). As mentioned above, other analogs of MTA which are substrates for MTA phosphorylase also inhibit the splitting of MTA by mammalian enzymes of this class, and in some instances may also become metabolically activated by MTA phosphorylase action to yield products that exert antineoplastic effects (cf. Savarese *et al.*, 1985; Williams-Ashman, 1985).

The widely distributed set of enzymes that can convert MTA to L-methionine enable many types of cells to synthesize spermidine and spermine *de novo* without any *net* consumption of L-methionine, although the amino acid is obviously required to initiate formation of decarboxylated AdoMet used for polyamine synthesis. In certain malignant cells that are apparently devoid of either MTA phosphorylase or the enzyme that catalyzes the conversion of MTR-1-P to 5-methylthioribulose-1-phosphate (for references see Williams-Ashman, 1985), the synthesis of polyamines (which is considerable in proliferating cancerous cells) must be dependent on exogenous methionine or endogenous transformation of AdoHcy to L-methionine. The efficacy of L-methionine salvage from MTA will depend not only on the activity of the requisite enzymes and the availability of glutamine or asparagine for enzymatic transamination of the intermediate KMTB to L-methionine, but also on the extent of operation of alternate pathways for KMTB metabolism, as considered by Williams-Ashman *et al.* (1982). The availability of drugs that selectively interfere with reactions involved in the recycling of MTA to L-methionine might have important chemotherapeutic implications. For this reason, the development of potent nonsubstrate inhibitors of MTA phosphorylase, or of the various enzymes required for the conversion of MTR-1-P to L-methionine, and which would not affect spermidine and spermine synthases or AdoHcy hydrolases in mammalian cells, would be a valuable contribution to medicinal chemistry.

REFERENCES

Artamonova, E.Y., Zavalova, L.L., Khomutov, R.M., and Khomutov, A.R. (1986). *Bioorg. Khim.* **12,** 206–212.

Backlund, P.S., and Smith, R.A. (1982). *Biochem. Biophys. Res. Commun.* **108,** 687–695.

Balint, R., and Cohen, S.S. (1985). *Virology* **144,** 194–203.

Batchelor, K.W., Smith, R.A., and Watson, N.S. (1986). *Biochem. J.* **233,** 311–312.

Bey, P., Danzin, C., VanDorsselaer, V., Mamont, P., Jung, M., and Tardif, C. (1978). *J. Med. Chem.* **21,** 50–55.

Bitonti, A.J., McCann, P.P., and Sjoerdsma, A. (1982). *Biochem. J.* **208,** 435–441.

Bitonti, A.J., Kelly, S.E., and McCann, P.P. (1984). *Life Sci.* **34,** 1513–1520.

Bitonti, A.J., Dumont, J.A., and McCann, P.P. (1986). *Biochem. J.* **237,** 685–689.

Bowman, W.H., Tabor, C.W., and Tabor, H. (1973). *J. Biol. Chem.* **248,** 2480–2486.

Carson, D.A., Willis, E.H., and Kamatani, N. (1983). *Biochem. Biophys. Res. Commun.* **112,** 391–397.

Carteni-Farina, M., Della Ragione, F., Ragosta, G., Oliva, A., and Zappia, V. (1979). *FEBS Lett.* **104,** 266–270.

Carteni-Farina, M., Cacciapuoti, G., Porcelli, M., Della Ragione, F., Lancieri, M., Geraci, G., and Zappia, V. (1984). *Biochim. Biophys. Acta* **805,** 158–164.

Corti, A., Dave, C., Williams-Ashman, H.G., Mihich, E., and Schenone, A. (1972). *Biochem. J.* **139,** 351–357.

Coward, J.K. (1982). *Annu Rep. Med. Chem.* **17,** 253–259.

Coward, J.K., Motola, N.M., and Moyer, J.D. (1977). *J. Med. Chem.* **20,** 500–505.

Coward, J.K., Anderson, O.L., and Tang, K.-C. (1983). *In* "Methods in Enzymology" (H. Tabor and C.W. Tabor, eds.), Vol. 94, pp. 286–294, Academic Press, New York.

Della Ragione, F., and Pegg, A.E. (1983). *Biochem. J.* **210,** 429–435.

Della Ragione, F., Porcelli, M., Carteni-Farina, M., Zappia, V., and Pegg, A.E. (1985). *Biochem. J.* **232,** 335–341.

Ferro, A.J., Barett, A., and Shapiro, S.K. (1978). *J. Biol. Chem.* **253,** 6021–6025.

Feurstein, B.O., Deen, D.F., and Marton, L.J. (1985). *Cancer Res.* **45,** 4950–4954.

Greenberg, M.L., and Cohen, S.S. (1985). *Plant Physiol.* **78,** 568–575.

Guranowski, A.B., Chiang, P.K., and Cantoni, G. (1981). *Eur. J. Biochem.* **114,** 293–299.

Hibasami, H., and Ito, H. (1981). *Gann* **72,** 512–516.

Hibasami, H., Borchardt, R.T., Chen, S.Y., Coward, J.K., and Pegg, A.E. (1980a). *Biochem. J.* **187,** 419–428.

Hibasami, H., Tanaka, M., Nagai, J., and Ikeda, T. (1980b). *FEBS. Lett.* **116,** 99–101.

Hibasami, H., Tanaka, M., and Nagai, J. (1982). *Biochem. Pharmacol.* **31,** 1649–1650.
Hibasami, H., Tsukada, T., and Nakashima, K. (1985). *In* "Polyamines: Basic and Clinical Aspects" (K. Imahori, F. Suzuki, O. Suzuki, and U. Bachrach, eds.), pp. 243–250. YNU Science Press, Utrecht.
Hietala, O.A., Laitinen, S.I., Laitinen, P.H., Lapinjoki, S.P., and Pajunen, A.E.I. (1983). *Biochem. Pharmacol.* **32,** 1581–1585.
Hietala, O.A., Lapinjoki, S.P., and Pajunen, A.E.I. (1984). *Biochem. Int.* **8,** 245–255.
Ito, H., Hibasami, H., Shimura, K., Nagai, J., and Hidaka, H. (1982). *Cancer Lett.* **15,** 229–235.
Johnson, E.M., and Taylor, A.S. (1980). *Biochem. Pharmacol.* **29,** 113–115.
Kallio, A., and Jänne, J. (1983). *Biochem. J.* **212,** 895–898.
Kamatani, N., and Carson, D.A. (1981). *Biochem. Biophys. Acta* **675,** 344–350.
Karvonen, E., Kauppinen, L., Partanen, T., and Pösö, H. (1985). *Biochem. J.* **231,** 165–169.
Khomutov, R.M., Zavalova, L.L., Siirku, V.I., Artamonova, E.Y., and Khomutov, A.R. (1983). *Bioorg. Khim.* **9,** 130–131.
Khomutov, R.M., Hyvönen, T., Karvonen, E., Kauppinen, L., Paalanen, T., Paulin, L., Eloranta, T., Pajula, R.-L., Anderson, L.C., and Pösö, H. (1985). *Biochem. Biophys. Res. Commun.* **130,** 596–602.
Kolb, M., and Barth, J. (1985). *Liebigs Ann. Chem.* 1036–1040.
Kolb, M., Danzin, C., Barth, J., and Claverie, N. (1982). *J. Med. Chem.* **25,** 550–556.
Kubota, M., Kamatani, N., and Carson, D.A. (1983). *J. Biol. Chem.* **258,** 7288–7291.
Kushad, M.M., Richardson, D.G., and Ferro, A.J. (1982). *Biochem. Biophys. Res. Commun.* **108,** 167–173.
Mattila, T., Honkanen-Buzalski, T., and Pösö, H. (1984). *Biochem. J.* **223,** 823–830.
Mihich, E. (1975). *Handb. Exp. Pharmacol.* **38/2,** 766–788.
Mitchell, J.L.A., Mahan, D.W., McCann, P.P., and Qasba, P. (1985). *Biochim. Biophys. Acta* **840,** 309–316.
Pajula, R.-L. (1983). *Biochem. J.* **215,** 669–676.
Pajula, R.-L., and Raina, A. (1979). *FEBS Lett.* **99,** 343–345.
Pajula, R.-L., Raina, A., and Eloranta, T.E. (1979). *Eur. J. Biochem.* **101,** 619.
Pankaskie, M., and Abdul-Monem, M.M. (1980a). *J. Pharm. Sci.* **69,** 1000–1004.
Pankaskie, M., and Abdul-Monem, M.M. (1980b). *J. Med. Chem.* **23,** 121–126.
Parks, R.E., Stoeckler, J.D., Cambor, C., Savarese, T.M., Crabtree, G.W., and Chu, S.-H. (1981). *In* "Molecular Actions and Targets for Chemotherapeutic Agents" (A.C. Sartorelli, J.S. Lazo, and J.R. Bertino, eds.), pp. 229–252. Academic Press, New York.
Paulin, L.G., Vehmaanperä, J., Nyjuanen, I., and Pösö, H. (1983). *Biochem. Biophys. Res. Commun.* **114,** 779–784.
Paulin, L.G., Brander, E.E., and Pösö, H. (1985). *Antimicrob. Agents Chemother.* **28,** 157–159.
Pegg, A.E. (1978). *J. Biol. Chem.* **253,** 539–542.
Pegg, A.E. (1983a). *In* "Methods in Enzymology" (H. Tabor and C.W. Tabor, eds.), Vol. 94, pp. 239–247. Academic Press, New York.
Pegg, A.E. (1983b) *In* "Methods in Enzymology" (H. Tabor and C.W. Tabor, eds.), Vol. 94, pp. 294–301. Academic Press, New York.
Pegg, A.E. (1984). *Cell Biochem. Funct.* **2,** 11–15.
Pegg, A.E. (1986). *Biochem. J.* **234,** 249–262.
Pegg, A.E., and Coward, J.K. (1981). *Adv. Polyamine Res.* **3,** 153–161.
Pegg, A.E., and Coward, J.K. (1985). *Biochem. Biophys. Res. Commun.* **133,** 82–89.
Pegg, A.E., and Jacobs, G. (1983). *Biochem. J.* **213,** 495–502.
Pegg, A.E., and Williams-Ashman, H.G. (1969). *Biochem. J.* **115,** 241–247.
Pegg, A.E., Borchardt, R.T., and Coward, J.K. (1981). *Biochem. J.* **194,** 79–89.
Pegg, A.E., Tang, K.-C., and Coward, J.K. (1982). *Biochemistry* **21,** 5082–5089.

Pegg, A.E., Bitonti, A.J., McCann, P.P., and Coward, J.K. (1983). *FEBS Lett.* **155,** 192–196.
Pegg, A.E., Erwin, B.G., and Persson, L. (1985). *Biochim. Biophys. Acta* **842,** 111–118.
Pegg, A.E., Coward, J.K., Talekar, R.R., and Secrist, J.A. (1986). *Biochemistry* **25,** 4091–4097.
Pösö, H., Paulin, L., and Brander, E. (1983). *Lancet* **2,** 1418.
Porta, R., Camardella, M., Gentile, V., and DeSantia, A. (1984). *J. Neurochem.* **42,** 321–325.
Porter, C.W., Dave, C., and Mihich, E. (1981). *In* "Polyamines in Biology and Medicine" (D.R. Morris and L.J. Marton, eds.), pp. 407–436. Dekker, New York.
Raina, A., Tuomi, K., and Pajula, R.-L. (1982). *Biochem. J.* **204,** 697–703.
Raina, A., Hyvönen, T., Eloranta, T., Voutilainen, M., Samejima, K., and Yamanoha, B. (1984). *Biochem. J.* **219,** 991–1000.
Savarese, J.M., Crabtree, G.W., and Parks, R.E. (1979). *Biochem. Pharmacol.* **30,** 189–199.
Savarese, J.M., Dexter, D.L., Parks, R.E., and Montgomery, J.A. (1983). *Biochem. Pharmacol.* **32,** 1907–1916.
Savarese, J.M., Chu, S.-H., Chu, M.-Y., and Parks, R.E. (1985). *Biochem. Pharmacol.* **34,** 361–367.
Samejima, K., and Nakazawa, Y. (1980). *Arch. Biochem. Biophys.* **201,** 241–246.
Samejima, K., and Yamanoha, B. (1982). *Arch. Biochem. Biophys.* **216,** 213–222.
Schlenk, F. (1983). *Adv. Enzymol. Rel. Areas Mol. Biol.* **54,** 196–265.
Seppänen, P., Fagerström, R., Alhonen-Hongisto, A., Elo, M., Lumme, P., and Jänne, J. (1984). *Biochem. J.* **221,** 483–488.
Shimuzu, S., Abe, T., Shiozaki, S., and Yamada, H. (1986). *Abstr. Int. Conf. Polyamines Life Sci. Lake Yamanaka, Japan* 193–194.
Shirahata, A., and Pegg, A.E. (1985). *J. Biol. Chem* **260,** 9583–9588.
Shirahata, A., and Pegg, A.E. (1986). *J. Biol. Chem.* **261,** 13833–13837.
Sindhu, R.K., and Cohen, S.S. (1984). *Plant Physiol.* **74,** 645–649.
Tabor, C.W., and Tabor, H. (1984). *Adv. Enzymol. Rel. Areas Mol. Biol.* **56,** 251–282.
Tang, K.-C., Pegg, A.E., and Coward, J.K. (1980). *Biochem. Biophys. Res. Commun.* **96,** 1371–137.
Tang, K.-C., Mariuzza, R., and Coward, J.K. (1981). *J. Med. Chem.* **24,** 1277–1284.
Trackman, P.C., and Abeles, R.H. (1981). *Biochem. Biophys. Res. Commun.* **103,** 1238–1244.
Trackman, P.C., and Abeles, R.H. (1983). *J. Biol. Chem.* **258,** 6717–6720.
Williams-Ashman, H.G. (1985). *In* "Recent Progress in Polyamine Research" (L. Selmeci, M.E. Brosnan, and N. Seiler, eds.), pp. 231–239. Akademiai Kiado, Budapest.
Williams-Ashman, H.G., and Pegg, A.E. (1981). *In* "Polyamines in Biology and Medicine" (D.R. Morris and L.J. Marton, eds.), pp. 43–73. Dekker, New York.
Williams-Ashman, H.G., and Schenone, A. (1972). *Biochem. Biophys. Res. Commun.* **46,** 288–295.
Williams-Ashman, H.G., and Seidenfeld, J. (1986). *Biochem. Pharmacol.* **35,** 1217–1225.
Williams-Ashman, H.G., Seidenfeld, J., and Galletti, P. (1982). *Biochem. Pharmacol.* **31,** 277–288.
Yamanoha, B., and Cohen, S.S. (1985). *Plant Physiol.* **78,** 784–790.
Yamanoha, B., and Samejima, K. (1980). *Chem. Pharm. Bull.* **28,** 2232–2234.
Yamanoha, B., Samejima, K., Nakajima, T., and Yasuhara, T. (1984). *J. Biochem.* 1273–1281.
Zappia, V., Galletti, P., Carteni-Farina, M., and Servillo, L. (1974). *Anal. Biochem.* **58,** 130–136.

3

Inhibition of Enzymes Oxidizing Polyamines

N. SEILER

Merrell Dow Research Institute
Strasbourg Centre
16, rue d'Ankara
F 67084 Strasbourg Cedex, France

I. INTRODUCTION

The first enzyme recognized to act specifically on spermidine and spermine is an amine oxidase especially abundant in the plasma of ruminants (Hirsch, 1953). Ironically, in spite of purification to homogeneity and crystallization (Yamada and Yasunobu, 1962; Mondovi *et al.*, 1983), extensive investigation of its occurrence in the animal kingdom (Blaschko, 1962; Blaschko and Bonney, 1962), and investigation of its mechanism of action (Mondovi and Finazzi Agro, 1982), a physiological function of this enzyme has not until now been established. In the course of the years a considerable number of oxidases capable of oxidizing diamines and/or polyamines have been found in animals, plants, and microorganisms, (Bachrach, 1973, 1981; Morgan 1980, 1985a; Smith, 1985).

The reactions involved in the biosynthetic formation of the polyamines were clarified in principle soon after the discovery of the key enzymes (Bachrach, 1973). In contrast, the principles of polyamine catabolism have only recently been understood to some extent. We recognize today two types of catabolic reactions: interconversion and terminal oxidation of the polyamines.

1. *Interconversion*. Spermine can be converted into spermidine, and spermidine into putrescine, along the so-called interconversion pathway,

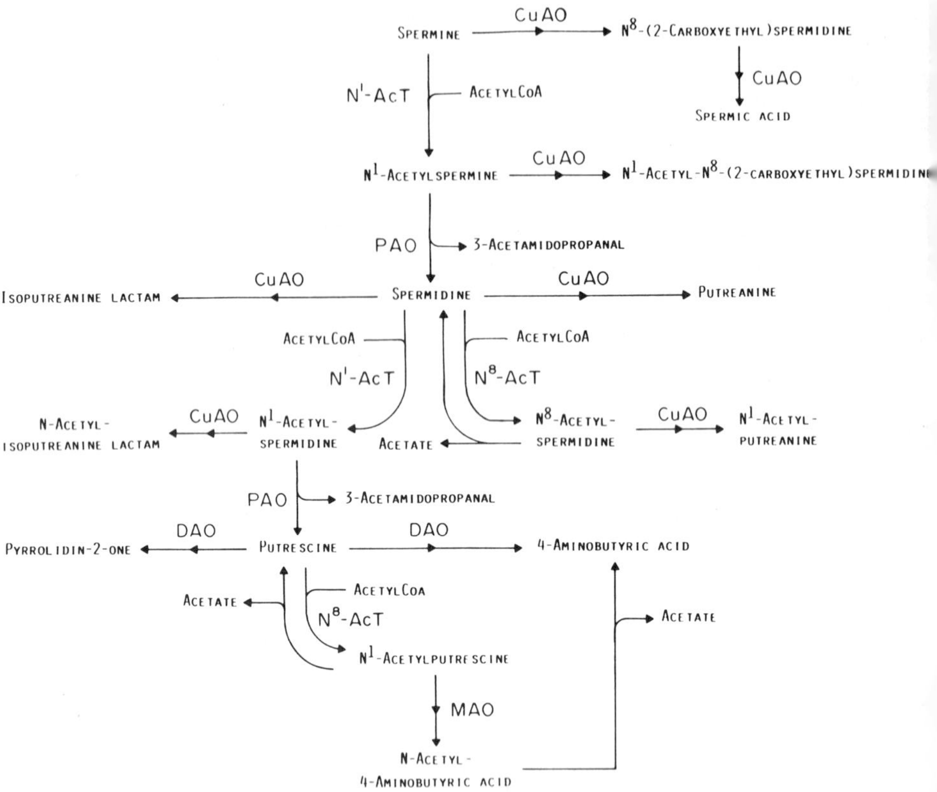

Fig. 1. Catabolic reactions of the polyamines. Abbreviations: MAO, monoamine oxidase (FAD dependent; EC 1.4.3.4.); DAO, diamine oxidase (EC 1.4.3.6.); CuAO, copper-containing amine oxidases (this designation implies serumamine oxidase, diamine oxidase, and tissue and serum enzymes which have not yet been characterized in detail; the transformation into amino acids requires in addition an aldehyde dehydrogenase or another enzyme which is able to oxidize the aldehydes formed by oxidative deamination to a carboxylic acid); PAO, polyamine oxidase (FAD dependent); N^1-AcT, acetyl-CoA: polyamine N^1-acetyltransferase (cytosolic); N^8-AcT, acetyl-CoA: spermidine N^8-acetyltransferase (nuclear; putrescine is also a substrate).

which is formally, but not mechanistically, the inversion of the biosynthetic reactions (Seiler *et al.*, 1981b; Pegg *et al.*, 1982). As is shown in Fig. 1, polyamine degradation is initiated by formation of N^1-acetylspermine and N^1-acetylspermidine, respectively. This step is catalyzed by a cytosolic acetyltransferase which uses acetyl CoA as cofactor (Della Ragione and Pegg, 1982). In the second step the acetamidopropyl moiety of monoacetylspermidine and -spermine is eliminated (Bolkenius and Seiler,

1981). This reaction is catalyzed by polyamine oxidase (Hölttä, 1977). Spermidine and putrescine, the products of the interconversion reaction, can be reutilized for the synthesis of spermine and spermidine, respectively (Antrup and Seiler, 1980; Seiler and Bolkenius, 1985; Seiler *et al.*, 1985c). As a result, putrescine, spermidine, and spermine can be converted into one another either by addition or by elimination of aminopropyl moieties, hence the designation interconversion pathway for this reaction sequence. It appears that cellular polyamine biosynthesis is not only controlled by regulation of the biosynthetic decarboxylases, ornithine decarboxylase and *S*-adenosylmethionine decarboxylase, but also by regulation of the rate of polyamine acetylation. This reaction seems to be enhanced whenever polyamines have to be removed from cells (Seiler *et al.*, 1981b; Mamont *et al.*, 1981; Matsui *et al.*, 1982). Since polyamine oxidase has been found in virtually all tissues (Seiler *et al.*, 1980), polyamine interconversion seems to be a general pathway.

In contrast with N^1-acetylspermine, N^1-acetylspermidine and N^8-acetylspermidine are normal excretory products (Seiler *et al.*, 1981d). This seems to indicate that the organism is not absolutely dependent on the action of polyamine oxidase, because the most important substrate of this enzyme can be eliminated as such.

2. *Terminal oxidation.* The second type of oxidative pathway of the polyamines has been called terminal oxidation (Seiler *et al.*, 1985a) in order to indicate that the terminal (primary) amino groups are oxidatively deaminated to aldehydes, which are further transformed into amino acids or, if the amino group is in the 4-position, into the corresponding γ-lactam (Fig. 1). The term terminal oxidative pathway also indicates that the compounds formed along these reactions are irreversibly eliminated from the polyamine metabolic cycle and cannot be retransformed into polyamines, if we disregard the possibility of incorporation of carbon atoms of these amino acids into polyamines in a very indirect manner via intermediary metabolism.

In contrast with polyamine oxidase, the enzymes involved in terminal polyamine oxidation are unevenly distributed in the vertebrate organism, with highest activities in small intestine (Shaff and Beaven, 1976; Seiler *et al.*, 1983). To date, the tissue enzymes are insufficiently characterized. There is no doubt that terminal polyamine oxidation is catalyzed by copper amine oxidases. It cannot be excluded that the classical diamine oxidase is actually responsible for this reaction, since virtually all intermediates of the interconversion pathway are substrates of diamine oxidases (Suzuki *et al.*, 1981; Gahl and Pitot, 1981, 1982). The reaction products of spermidine and spermine (see Fig. 1) have been detected as normal urinary constituents (Nakajima *et al.,* 1980; Seiler and Knödgen, 1983) and there is evidence that N^1-acetylspermidine can be degraded to *N*-(3-acetamidopropyl) pyrrolidin-2-one (*N*-acetylisoputreanine lactam) (Seiler *et al.,* 1982, 1983).

Neither polyamine biosynthesis nor polyamine catabolism follows simple reaction schemes. Specific inhibitors of biosynthetic enzymes such as α-difluoromethylornithine (Bey, 1978) have had a considerable influence on the development of our knowledge about the function of the polyamines (Heby and Jänne, 1981; Koch-Weser *et al.,* 1981). Inhibitors of the catabolic enzymes will most probably also be of importance in the elucidation of the significance of the various pathways. Their mode of action and their most conspicuous effects on polyamine metabolism in vertebrates will be discussed to some detail in the following sections.

II. COPPER AMINE OXIDASES

A. Enzymes

There is considerable confusion with regard to the identity or nonidentity of the copper amine oxidases that have been isolated from tissues and blood of various species during the last four decades. Full agreement exists only with regard to two enzymes: serum enzyme from ruminant blood and diamine oxidase from hog kidney.

1. Bovine Serum Oxidase

The pure enzyme has an M_r of about 180,000 and appears to contain two Cu^{2+} atoms (Mondovi *et al.,* 1983) and, as prosthetic group, a covalently bound pyrroloquinoline quinone (Lobenstein-Verbeek *et al.,* 1984) (Fig. 2). It is composed of two equivalent subunits (Achee *et al.,* 1968). The bovine enzyme attacks exclusively the primary amino group of the aminopropyl moieties of spermidine and spermine, as is formulated in Fig. 3 (Tabor *et al.,* 1964), while putrescine is not a substrate (Mondovi *et al.,* 1983). Adult bovine serum, in contrast with fetal calf serum, contains no diamine oxidase-like activity (Gahl and Pitot, 1962).

Blaschko and his collaborators have systematically studied the activity and properties of polyamine-oxidizing serum enzymes in a great variety of species (Blaschko, 1962; Blaschko and Bonney, 1962). A functional explanation for the fact that ruminants have extremely high activities of this enzyme whereas other species, for example rodents, have very low activities, has not yet been found. It is not unlikely that the enzyme protects the body from alimentary polyamines. However, experimental data on the

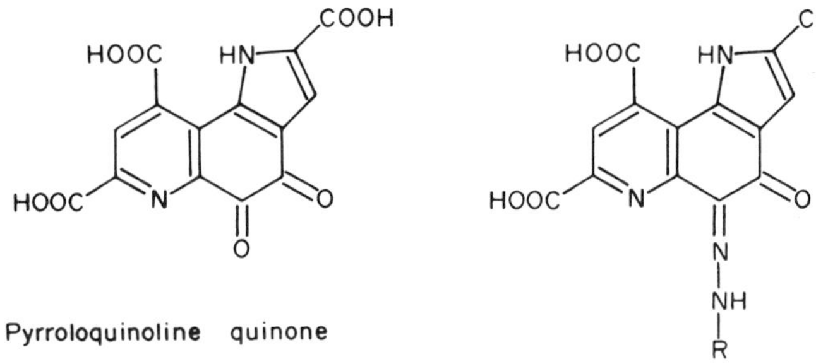

Fig. 2. Pyrroloquinoline quinone, the cofactor of serum spermine oxidase, and the reaction product of pyrroloquinoline quinone with an hydrazine.

uptake of polyamines from the gastrointestinal tract into the circulation do not exist.

In addition to the bovine serum amine oxidase a number of other polyamine-oxidizing serum enzymes are known to exist but are inadequately characterized. In human pregnancy serum, for example, an oxidase has been found which deaminates both polyamines and diamines (Gahl *et al.*, 1982). Unlike in fetal calf serum, the putrescine-oxidizing activity could not be separated from the spermidine-oxidizing activity in human pregnancy serum.

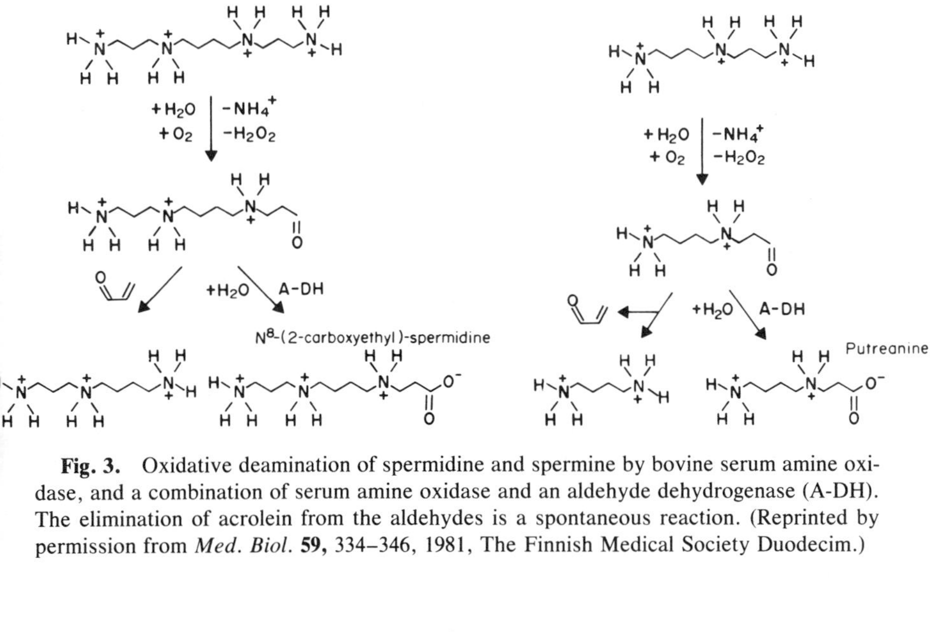

Fig. 3. Oxidative deamination of spermidine and spermine by bovine serum amine oxidase, and a combination of serum amine oxidase and an aldehyde dehydrogenase (A-DH). The elimination of acrolein from the aldehydes is a spontaneous reaction. (Reprinted by permission from *Med. Biol.* **59,** 334–346, 1981, The Finnish Medical Society Duodecim.)

2. *Diamine Oxidase*

Copper-containing amine oxidases which preferentially oxidize diamines (putrescine, cadaverine) and histamine are termed diamine oxidases. Histaminase is now often used as a synonym (Mondovi *et al.*, 1975), although there are enzymes such as pig plasma benzylamine oxidase for which histamine is a good substrate, whereas putrescine and cadaverine are not oxidized (Buffoni, 1966). Moroever, the temperature inactivation curves and the inhibitory effect of $AgNO_3$ on diamine oxidase from hog kidney differed greatly when using cadaverine and histamine as substrates (Suetsugu, 1974).

Gorkin and his collaborators have demonstrated (Gorkin *et al.*, 1971) that treatment of diamine oxidase from hog kidney with reducing agents partially inhibits its enzymatic activity with putrescine as substrate, while inducing its ability to deaminate monoamines. Conversely, by treatment with Cu^{2+} in aerobic conditions, monoamine oxidase can be transformed into an enzyme which oxidizes diamines. This suggests that changes in substrate specificity may occur during purification procedures, which may add to the present confusion about the identity or nonidentity of the diamine oxidase activities which were isolated from different sources (Buffoni, 1966; Zeller, 1963a; Kapeller-Adler, 1970).

The enzyme isolated from pig kidney has a minimum molecular weight of 90,000, calculated on the basis of its copper content, but the actual molecular weight as obtained by hydrodynamic methods does not fit any multiples of this figure and depends on the experimental conditions (Yasunobu *et al.*, 1976; Pionetti, 1974; Kluetz and Schmidt, 1977).

Traditionally diamine oxidase is considered as a cytoplasmic enzyme (Zeller, 1963b), but it has recently been demonstrated that 60% of the enzyme in rabbit kidney cortex is bound to the membranes of the endoplasmic reticulum (Sartori *et al.*, 1983).

There is photometric evidence that hog kidney diamine oxidase also has pyrroloquinoline quinone as its prosthetic group (Ameyama *et al.*, 1984). Therefore, it will be necessary to revise all previous mechanistic considerations.

Zeller was the first to point out that hog kidney diamine oxidase attacks cadaverine and putrescine rapidly, spermidine more slowly, and spermine hardly at all (Zeller *et al.*, 1939), but crude kidney preparations from pork, guinea pig, and sheep displayed a minute capacity for deaminating spermine (Hirsch, 1953; Barsky, 1959). Guarnieri *et al.* (1977) described an electrophoretically separated protein band from the same source which more actively oxidized spermidine than it did putrescine.

The oxidation of putrescine, spermidine, and spermine is also reported for a mouse liver enzyme preparation (Syatkin and Galaev, 1977), and an amine oxidase of M_r 182,000 has been purified to homogeneity from human seminal plasma (Hölttä *et al.*, 1975). The following K_m values were determined for the latter enzyme: putrescine, 13 μM; spermidine, 0.56 mM; and spermine, 0.1 mM.

Diamine oxidase activities are found in most organisms but occur only in selected tissues (Shaff and Beaven, 1976), with highest activity in placenta and small intestines and very low activity in vertebrate brains.

Much attention has been paid to the localization of diamine oxidase in intestine (Shakir *et al.*, 1977; Dagostini *et al.*, 1984; Bieganski *et al.*, 1983), because functional interrelations are suspected between this enzyme and the enormous cell turnover in intestinal mucosa. Interestingly, diamine oxidase (Shakir *et al.*, 1977) and ornithine decarboxylase (Porter *et al.*, 1980) show a similar distribution pattern in this organ, but an inverse relationship in activity during postnatal development. The activity of diamine oxidase increases at the time ornithine decarboxylase activity decreases (Jänne and Hölttä, 1973).

Further evidence for interrelationships between enhanced growth and diamine oxidase activity comes from observations of elevated diamine oxidase activity in numerous tumors and tumor cell lines (Borglin and Willert, 1962; Baylin *et al.*, 1970, 1975; Lin *et al.*, 1975; Kusche *et al.*, 1980; Perin *et al.*, 1981; Thomassek *et al.*, 1982) and enhanced diamine oxidase activities in tissues with enhanced normal growth, such as regenerating liver and hypertrophic heart or kidney (Perin *et al.*, 1981, 1983; Desiderio *et al.*, 1982; Sessa *et al.*, 1982). These findings lead to the suggestion that diamine oxidase activity is involved in the regulation of putrescine levels in growing tissue (Perin *et al.*, 1984).

B. Inhibitors

A common characteristic of all copper amine oxidases is their sensitivity against copper chelating agents and against carbonyl reagents, for which semicarbazide and aminoguanidine are prototypes. A collection of this type of compound was summarized by Buffoni (1966). Into this category of compounds belong methylglyoxal bis(guanylhydrazone) and its analogs, which are also effective inhibitors of *S*-adenosylmethionine decarboxylase (Pegg, 1978).

There are also numerous substrate analogs which act as inhibitors (Bardsley and Ashford; 1972). Among these, 1,4-diamino-2-butanone and 1,5-diamino-3-pentanone deserve special mention because of their close structural relationship with putrescine and cadaverine (Macholan, 1974).

Although it has not been firmly established, it is likely that all copper amine oxidases have pyrroloquinoline quinone as cofactor, and that carbonyl reagents and substrate analogs react with the cofactor as indicated in Fig. 2 for a hydrazine, and thus prevent binding of the substrate.

A third category of inhibitors are enzyme-activated irreversible inhibitors (suicide inactivators) (Abeles and Maycok, 1976; Tobes *et al.*, 1983) such as 1,4-diaminobutyne-2,β-bromoethylamine and glycine phenyl esters. Disregarding aminoacetonitrile, these compounds have been mainly used in investigations of mechanisms.

There are considerable differences with respect to specificity. For example, 3-hydroxybenzyloxyamine is *in vitro* a much more potent inhibitor of bovine serum amine oxidase than is aminoguanidine (Seiler *et al.*, 1981c), while both compounds are very potent inhibitors of diamine oxidase (Zeller, 1963a; Gaugas, 1981). Therefore, these inhibitors can be used to distinguish between the two enzymes.

Very few inhibitors of the copper amine oxidases have been used *in vivo*, and only aminoguanidine seems to have been administered to humans (Lindell *et al.*, 1960; Baylin *et al.*, 1970, 1972). As will be shown below, aminoguanidine is suitable to inhibit all known oxidative deaminations of the polyamines and their derivatives, and since it seems nontoxic, it has been the preferred tool in the *in vivo* studies of oxidative polyamine catabolism.

C. Biological Consequences of Inhibition of Copper Amine Oxidases

1. Putrescine Catabolism

The formation of Δ^1-pyrroline from putrescine (Fig. 4) and of Δ^1-tetrahydropyridine from cadaverine has been studied *in vitro* in great detail and is used in the method of Okuyuma and Kobayashi, published in 1961, for the determination of diamine oxidase activity. However, the metabolic fate of exogenously administered, compared with endogenous, putrescine was clarified considerably later.

If trace amounts of labeled putrescine are administered to rodents, about 40% of the labeled carbon appears within 2 hr in the expired air (Jänne, 1967; Seiler and Eichentopf, 1975). Formation of $^{14}CO_2$ from labeled putrescine is prevented completely by administration of aminoguanidine and also by inhibitors of 4-aminobutyrate: 2-oxoglutarate aminotransferase. The latter treatment causes the accumulation of labeled 4-aminobutyric acid in tissue and enhanced urinary excretion of this amino acid (Seiler and Eichentopf, 1975). These findings were interpreted as follows: Putrescine is oxidatively deaminated, forming 4-aminobutyraldehyde, which in the absence of an aldehyde dehydrogenase is spontaneously converted into Δ^1-pyrroline. In tissues, however, 4-aminobutyric acid is formed (Fig. 4) which may be further metabolized via the tricarboxylic acid cycle. These findings have been confirmed (Sourkes and Missala 1981; Fogel *et al.*, 1981). A 60-mg/kg dose of pargyline prevented partially $^{14}CO_2$ formation from [^{14}C]putrescine. This was initially confusing, because it seemed to indicate a role of monoamine oxidase in putrescine catabolism. However, it is clear now that a metabolite formed from pargyline *in vivo* causes the inhibition of aldehyde dehydrogenase(s) (Lebsack and Anderson, 1979; Shirota *et al.*, 1979) and prevents in this way the conversion of 4-aminobutyraldehyde into 4-aminobutyric acid. In addition to 4-aminobutyrate, putrescine is also converted into pyrrolidin-2-one (Fig. 4) (Lundgren *et al.*, 1980).

After a single 0.2-mmol/kg dose of aminoguanidine (25 mg/kg aminoguanidine sulfate), putrescine levels in small intestine of mice increased by about 45% within 1 hr and remained elevated during the following 24

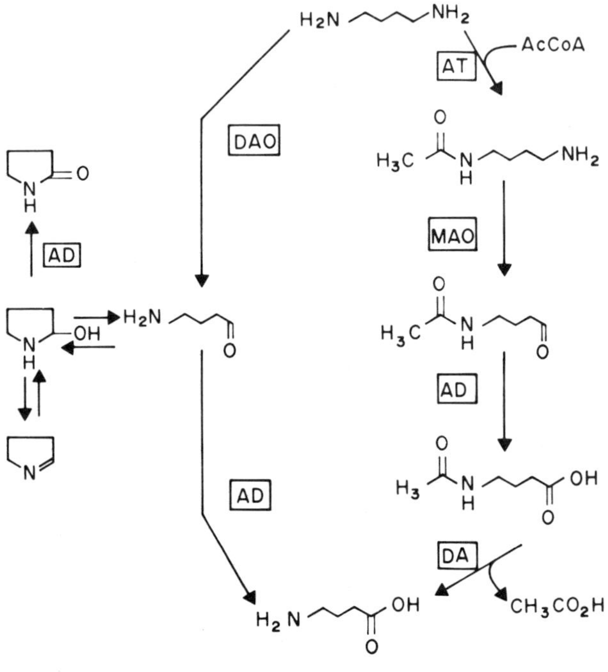

Fig. 4. Pathways of oxidative deamination of putrescine. Abbreviations: DAO, diamine oxidase (EC 1.4.3.6); MAO, monoamine oxidase (EC 1.4.3.4.); AD, aldehyde dehydrogenase; DA, deacetylase (hydrolase); AT, acetyl-CoA : spermidine N^8-acetyltransferase (nuclear; putrescine is a substrate of this enzyme). [Reprinted by permission from "Structure and Function of Amine Oxidases" (B. Mondovi, ed.), 1986, CRC Press, Inc.]

hr. No other tissue accumulated putrescine levels. Spermidine concentration was significantly elevated in intestines at 24 hr. Precisely the same results were obtained after administration of 0.5 mmol/kg aminoacetonitrile (Seiler *et al.*, 1983), an inhibitor of copper amine oxidases which has been reported to penetrate into brain (Riceberg *et al.*, 1975). It appears from these observations that the small intestine is the major site of *in vivo* putrescine catabolism by diamine oxidase. Fogel *et al.* (1981) arrive at the same conclusion from the tissue distribution of labeled putrescine catabolites.

In contrast with normal kidney, in kidney undergoing hypertrophy, putrescine accumulates to a significant extent if diamine oxidase is inhibited (Desiderio *et al.*, 1984). This finding again supports the notion that this enzyme may be involved in the limitation of putrescine levels in growing tissue.

Comparison of the urinary polyamine patterns of rats before and after treatment with aminoguanidine (Fig. 5) shows that about 70% of the putrescine that is formed in the body and needs to be eliminated is oxidized by diamine oxidase, and only 30% appears in the urine (Seiler *et al.*, 1985a,b). The excreted proportion of putrescine is even smaller in situations where there is enhanced diamine oxidase activity: it can be calculated from the values reported by Rojansky *et al.* (1979), who showed that rats excrete, around day 20 of pregnancy, only about 13% of the total amount of putrescine destined to be eliminated from the body.

Recent data obtained from two male human subjects taking 50 mg aminoguanidine four times a day for 2 days also showed enhancement of urinary putrescine excretion by 56 and 192%, respectively (Chayen *et al.*, 1985), demonstrating that *in vivo* metabolism of putrescine in humans is not basically different from that in experimental animals. Urinary cadaverine excretion is similarly enhanced following inhibition of diamine oxidase (Seiler *et al.*, 1985a,b; Chayen *et al.*, 1985).

In tumor-bearing mice treated with a combination of α-difluoromethylornithine (Koch-Weser *et al.*, 1981) and methylglyoxal bis(guanylhydrazone), the polyamine pattern of the tumor cells normalized and a considerable accumulation of cadaverine in the leukemia cells was noticed at the same time, due to the capacity of the diguanide to inhibit diamine oxidase (Pegg and McGill, 1978). However, in spite of the normalization of the tumor cell polyamine pattern, the combination of the two drugs produced a growth inhibitory effect which was not achieved with either of the compounds alone (Kallio and Jänne, 1983).

No experimental data exist for effects on putrescine metabolism of long-term administration of aminoguanidine or other inhibitors of diamine oxidase.

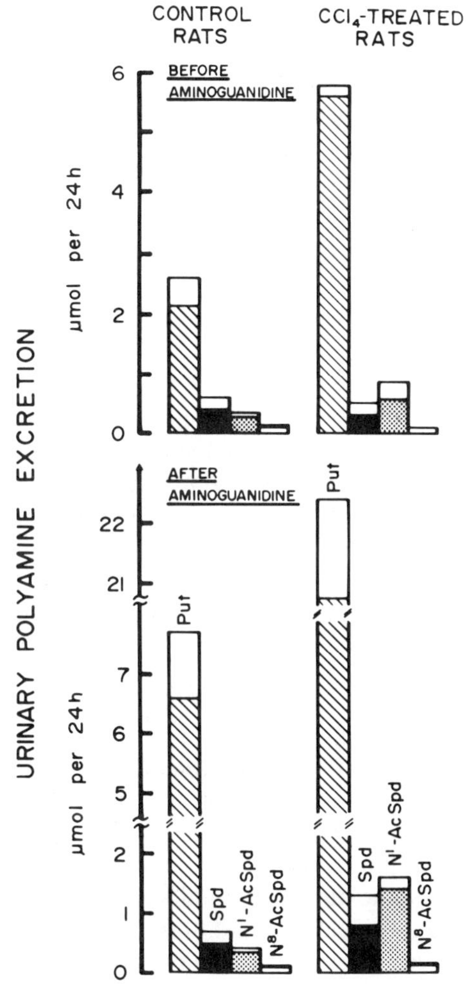

Fig. 5. The effect of treatment with aminoguanidine and carbon tetrachloride on the urinary excretion of polyamines by rats. Groups of three rats received either physiological, saline, or 100 mg/kg aminoguanidine sulfate, or 2 ml/kg carbon tetrachloride, or a combination of these drugs. The bars indicate average daily excretion during the 24 hr following treatment ± SD. Abbreviations: Put, putrescine; Spd, spermidine; N^1-AcSpd, N^1-acetylspermidine; N^8-AcSpd, N^8-acetylspermidine. [Reprinted by permission from "Structure and Function of Amine Oxidases" (B. Mondovi, ed.), 1986, CRC Press, Inc.]

2. *Polyamine Catabolism*

It has been stated that polyamines and their acetyl derivatives are *in vitro* substrates of diamine oxidases from different sources (Suzuki *et al.*, 1981; Gahl and Pitot, 1981, 1982). The following observations suggest that *in vivo,* diamine oxidase or closely related enzymes with the same sensitivity to aminoguanidine are responsible for the oxidation of polyamines.

a. Exogenous Polyamines. Administration of spermidine to mice is followed by the accumulation of putreanine and isoputreanine lactam in liver, kidney, and small intestines. Spermine administration causes the accumulation of N^8-(2-carboxyethyl)spermidine (Seiler *et al.*, 1981c, 1982). These observations suggest that the tissue enzymes involved in the oxidative deamination of the polyamines are attacking both primary amino groups, in contrast with the serum amine oxidase. However, it has not been clarified whether one or several enzymes are involved in these reactions. Since tissues obviously contain suitable aldehyde dehydrogenases, conversion of spermidine into putreanine and isoputreanine lactam [*N*-(3-aminopropyl)-2-pyrrolidinone] can be formulated as shown in Fig. 6. Isoputreanine is the hydrolysis product of the lactam. N^1-Acetylspermidine is accordingly converted into *N*-(3-acetamidopropyl)-2-pyrrolidinone (Seiler *et al.*, 1982).

After an intraperitoneal dose of 1 mmol/kg spermidine the rate of putreanine accumulation is constant for more than 30 min. The amount of putreanine in liver 30 min after spermidine administration is therefore a suitable measure for the *in vivo* rate of spermidine oxidation. With increasing doses of aminoguanidine and aminoacetonitrile, conversion of spermidine into putreanine is increasingly inhibited, as shown in Fig. 7. In the case of aminoguanidine only 0.16 μmol/kg was needed to achieve 50% inhibition. This amount is similar to that (0.07 μmol/kg) required to reduce diamine oxidase activity in rat intestines by 50% (Burkard *et al.*, 1960).

b. Endogenous Polyamines. First evidence for the fact that spermidine, in addition to putrescine, is attacked by an aminoguanidine-sensitive oxidase was reported by Rojansky *et al.* (1979), who demonstrated augmented excretion of spermidine by pregnant rats which were pretreated with aminoguanidine.

Spermidine and N^1-acetylspermidine excretion is enhanced after inhibition of the copper amine oxidases, whereas excretion of N^8-acetylspermidine seems not to be affected by this treatment (Seiler *et al.*, 1983, 1985a,b). A logical consequence of the above described metabolic interrelationships is the decreased excretion of isoputreanine lactam observed under these conditions (Seiler and Knödgen, 1983).

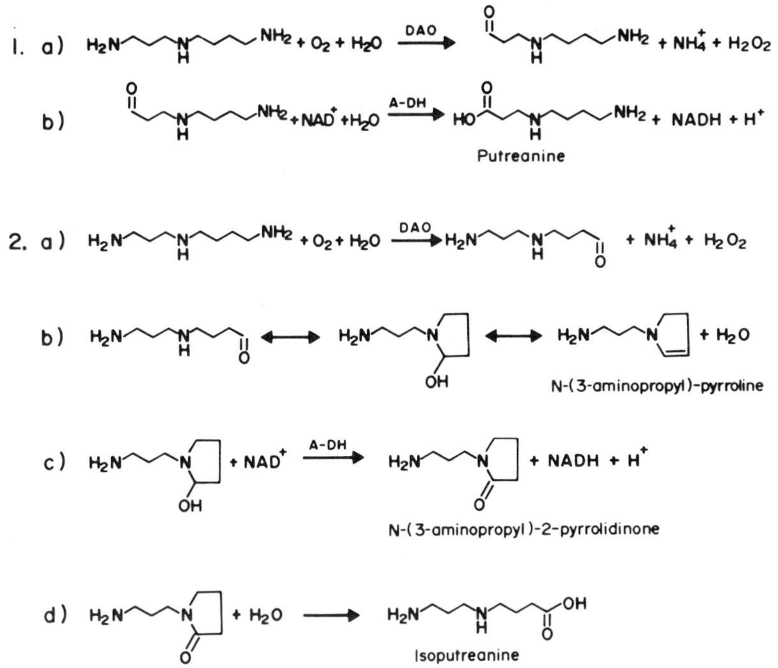

Fig. 6. Reactions involved in putreanine and isoputreanine formation from spermidine. Conversion of *N*-(3-aminopropyl)pyrrolidin-2-one into isoputreanine (step 2d) is achieved by acid hydrolysis. Abbreviations: DAO, diamine oxidase (EC 1.4.3.6) or a closely related aminoguanidine-sensitive enzyme; A-DH, aldehyde dehydrogenase or an oxidase capable of converting the aldehydes into carboxylic acids and the lactam, respectively. (Reprinted by permission from *Biochem. J.* **208,** 189–197, 1982, The Biochemical Society, London.)

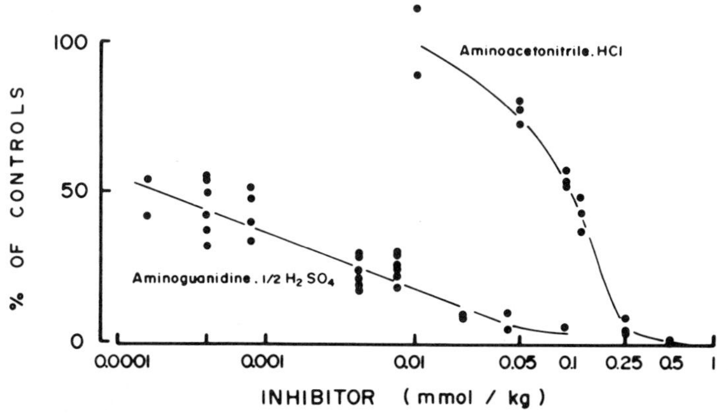

Fig. 7. Inhibition of the conversion of exogenous spermidine into putreanine in mouse liver by aminoguanidine and aminoacetonitrile. The inhibitors were given intraperitoneally 1 hr before ip administration of 1 mmol/kg spermidine·3HCl. Each point represents the result obtained from a separate animal. (Reprinted by permission from *Adv. Polyamine Res.* **4,** 135–154, 1983, Raven Press.)

The proportion of spermidine and N^1-acetylspermine which is oxidatively deaminated by copper amine oxidases is normally small compared with the excreted amount (Fig. 5), but this proportion may increase considerably in certain physiological and pathological conditions. Pregnancy has already been mentioned. Another situation for the enhanced oxidative deamination of both spermidine and N^1-acetylspermidine is given in the following example:

Intoxication of mice or rats with 2 ml/kg of carbon tetrachloride causes enhanced formation of N^1-acetylspermidine and N^1-acetylspermine in the liver (Seiler *et al.*, 1981e; Abdel-Monem and Merdink, 1981). Putrescine concentrations also increase, whereas spermidine and spermine concentrations decrease during the first 24 hr. These changes are reflected in the urinary polyamine pattern (Fig. 5) (Seiler *et al.*, 1983). Putrescine and N^1-acetylspermidine excretion is signficantly enhanced under these conditions. If aminoguanidine was administered prior to liver intoxication with carbon tetrachloride, a doubling of spermidine excretion and about a fourfold increase of N^1-acetylspermidine concentration in urine are observed. These observations are compatible with a function of the amine oxidases in the detoxification of excessive amounts of the polyamines and their derivatives.

There is no doubt that the same principles of catabolism are also active in humans. In contrast with rodents, humans excrete polyamines nearly exclusively in the form of their monoacetyl derivatives (Seiler *et al.*, 1981d). The observed enhancement of spermidine in urine hydrolysates after aminoguanidine ingestion (Chayen *et al.*, 1985) most probably reflects, therefore, the enhanced excretion of N^1-acetylspermidine.

Spermine is excreted by rodents only in trace amounts; in contrast, the amounts excreted by humans in nonconjugated form are easily measurable and are usually higher in males than in females (Seiler *et al.*, 1981d). Treatment with aminoguanidine did not significantly enhance spermine excretion in rodents (Seiler *et al.*, 1985a,b) and gave inconsistent results in humans (Chayen *et al.*, 1985). Not even the intoxication of mice with carbon tetrachloride, which resulted in the accumulation of N^1-acetylspermine in the liver (Seiler *et al.*, 1981e), caused an enhancement of spermine or spermine conjugate excretion. On the other hand, it is known that typical reaction products of the oxidative deamination of spermine, N^8-(2-carboxyethyl)spermidine, and spermic acid [*N*,*N*′-bis(2-carboxyethyl)-1,4-butanediamine] are formed from exogenous polyamines and are found in urine (Nakajima *et al.*, 1980; Van den Berg *et al.*, 1984). This apparent dichotomy needs resolution. Determination of the spermine-derived amino acids in urine before and after treatment with aminoguanidine is probably the most direct way to obtain evidence for the physiological significance of oxidative deamination of spermine.

c. Putreanine in Brain. Putreanine (Fig. 6) is a normal tissue constituent. Although it is formed from exogenous spermidine in peripheral organs and is excreted (Nakajima *et al.*, 1980; Seiler *et al.*, 1981c; 1982; Van den Berg *et al.*, 1984), brain is the only organ with signficant endogenous putreanine concentrations (Kakimoto *et al.*, 1969). Its late appearance during brain development (Kremzner and Sturman, 1979; Seiler *et al.*, 1985d) has not yet been correlated with the appearance of a suitable amine oxidase. Treatment of mice with aminoguanidine or aminoacetonitrile for 5 days produced no significant change in brain putreanine levels (Seiler *et al.*, 1983). This may indicate either the inadequacy of the presently available inhibitors of copper amine oxidases to inhibit the oxidases in brain, a very slow turnover rate of putreanine, or an alternative pathway for putreanine formation in brain. The unique occurrence of putreanine in brain may indicate a specific function. Therefore, it seems of considerable importance to direct more attention to this amino acid.

III. FLAVIN-DEPENDENT ENZYMES IN POLYAMINE CATABOLISM

A. Monoamine Oxidase

The activity of diamine oxidase in vertebrate brains is low (Burkard *et al.*, 1963), but putrescine is rapidly converted within the brain into 4-aminobutyrate (Seiler *et al.*, 1971). In order to resolve this discrepancy, it was postulated that putrescine can undergo a second sequence of metabolic reactions, which are shown in Fig. 4. The difference between the diamine oxidase-catalyzed and the monoamine oxidase-catalyzed pathway is that the latter requires monoacetylation of putrescine, monoacetylputrescine being a substrate of monoamine oxidase (Seiler and Al-Therib, 1974a). In addition, the following arguments are in favor of the proposed scheme:

1. Monoacetylputrescine is a natural constituent of vertebrate brains (Perry *et al.*, 1967; Seiler *et al.*, 1973).

2. Brain (and other tissues) contain an acetyltransferase which is capable of acetylating putrescine (Seiler and Al-Therib, 1974b). This enzyme also possesses spermidine N^8-acetyltransferase activity (Blankenship and Walle, 1978).

3. Brain homogenates and brain cell suspensions do not effectively convert putrescine into 4-aminobutyric acid, unless acetylation is favored by addition of acetyl CoA to the reaction mixture (Seiler *et al.*, 1979).

4. Pargyline inhibits in mouse brain the conversion of putrescine and of monocetylputrescine into 4-aminobutyric acid to an equal extent (Seiler and Al-Therib, 1974a).

The oxidative deamination of putrescine in brain is, under physiological conditions, a minor process. Therefore, inhibition of monoamine oxidase, even for an extended period of time, is unlikely to have any effects on polyamine metabolism.

B. Polyamine Oxidase

The enzyme which has been shown to catalyze the oxidative cleavage of N^1-acetylspermidine and N^1-acetylspermine (Fig. 1) has been purified to homogeneity from rat liver (Hölttä, 1977). Its localization in peroxisomes has been established for liver but not for other organs.

Polyamine oxidase activity is found in all tissues, including tumor cells, usually at relatively high activity (Table I) (Seiler *et al.*, 1980). Its ubiquitous occurrence is in agreement with the assumption that it is involved in normal polyamine turnover.

The molecular weight as obtained by density gradient centrifugation, gel filtration, and gel electrophoresis is 60,000. The enzyme has a tightly bound FAD and there is evidence that Fe^{2+} may be a metal cofactor (Hölttä, 1983). In contrast with the acetyltransferase (Pegg and Erwin, 1985; Pegg *et al.*, 1985), polyamine oxidase seems not to be an inducible enzyme.

Although the acetyl derivatives of spermidine and spermine are the natural substrates (Bolkenius and Seiler, 1981), the enzyme tolerates other acyl residues as well. Even monodansylspermine is oxidatively cleaved to form spermidine and *N*-dansyl-3-aminopropionaldehyde. Acylation of the aminopropyl moiety and thus removal of the positive charge of N^1 obviously enhances the affinity and the maximum reaction rate, as determined from the comparison of the kinetic data of N^1-acetylspermine and spermine and of N^1-acetylspermidine and spermidine (Table II).

The partial purification of a polyamine oxidase from human pregnancy serum was recently reported (Morgan, 1985b). The enzyme preparation contained a major electrophoretically separable band (M_r 67,000) which contained copper and iron. Since it was capable of cleaving N^1,N^{12}-diacetylspermine to N^1-acetylspermidine, it is likely that this serum enzyme is identical with or similar to the intracellular polyamine oxidase.

Disregarding the fact that it is not localized in mitochondria, polyamine oxidase in many ways resembles monoamine oxidase. The reaction mechanism has not been studied, but one may assume that the reaction starts,

TABLE I

Polyamine Oxidase Activity in Rat Tissues

Organ	Enzyme activity[a]
Pancreas	22.9 ± 2.2 (3)
Liver	12.1 ± 1.7 (14)
Spleen	9.1 ± 1.6 (7)
Kidney	8.9 ± 1.3 (6)
Small intestine	6.5 ± 0.9 (6)
Testes	5.9 ± 0.5 (5)
Thymus	5.6 ± 1.2 (3)
Ventral prostate	4.1 ± 0.5 (3)
Brain	3.8 ± 1.3 (6)
Lung	2.6 ± 1.1 (4)
Heart	1.2 ± 0.2 (3)
Skeletal muscle	0.6 ± 0.2 (1)
Serum	0.4 ± 0.02 (2)

[a] As micromoles N^1-acetylspermidine formed from 0.56 m*M* N^1,N^{12}-diacetylspermine per gram of wet tissue per hour; number of animals is given in parentheses. Data from Seiler *et al.* (1980).

as is the case for monoamine oxidase, by the elimination of a hydrogen atom which is adjacent to the amino group. Both enzymes require molecular oxygen as electron acceptor, both produce hydroperoxide, and both have the same cofactors. It is therefore not surprising that both enzymes are inhibited by pargyline.

TABLE II

Substrate Properties of Polyamines and Some Acetylderivatives

Substrate	Reaction product	K_m (μM)	V_{max} (nmol/min/mg protein)
N^1-Acetylspermine	Spermidine	0.6	2.8
N^1,N^{12}-Diacetylspermine	N^1-Acetyl-spermidine	5.0	4.0
N^1-Acetylspermidine	Putrescine	14	2.5
Spermine	Spermidine	40	0.1
Spermidine	Putrescine	—[a]	—[a]

[a] Below detection limit of the analytical method. Data from Bolkenius and Seiler (1981).

C. Inhibitors of Tissue Polyamine Oxidase

1. Inhibitors Acting on Cofactors

Quinacrine [N^4-(6-chloro-2-methoxy-9-acridinyl)-N^1,N^1-diethyl-1,4,-pentanediamine] is an FAD analog. At 0.1 mM it inhibits polyamine oxidase activity.

In contrast with copper chelators, the iron chelators, α,α-dipyridyl, 8-hydroxyquinoline, and *o*-phenanthroline are effective inhibitors (Hölttä, 1977). However, neither quinacrine nor the iron chelators are specific for polyamine oxidase and are therefore of little use in the elucidation of its physiological role.

2. Enzyme-Activated Irreversible Inhibitors

Hölttä (1977) had already pointed out that pargyline (*N*-methyl-*N*-benzylpropargylamine), the well-known inhibitor of monoamine oxidase, is an inhibitor of polyamine oxidase, although a weak one. Taking this lead and the considerations of Krantz *et al.* (1979) concerning the reaction of propargylic and allenic amines with monoamine oxidase, a series of putrescine derivatives substituted on nitrogen with the same unsaturated groups were prepared (Bey *et al.*, 1985). All these compounds were inhibitors of polyamine oxidase, but with greatly differing potencies. In the case of monoamine oxidase, propargylic amines are more potent than allylic and allenic amines (Krantz *et al.*, 1979). The situation is exactly opposite in the case of polyamine oxidase.

The two most potent compounds of the series, *N*-methyl-*N*′-(2,3-butadienyl)-1,4-butanediamine (MDL 72521) and *N*,*N*′-bis(2,3-butadienyl)-1,4-butanediamine (MDL 72527) (Fig. 8) were studied in some detail.

Incubation of the compounds with a partially purified polyamine oxidase from pig liver resulted in a time- and dose-dependent loss of enzyme activity that followed pseudo first-order kinetics for at least two half-lives. Exhaustive dialysis or incubation with the substrate, N^1,N^{12}-diacetylspermine, of the inactivated enzyme did not restore activity. Preincubation of polyamine oxidase with inhibitors in the presence of the substrate provided protection against inactivation, demonstrating that substrate and inhibitors were competing for the enzyme's active site. The minimum half-life ($\tau_{1/2}$) of the enzyme and the apparent dissociation constant K_i under saturating conditions are shown in Fig. 8. The kinetic and inhibitory characteristics are compatible with the assumption that the butadienyl-1,4-butanediamines are enzyme-activated irreversible inhibitors (suicide inactivators) of polyamine oxidase.

In accordance to this postulated mode of action the two compounds show high selectivity for polyamine oxidase. They do not inhibit any

	K_i (μM)	$\tau_{1/2}$ (min)
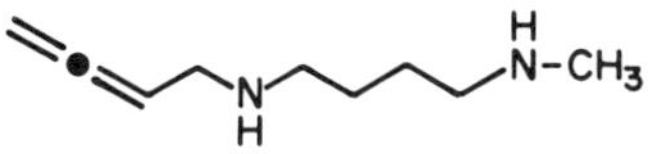 MDL 72521	0.3	0.5
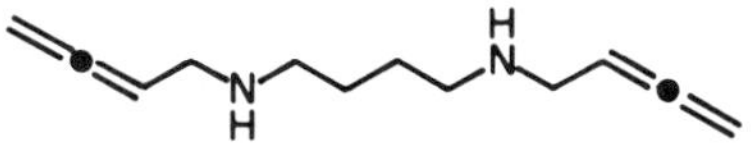 MDL 72527	0.1	2.2

Fig. 8. Structural formulas and kinetic parameters of the polyamine oxidase inhibitors *N*-methyl-*N'*-(2,3-butadienyl)-1,4-butanediamine (MDL 72521) and *N,N'*-bis(2,3-butadienyl)-1,4-butanediamine (MDL 72527). Partially purified polyamine oxidase from pig liver was mixed at time zero with a solution of the inhibitor in borate buffer (pH 9) and the solution was incubated at 30° C. At different time intervals 10-μl aliquots were transferred to 200 μl of assay medium containing 0.56 m*M* N^1,N^{12}-diacetylspermine as substrate, and the residual enzyme activity was determined by determination of H_2O_2 formation. The apparent dissociation constant (K_i) and the enzyme half-life under saturating conditions ($\tau_{1/2}$) were determined according to the method of Kitz and Wilson (1962). [Data from Bey *et al.* (1985).]

other enzyme involved in polyamine metabolism, except N^8-acetylspermidine deacetylase, for which they are reversible inhibitors. However, as has been shown (Bolkenius *et al.*, 1985), even repeated administrations of the inhibitors to experimental animals in doses which inactivate polyamine oxidase almost completely did not produce enhanced N^8-acetylspermidine concentrations in tissues and urine.

IV. BIOLOGICAL CONSEQUENCES OF POLYAMINE OXIDASE INACTIVATION

A. Metabolic Effects on Cultured Cells

Very little is known of the biological consequences of specific inhibition of polyamine oxidase in cell lines. As a matter of fact, the only experiments have been with HTC cells (Mamont *et al.*, 1984; Seiler *et al.*, 1985c) a tumor cell line originating from a Morris hepatoma. These cells have a relatively high polyamine oxidase activity which is independent of the rate of cell growth (Seiler *et al.*, 1980).

Addition of MDL 72521 inactivated polyamine oxidase in HTC cells completely and irreversibly if added to the culture medium at 10 μM or greater concentrations. Following polyamine oxidase inactivation the intracellular concentrations of N^1-acetylspermidine and N^1-acetylspermine increased with time, from undetectable levels at day 0 to 90 and 75 pmol per 10^6 cells, respectively, on day 3 of cell culture. N^8-Acetylspermidine concentration remained unmeasurably low under these conditions, and spermine and spermidine concentrations were not significantly altered. These findings are strong evidence for the assumption that the N^1-acetyl derivatives of spermidine and spermine, but not the nonderivatized polyamines, are the natural substrates of polyamine oxidase.

Cell growth rates were not affected after the inactivation of polyamine oxidase and the consecutive accumulation of the monoacetyl derivatives of spermidine and spermine. Thus it is apparent that polyamine oxidase is not of vital importance for the cells, and even considerable intracellular accumulations of N^1-acetylspermidine and N^1-acetylspermine can be tolerated without any obvious disadvantage.

B. *In Vivo* Effects in Rodents

1. In Vivo Potency of MDL 72521 and MDL 72527

Administration of single doses of MDL 72521 and 72527 to rats and mice inactivated polyamine oxidase in all tissues in a time- and dose-dependent manner. The two compounds were equipotent in the inactivation of polyamine oxidase in peripheral tissues. However, presumably due to its higher lipophilicity, *N,N'*-bis(2,3-butadienyl)-1,4-butanediamine (MDL 72527) is a more efficient inhibitor in brain than MDL 72521: for inactivation of 50% of the enzyme in mouse brain an ip dose of 0.25 mg/kg MDL 72527 is required; in order to achieve the same effect with MDL 72521, a 20-fold greater dose (5 mg/kg) is needed. For complete inactivation of the enzyme for several hours in all organs of rat and mice, the ip administration of 20 mg/kg of MDL 72527 is sufficient (Bolkenius *et al.*, 1985). Inactivation is maximal within 30 min, and polyamine oxidase activity remains for about 24 hr below the detection limit of the assay. Thereafter one observes a gradual reappearance of polyamine oxidase activity. In liver 50% of the normal enzyme activity is observed after about 6 days. In brain enzyme activity remains below normal for more than 2 weeks (Bolkenius *et al.*, 1985). After administration of MDL 72521 recovery of polyamine oxidase activity is faster ($\tau_{1/2}$ = 2–3 days). Evidence for a biological half-life of polyamine oxidase in the order of several days, i.e., a slow turnover, has previously been obtained in mouse skin by inhibition of protein biosynthesis with cycloheximide (Seiler *et al.*, 1980).

2. Tissue Polyamine Levels after Inactivation of Polyamine Oxidase

If polyamine oxidase is inactivated, one expects an increase in the concentrations of the substrates of this enzyme and a decrease in the levels of the reaction products which is proportional to their rate of formation. These expectations are indeed fulfilled. Not only are the concentrations of N^1-acetylspermidine and N^1-acetylspermine elevated, dependent on the dose of inactivator, but putrescine and spermidine concentrations decrease at the same time, although a statistically significant decrease in spermidine concentrations of about 20% has been observed only in the brain and testes of mice after daily treatment with 20 mg/kg MDL 72727 for 6 weeks (Bolkenius *et al.*, 1985). The most striking result of this experiment was the fact that the animals with a totally inactivated polyamine oxidase showed normal weight gain and a normal behavior, suggesting that polyamine oxidase is not essential, despite its obvious metabolic function in polyamine catabolism.

The greatest changes in polyamine concentrations were observed in brain. A daily dose of 20 mg/kg of MDL 72527 produced a linear increase of N^1-acetylspermidine (1.1 nmol/g/hr) and N^1-acetylspermine (0.05 nmol/g/hr) concentrations for about 72 hr. Putrescine concentration decreased during the first 8 hr with the rate of N^1-acetylspermidine accumulation; its subsequent decrease was much slower. At 24 hr a level of about 30% of the control value was reached. This suggests that in brain normally only about 30% of the putrescine is derived from ornithine by decarboxylation, whereas about 70% is formed by degradation of spermidine (Seiler and Bolkenius, 1985). This portion of putrescine is reused for spermidine biosynthesis, in agreement with the previously published hypothesis of polyamine reutilization (Antrup and Seiler, 1980).

After a single dose of MDL 72527 brain putrescine levels were somewhat lower (Seiler and Bolkenius, 1985) than after long-term treatment with this drug (Bolkenius *et al.*, 1985). Therefore, it cannot be excluded that a small increase of ornithine decarboxylase activity partially compensates for the decreased formation of putrescine via the interconversion pathway, although ornithine decarboxylase activity was not measurably enhanced in liver or brain during chronic administration of the polyamine oxidase inactivator (Bolkenius *et al.*, 1985), nor were the activities of other enzymes of polyamine metabolism (*S*-adenosylmethionine decarboxylase; acetyl-CoA : polyamine N^1-acetyltransferase; spermidine and spermine synthase) measurably altered.

Polyamines do not penetrate the blood–brain barrier very well (Shin *et al.*, 1985), and since N^1-acetylspermidine accumulates linearly in brain for 3 days from nondetectable levels to a final concentration of 60 nmol/g, it can be assumed that its elimination from brain is slow compared with its formation. Moreover, in brain no rapid degradative processes exist for this compound. Therefore, one may assume that the rate of N^1-acetylspermidine accumulation in brain is not only a direct measure for the rate of spermidine degradation, but also a measure for polyamine turnover. From the experimental data, an average half-life of 8 days can be calculated for spermidine (Seiler and Bolkenius, 1985).

Similar considerations might be valid for N^1-acetylspermine. However, since this compound accumulates very gradually, even slow elimination processes have a great effect on the data derived from N^1-acetylspermine concentration measurements.

From this observation *in vivo* of the consequences of polyamine oxidase inactivation, it is evident that the compounds are useful in recognizing enhanced formation of N^1-acetylspermidine and N^1-acetylspermine. The method is complementary to the *ex vivo* determination of the activity of acetyl-coA : polyamine N^1-acetyltransferase. Obviously, quantitative determination of *in vivo* rates of polyamine acetylation can only be made in metabolically closed systems, of which the brain is an example. Semi-quantitative data, however, can be obtained in other tissues as well. It had previously been shown, for example, that treatment of rats with secretin (Danzin *et al.*, 1982) or an inhibitor of cyclic nucleotide phosphodiesterase such as 3-isobutyl-1-methylxanthine (IBMX) (Danzin *et al.*, 1983) causes a dramatic transient increase of putrescine concentration in the pancreas. Pretreatment of the rats with the polyamine oxidase inhibitor MDL 72527 demonstrated that the formation of putrescine after IBMX was clearly due to induction of the interconversion pathway: N^1-acetylspermidine levels increased in pancreas from 47 ± 6 to 385 ± 120 nmol/g (Danzin *et al.*, 1984; Seiler *et al.*, 1985c). In these examples the induced activity of the interconversion pathway is so enormous that even the accumulation of β-alanine, the product of the aminopropyl moiety of spermidine and spermine, could be directly demonstrated; β-alanine accumulation was abolished by inactivation of polyamine oxidase, as predicted.

3. *Polyamine Oxidase Inactivation and Urinary Polyamine Excretion*

The relatively small increase of N^1-acetylspermidine and N^1-acetylspermine concentrations in tissues other than brain after inactivation of polyamine oxidase is in principle open to three different interpretations: (1) Polyamine degradation along the interconversion pathway is slow. (2) The substrates of polyamine oxidase can be degraded by other enzymes. (3) Acetylpolyamine excretion is enhanced.

As we know from other chapters of this review, acetylpolyamines are both normal urinary excretory products and substrates of copper amine oxidases. With the aid of aminoguanidine and a polyamine oxidase inhibitor it is possible to obtain relatively detailed insights into the significance of the various pathways of polyamine catabolism and other dynamic aspects of polyamine metabolism, if one accepts the premise that the amount of polyamines accumulated in tissues and excreted in the feces is negligible compared with the amount excreted in the urine. We know that the first assumption is correct; however, we have no information on fecal polyamine excretion. Since we must assume that the major portion of fecal polyamines is of alimentary and bacterial origin, it is not possible to obtain meaningful data from polyamine determinations in the feces. The following discussion is therefore hampered by this lack of information.

Figure 9 summarizes results of average daily polyamine excretions by rats after daily administration of aminoguanidine (25 mg/kg, ip) or MDL 72521 (20 mg/kg, ip) or a combination of the two inhibitors in doses

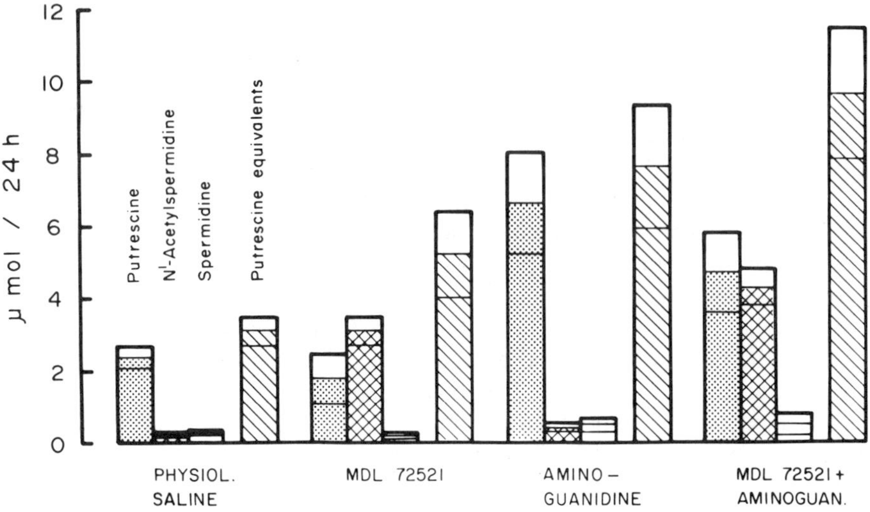

Fig. 9. Average daily polyamine excretion by rats during days 3–5 of treatment. There were three rats per treatment group. MDL 72521 (20 mg/kg, ip) or aminoguanidine sulfate (25 mg/kg, ip) or the combination of the two drugs was given on 5 consecutive days. The bars indicate average daily excretion ± SD. Putrescine equivalents: sum of the amounts of putrescine, spermidine, N^1-acetylspermidine, N^8-acetylspermidine, and spermine. [Data from Seiler *et al.* (1985a).]

sufficient to inactivate both oxidative pathways. Cadaverine as a bacterial product has not been taken into consideration. Since the treatment with the inhibitors did not affect the excretion of N^8-acetylspermidine or spermine, these compounds were considered only in the calculation of total polyamine excretion. Conjugates of spermine as major urinary constituents are not known. N^1-Acetylspermine is not found in urine, even if its formation is enhanced by intoxication with carbon tetrachloride (Seiler *et al.*, 1981e).

Putrescine is the major urinary excretion product of rats. Inhibition of polyamine oxidase alone enhances exclusively the excretion of N^1-acetylspermidine. Inhibition of the copper amine oxidases with aminoguanidine increases excretion of putrescine, spermidine, and N^1-acetylspermidine, as has been stated above. From the urinary polyamine pattern after inactivation of both oxidative pathways the following conclusions can be inferred:

1. About 30% of the putrescine destined to be eliminated is formed by degradation of spermidine, while 70% is formed by decarboxylation of ornithine. Part of the putrescine called "ornithine derived" may originate from the gastrointestinal bacterial flora. The quantitative significance of the latter source is unknown; however, an upper limit could be estimated. Rats recieved a 3% solution of α-difluoromethylornithine for several days in order to inhibit ornithine decarboxylase as completely as possible. Total urinary polyamine excretion was reduced under these conditions by more than 80% (Seiler *et al.*, 1985a). Since α-difluoromethylornithine does not affect ornithine decarboxylase of *E. coli* (Kallio and McCann, 1981), it follows that the bacterial putrescine is less than 20% of the total urinary polyamines.
2. Inhibition of polyamine oxidase causes an increase in total polyamine excretion of about 2 μmol per 24 hr, irrespective of whether or not the copper amine oxidases have been inhibited by treatment with aminoguanidine (Seiler *et al.*, 1985a). The explanation for this finding is as follows: A part of the putrescine formed by spermidine degradation is normally used for spermidine biosynthesis. If polyamine oxidase is inactivated, this portion of putrescine cannot enter the synthetic reactions and is therefore excreted in the form of N^1-acetylspermidine, in addition to the amount which is normally excreted. In other words, a normal rat reutilizes about 2 μmol of putrescine per day for spermidine biosynthesis, this putrescine having been formed by spermidine degradation somewhere in the organism. No data exist presently which would allow us to calculate the amount of ornithine-derived putrescine needed per day for the maintenance of normal polyamine levels in the rat.

V. CONCLUSIONS

Specific inhibitors of copper amine oxidases and of polyamine oxidase have only very recently been used in the study of polyamine metabolism *in vivo*. Therefore, only a few examples are available which demonstrate biological consequences of the inhibition of these enzymes. However, even from the few published observations it is obvious that aminoguanidine and the polyamine oxidase inhibitors are excellent scientific tools which allow us to obtain insights into the quantitative significance of the two major pathways of polyamine catabolism and into dynamic aspects of polyamine metabolism. Furthermore, it is possible now to demonstrate directly the extent of putrescine reutilization in an intact animal, and we have for the first time a relatively simple method for the determination of spermidine turnover in a closed metabolic system. This method will be useful, for instance, in the study of polyamine turnover in brain under a variety of physiological and pathophysiological states, and under the influence of drugs.

It has been shown both in rodents (Seiler *et al.*, 1985a,b) and in humans (Chayen *et al.*, 1985) that normally a very significant proportion of the polyamines does not appear in the urine because of catabolism. The excreted amount is strongly dependent on the activity of the catabolic reactions, which may change due to disease or due to therapy with, for instance, cytotoxic agents, ornithine decarboxylase inhibitors (α-difluoromethylornithine), or drugs that affect diamine oxidase activity [methylglyoxal bis(guanylhydrazone)]. Administration of aminoguanidine could be the means to improve the diagnostic value of polyamine determinations in urine, because complete inhibition of terminal polyamine catabolism allows determination of the actual amount of polyamines disposed of by an organism (Seiler *et al.*, 1985a,b).

We are familiar now with the major principles of *in vivo* polyamine catabolism, although not with many details. After administration of deuterated polyamines to rats and subsequent determination of the deuterated metabolites in urine, 1,3-diaminopropane and *N*-(2-carboxyethyl)-4-amino-*n*-butyric acid were found in addition to the acetylated polyamines and the previously described monocarboxylic acids which derive from spermidine and spermine (Van den Berg *et al.*, 1984). *N*-(2-Carboxyethyl)-4-amino-*n*-butyric acid is formally the result of the oxidative deamination of both primary amino groups of spermidine. It will now be necessary to search for the enzymes responsible for the formation of these metabolites. It can be expected that in this search inhibitors of polyamine-oxidizing enzymes will again be of importance.

One of the greatest surprises of the work with specific inhibitors of polyamine-oxidizing enzymes was the lack of conspicuous pharmacological effects, even after prolonged treatment of the experimental animals with doses of the compounds sufficiently high to completely inactivate the respective enzymes. This lack of toxic side effects increases the value of the compounds as scientific tools. At the same time we may be tempted to conclude from these findings that oxidative polyamine catabolism, although of major quantitative significance under physiological conditions, is a dispensable way of polyamine inactivation, because urinary excretion is capable of taking over the functional role of oxidation. However, our present experience with inhibitors of polyamine catabolism is not extensive enough to allow us to obtain a definitive answer to this important question.

REFERENCES

Abdel-Monem, M.M., and Merdink, J.L. (1981). *Life Sci.* **28,** 2017–2023.

Abeles, R.H., and Maycock, A.L. (1976). *Acc. Chem. Res.* **9,** 313–319.

Achee, F.M., Chervenka, C., Smith, R.A., and Yasunobu, K.T. (1968). *Biochemistry* **7,** 4329–4336.

Ameyama, M., Hayashi, M., Matsushita, K., Shinagawa, E., and Adachi, O. (1984). *Agric. Biol. Chem.* **48,** 561–565.

Antrup, H., and Seiler, N. (1980). *Neurochem. Res.* **5,** 123–143.

Bachrach, U. (1973). "Function of Naturally Occurring Polyamines," pp. 96–106. Academic Press, New York.

Bachrach, U. (1981). *In* "Polyamines in Biology and Medicine" (D.R. Morris and L.J. Marton, eds.), pp. 151–168. Dekker, New York.

Bardsley, W.G., and Ashford, J.S. (1972). *Biochem. J.* **128,** 253–263.

Barsky, J. (!959). Doctoral thesis, North Western University, Evanston, Illinois, *Diss. Abstr.* **19,** 1191 (1958–1959).

Baylin, S.B., Beaven, M.A., Engelman, K., and Sjoerdsma, A. (1970). *N. Engl. J. Med.* **283,** 1239–1244.

Baylin, S.B., Beaven, M.A., Buja, L.M., and Keiser, M.R. (1972). *Am. J. Med.* **53,** 723–733.

Baylin, S.B., Abeloff, M.D., Wieman, K.C., Tomford, J.W., and Ettinger, D.S. (1975). *N. Engl. J. Med.* **293,** 1286-1290.

Bey, P. (1978). *In* "Enzyme-Activated Irreversible Inhibitors" (N. Seiler, M.J. Jung, and J. Koch-Weser, eds.), pp. 27–41. Elsevier, Amsterdam.

Bey, P., Bolkenius, F.N., Seiler, N., and Casara, P. (1985). *J. Med. Chem.* **28,** 1–2.

Bieganski, T., Kusche, J., Lorenz, W., Hesterberg, R., Stahlknecht, C.D., and Feussner, K.D. (1983). *Biochim. Biophys. Acta* **756,** 196–203.

Blankenship, J., and Walle, T. (1978). *Adv. Polyamine Res.* **2,** 97-110.

Blaschko, H. (1962). *Adv. Comp. Physiol. Biochem.* **1,** 67–116.

Blaschko, H., and Bonney, R. (1962). *Proc. R. Soc. London Ser. B* **156,** 268–279.

Bolkenius, F.N, and Seiler, N. (1981). *Int. J. Biochem.* **13,** 287–292.

Bolkenius, F.N., and Seiler, N. (1985). *Biochim. Biophys. Acta* **838,** 69–76.

Borglin, N.E., and Willert, B. (1962). *Cancer* **15,** 271–275.

Buffoni, F. (1966). *Pharmacol. Rev.* **18,** 1163–1199.
Burkard, W.P., Gey, K.F., and Pletscher, A. (1960). *Biochem. Pharmacol.* **3,** 249–255.
Burkard, W.P., Gey, K.F., and Pletscher, A. (1963). *J. Neurochem.* **10,** 183–186.
Chayen, R., Goldberg, S., and Burke, M. (1985). *Isr. J. Med. Sci.* **21,** 543–545.
Dagostini, L., Dargenio, G., Ciacci, C., Daniele, B., Macchia, V., and Mazzaca, G. (1984). *Enzyme* **31,** 217–220.
Danzin, C., Bolkenius, F.N., Claverie, C., Wagner, J., and Grove, J. (1982). *Biochem. Biophys. Res. Commun.* **109,** 1234–1239.
Danzin, C., Claverie, N., Wagner, J., Bolkenius, F.N., and Grove, J. (1983). *Life Sci.* **33,** 2173–2178.
Danzin, C., Claverie, N., and Wagner, J. (1984). *Int. Conf. Polyamines, Budapest* Abstr. No. 57.
Della Ragione, F., and Pegg, A.E. (1982). *Biochemistry* **21,** 6152–6158.
Desiderio, M.A., Sessa, A., and Perin A. (1982). *Biochim. Biophys. Acta* **714,** 243–249.
Desiderio, M.A., Sessa, A., and Perin A. (1984). *In* "Advances in Polyamines in Biomedical Science" (C.M. Caldarera and U. Bachrach, eds.), pp. 141–144. CLUB, Bologna.
Fogel, A., Bieganski, T., Schayer, R.W., and Maslinski, C. (1981). *Agents Actions* **11,** 679–684.
Gahl, W.A., and Pitot, H.C. (1981). *Life Sci.* **29,** 2177–2179.
Gahl, W.A., and Pitot, H.C. (1982). *Biochem. J.* **202,** 603–611.
Gahl, W.A., Vale, A.M., and Pitot, H. (1982). *Biochem. J.* **201,** 161–166.
Gaugas, J.M. (1981). *Biochem. Pharmacol.* **30,** 3121–3122.
Gorkin, V.Z., Akopyan, Zh.I., Veryorskina, I.V., Stesina, L.N., and Abdel Samed, M.M. (1971). *Biochem. J.* **121,** 31P-32P.
Guarnieri, C., Clo, C., Tantini, B., and Caldarera, C.M. (1977). *Boll. Soc. Ital. Biol. Sper.* **53,** 2146–2151.
Heby, O., and Jänne, J. (1981). *In* "Polyamines in Biology and Medicine" (D.R. Morris and L.J. Marton, eds.), pp. 243–310. Dekker, New York.
Hirsch, J.G. (1953). *J. Exp. Med.* **97,** 345–355.
Hölttä, E. (1977). *Biochemistry* **16,** 91–100.
Hölttä, E. (1983). *In* "Methods in Enzymology" (H. Tabor and C.W. Tabor, eds.), Vol. 94, pp. 306–311. Academic Press, New York.
Hölttä, E., Pulkinen, P., Elfving, K., and Jänne, J. (1975). *Biochem. J.* **145,** 373–378.
Jänne, J. (1967). *Acta Physiol. Scand. Suppl.* **300,** 1–71.
Jänne, J., and Hölttä, E. (1973). *Acta Chem. Scand.* **27,** 2399–2404.
Kakimoto, J., Nakajima, T., Kumon, A., Matsuoka, J., Imaoka, N., and Sano, T. (1969). *J. Biol. Chem.* **244,** 6003–6006.
Kallio, A., and Jänne, J. (1983). *Biochem. J.* **212,** 895–898.
Kallio, A., and McCann, P.P. (1981). *Biochem. J.* **200,** 69–75.
Kapeller-Adler, R. (1970). "Amine Oxidases and Methods for their Study." Wiley, New York.
Kitz, R., and Wilson, I.B. (1962). *J. Biol. Chem.* **237,** 3245–3249.
Kluetz, M.D., and Schmidt, P.G. (1977). *Biochem. Biophys. Res. Commun.* **76,** 40–45.
Koch-Weser, J., Schechter, P., Bey, P., Danzin, C., Fozard, J.R., Jung, M.J., Mamont, P.S., and Seiler, N. (1981). *In* "Polyamines in Biology and Medicine" (D.R. Morris and L.J. Marton, eds.), pp. 437–453. Dekker, New York.
Krantz, A., Kokel, B., Sachdeva, Y.P., Salach, J., Claesson, A., and Sahlberg, C. (1979). *In* "Drug Action and Design: Mechanism-Based Enzyme Inhibitors" (T.I. Kalman, ed.), pp. 145–174. Elsevier, Amsterdam.
Kremzner, L.T., and Sturman, J.A. (1979). *J. Neurochem.* **33,** 1115–1117.

Kusche, J., Bieganski, T., Hesterberg, R., Stahlknecht, C.-D., Feussner, K.D., Stahlenberg, I., and Lorenz, W. (1980). *Agents Actions* **10,** 110–113.
Lebsack, M.E., and Anderson, A.D. (1979). *Res. Commun. Chem. Pathol. Pharmacol.* **26,** 263–275.
Lin, C.-W., Orcutt, M.L., and Stolback, L.L. (1975). *Cancer Res.* **35,** 2762–2765.
Lindell, S.E., Nilsson, K., Roos, B.E., and Westling, H. (1960). *Br. J. Pharmacol.* **15,** 351–355.
Lobenstein-Verbeek, C.L., Jongejan, J.A., Frank, J., and Duine, J.A. (1984). *FEBS Lett.* **170,** 305–309.
Lundgren, D.W., Lloyd, H.A., and Hankins, J. (1980). *Biochem. Biophys. Res. Commun.* **97,** 667–672.
Macholan, L. (1974). *Coll. Czech. Chem. Commun.* **39,** 653–661.
Mamont, P.S., Seiler, N., Siat, M., Joder-Ohlenbusch, A.-M., and Knödgen, B. (1981). *Med. Biol.* **59,** 347–353.
Mamont, P.S., Bolkenius, F., Seiler, N., Bey, P., and Kolb, M. (1984). *Int. Conf. Polyamines, Budapest* Abstr. No. 6.
Matsui, I., Pösö, H., and Pegg, A.E. (1982). *Biochim. Biophys. Acta* **719,** 199–207.
Mondovi, B., and Finazzi-Agro, A. (1982). *In* "Structure and Function Relationships in Biochemical Systems" (F. Bossa, E. Chiancone, A. Finazzi Agro, and R. Storm, eds.), pp. 141–153. Plenum, New York.
Mondovi, B., Scioscia-Santoro, A., Rotilio, G., Costa, M.T., and Finazzi-Agro, A. (1975). *Agents Actions* **5,** 460.
Mondovi, B., Turini, P., Befani, O., and Sabatini, S.(1983). *In* "Methods in Enzymology" (H. Tabor and C.W. Tabor, eds.), Vol. 94, pp. 314–318. Academic Press, New York.
Morgan, D.M.L. (1980). *In* "Polyamines in Biomedical Research" (J.M. Gaugas, ed.), pp. 285–302. Wiley, New York.
Morgan, D.M.L. (1985a). *Biochem. Soc. Trans.* **13,** 322–326.
Morgan, D.M.L. (1985b). *Biochem. Soc. Trans.* **13,** 351–352.
Nakajima, T., Noto, T., and Kato, N. (1980). *Physiol. Chem. Phys.* **12,** 401–410.
Okuyuma T., and Kobayashi Y. (1961). *Arch. Biochem. Biophys.* **95,** 242–250.
Pegg, A.E., and Erwin, B.G. (1985). *Biochem. J.* **231,** 285–289.
Pegg, A.E., and McGill, S. (1978). *Biochem. Pharmacol.* **27,** 1625–1629.
Pegg, A.E., Seely, J.E., Pösö, H., Della Ragione, F., and Zagon, I.S. (1982). *Fed. Proc., Fed. Am. Soc. Exp. Biol.* **41,** 3065–3072.
Pegg, A.E., Erwin, B.G., and Persson, L. (1985). *Biochim. Biophys. Acta* **842,** 111–118.
Perin, A., Sessa, M., and Desiderio, M.A. (1981). *Adv. Polyamine Res.* **3,** 397–407.
Perin, A., Sessa, A., and Desiderio, M.A. (1983). *Biochim. Biophys. Acta* **755,** 344–351.
Perin, A., Desiderio, M.A., and Sessa, A. (1984). *Ital. J. Biochem.* **33,** A50–A52.
Perry, T.L., Hansen, S., and MacDougall, L. (1967). *J. Neurochem.* **14,** 775–782.
Pionetti, J.M. (1974). *Biochem. Biophys. Res. Commun.* **58,** 495–498.
Porter, C.W., Dworaczyk, D., Ganis, B., and Weiser, M.M. (1980). *Cancer Res.* **40,** 2330–2335.
Riceberg, L.J., Simon, M., Van Vunakis, H., and Abeles, R.H. (1975). *Biochem. Pharmacol.* **24,** 119–125.
Rojansky, N., Neufeld, E., and Chayen, R. (1979). *Biochim. Biophys. Acta* **586,** 1–9.
Sartori, C., Bargagli, A.M., and Argentu-Ceru, M.P. (1983). *Biochim. Biophys. Acta* **758,** 58–69.
Seiler, N., and Al-Therib, M.J. (1974a). *Biochem. J.* **144,** 29–35.
Seiler, N., and Al-Therib, M.J. (1974b). *Biochim. Biophys. Acta* **354,** 206–212.
Seiler, N., and Bolkenius, F.N. (1985). *Neurochem. Res.* **10,** 529–544.

Seiler, N., and Eichentopf, B. (1975). *Biochem. J.* **152,** 201–210.
Seiler, N., and Knödgen, B. (1983). *Int. J. Biochem.* **15,** 907–915.
Seiler, N., Wiechmann, M., Fischer, H.A., and Werner, G. (1971). *Brain Res.* **28,** 317–325.
Seiler, N., Al-Therib, M.J., and Knödgen, B. (1973). *Hoppe-Seyler's Z. Physiol. Chem.* **354,** 589–590.
Seiler, N., Schmidt-Glenewinkel, T., and Sarhan, S. (1979). *J. Biochem. (Tokyo)* **86,** 277–278.
Seiler, N., Bolkenius, F.N., Knödgen, B., and Mamont, P. (1980). *Biochim. Biophys. Acta* **615,** 480–488.
Seiler, N., Bolkenius, F.N., and Sarhan, S. (1981a). *Int. J. Biochem.* **13,** 1205–1214.
Seiler, N., Bolkenius, F.N., and Rennert, O.M. (1981b). *Med. Biol.* **59,** 334–346.
Seiler, N., Knödgen, B., Gittos, M.W., Chan, W.-Y., Griesmann, G., and Rennert, O.M. (1981c). *Biochem. J.* **200,** 123–132.
Seiler, N., Koch-Weser, J., Knödgen, B., Richards, W. Tardif, C., Bolkenius, F.N., Schechter, P., Tell, G., Fozard, J., Bachrach, U., and Grosshans, E. (1981d). *Adv. Polyamine Res.* **3,** 197–211.
Seiler, N., Bolkenius, F.N., Knödgen, B., and Haegele, K. (1981e). *Biochim. Biophys. Acta* **676,** 1–7.
Seiler, N., Knödgen, B., and Haegele, K. (1982). *Biochem. J.* **208,** 189–197.
Seiler, N., Knödgen, B., Bink, G., Sarhan, S., and Bolkenius, F. (1983). *Adv. Polyamine Res.* **4,** 135–154.
Seiler, N., Bolkenius, F.N., and Knödgen, B. (1985a). *Biochem. J.* **225,** 219–226.
Seiler, N., Knödgen, B., and Bartholeyns, J. (1985b). *Anticancer Res.* **5,** 371–378.
Seiler, N., Bolkenius, F.N., Bey, P., Mamont, P.S., and Danzin, C. (1985c). *In* "Recent Progress in Polyamine Research" (L. Selmeci, M.E. Brosnan, and N. Seiler, eds.), pp. 305–319. Akademiai Kiado, Budapest.
Seiler, N., Sarhan, S., and Knödgen, B. (1985d). *Int. J. Dev. Neurosci.* **3,** 317–322.
Sessa, A., Desiderio, M.A., and Perin, A. (1982). *Biochim. Biophys. Acta* **698,** 11–14.
Shaff, R.E., and Beaven, M.A. (1976). *Biochem. Pharmacol.* **25,** 1057–1062.
Shakir, K.M., Margolis, S., and Baylin, S.B. (1977). *Biochem. Pharmacol.* **26,** 2343–2347.
Shin, W.W., Fong, W.-F., and Wong, P.C.-L. (1985). *J. Neurochem.* **44,** 1056–1059.
Shirota, F.N., Master, E.G., and Nagasawa, H.T. (1979). *J. Med. Chem.* **22,** 463–464.
Smith, T.A. (1985). *Biochem. Soc. Trans.* **13,** 319–322.
Sourkes, T.L., and Missala, K. (1981). *Agents Actions* **11,** 19–27.
Suetsugu, T. (1974). *Jpn. J. Pharmacol.* **24,** 141–149.
Suzuki, O., Matsumoto, T., Oya, M., and Katsumata, Y. (1981). *Biochim. Biophys. Acta* **677,** 190–193.
Syatkin, S.P., and Galaev, Y.V. (1977). *Biochemistry SSR* **42,** 783–786.
Tabor, C.W., Tabor, H., and Bachrach, U. (1964). *J. Biol. Chem.* **239,** 2194–2203.
Thomassek, N., Quash, G., and Dore, J.F. (1982). *Br. J. Cancer* **46,** 58–66.
Tobes, M.C., Griffith, R.K., Dipietro, R.A., and Zoorob, H.H. (1983). *Fed., Proc. Fed. Am. Soc. Exp. Biol.* **42,** 2042.
Van den Berg, G.A., Elzinga, H., Nagel, G.T., Kingsma, A.W., and Muskiet, F.A.J. (1984). *Biochim. Biophys. Acta* **802,** 175–187.
Yamada, H., and Yasunobu, K.T. (1962). *J. Biol. Chem.* **237,** 1511–1516.
Yasunobu, K.T., Ishizaki, H., and Minamiura, N. (1976). *Mol. Cell. Biochem.* **13,** 3–29.
Zeller, E.A. (1963a). *In* "The Enzymes" (P.D. Boyer, H. Lardy, and K. Myrback, eds.), Vol. 8, pp. 313–335. Academic Press, New York.
Zeller, E.A. (1963b). *In* "Metabolic Inhibitors" (R.M. Hochster and J.H. Quastel, eds.), Vol. 2, pp. 53–78. Academic Press, New York.
Zeller, E.A., Schär, B., and Staehlin, S. (1939). *Helv. Chim. Acta* **22,** 837–850.

4

Molecular and Cellular Functions of the Polyamines

LAURENCE J. MARTON*† AND DAVID R. MORRIS‡

** The Department of Laboratory Medicine and*
† The Brain Tumor Research Center
Department of Neurological Surgery
School of Medicine, University of California
San Francisco, California 94143

‡ The Department of Biochemistry
University of Washington School of Medicine
Seattle, Washington 98195

I. INTRODUCTION

The phrase "the function of the polyamines is unknown," or some variation thereof, is frequently encountered. It is a cautious misstatement. Many actions of the polyamines are known and significant progress has been made toward an understanding of the molecular mechanisms that relate these actions to cellular functions. To appreciate fully the possible molecular consequences of polyamine depletion, it is essential to have a detailed understanding of the known interactions of these compounds with other molecules and of the molecular mechanisms involved. An understanding of the functions of the polyamines must include key factors such as the exquisite biosynthetic pathways that regulate their intracellular levels, the metabolic interconversion of polar, nonacetylated polyamines into less polar, acetylated derivatives, and, very importantly, the realization that the positive charges of the protonated polyamines are distributed over the length of the molecules and are not the point charges of inorganic ions such as Ca^{2+} and Mg^{2+}. The subcellular compartmentalization of these compounds, clearly a very important factor that must be understood for a more precise definition of functional activity of polyamines, is not known with any certainty.

Because the protonated forms of intracellular polyamines are cations, their function is frequently compared to those of inorganic cations. The essential participation of the polyamines is frequently dismissed in cellular functions that are supported by both polyamines and inorganic cations. It should be kept in mind that polyamines may participate with inorganic cations in physiological processes that maintain normal structure and function in cells. The relative importance of these ions will not be known until accurate measurements of intracellular concentrations with subcellular locations for both polyamines and the inorganic cations are available. An inherent benefit of the involvement of polyamines in cellular processes is that the polarity and concentration of polyamines within cells are controlled by mechanisms in addition to those that control inorganic cations intracellularly.

Despite these facts, specific functions for the polyamines that cannot be duplicated by other cations at physiological concentrations are being sought and are being found. These specific interactions should not be considered as the only functions of importance, however; polyamines contribute significantly to the overall cationic balance of the cell.

II. NUCLEIC ACID STRUCTURE

The early historical associations between the polyamines and nucleic acids have been well reviewed (Bachrach, 1973; Cohen, 1971; Stevens, 1970; Tabor and Tabor, 1972; Tabor and Tabor, 1976) and may be briefly summarized as follows: (1) Polyamines have been found associated with nucleic acids isolated from viral, bacterial, and mammalian cells. (2) In cell-free systems, polyamines are able to precipitate nucleic acids, increase their melting point (double-stranded nucleic acids), and protect against enzymatic degradation, X-irradiation, and mechanical shearing. (3) Polyamines promote the aggregation of ribosomal subunits and protect against their dissociation.

The results of these studies provide little or no information on the nature of the association between these molecules. Obviously, interactions may be totally nonspecific, very specific (site related), or a combination of the two. We will review the evidence for these possibilities from both early and more recent studies.

Seymour S. Cohen's pioneering studies of the association of polyamines with transfer RNA (tRNA) (Cohen, 1971; Pochon and Cohen,

1972; Sakai and Cohen, 1973) are excellent examples of how carefully conducted biochemical experiments can predict structure–function relationships. By isolating tRNA from both bacterial and mammalian cells at low ionic strength, a molar relationship between tRNA and polyamines could be established for specific growth conditions. The finding that the specific polyamine(s) bound to tRNA under various conditions did not merely reflect the concentration of polyamines in the cell under those conditions suggested that there was some specificity of binding. Even when an excess of polyamines was present, the molar ratio between the polyamines and tRNA never exceeded 2–3 : 1. Cohen *et al.* cautiously interpreted their studies and predicted that polyamines might act to stabilize the cloverleaf structure of tRNA. Crystallographic studies of tRNA now indicate that two spermine molecules, the polyamine used to facilitate the crystallization, are found at very specific sites in the tRNA molecule and appear to stabilize the secondary and tertiary structures (Quigley *et al.*, 1978). One molecule is found in the major groove of the double helix and extends into the anticodon stem from one end of the D stem. It appears to bond to four phosphate oxygens in such a way that the opposite sides of the major groove are drawn together; the tRNA strands in this region are approximately 3 Å closer together than they would be normally. This spermine binding helps produce a 25° deviation from colinearity between the axes of the anticodon and D stems. This is strikingly similar to the description of a DNA major groove binding model discussed below. The other spermine in the tRNA crystal "is near the variable loop and curls around phosphate 10 in a region where the polynucleotide chain takes a sharp turn" (Quigley *et al.*, 1978).

A number of studies of the structure of specific tRNAs in solution have given different results regarding the stoichiometry and nature of binding of polyamines. It is very possible that these differences might reflect different conformations, no matter how small the differences are, in specific tRNAs. McMahon and Erdman (1982) found evidence for electrostatic binding only without cooperative ligand binding. Nöthig-Laslo *et al.* (1981) showed that spermine stabilized the conformation of the anticodon loop and Nilsson *et al.* (1983) reported evidence that spermine stabilized one particular conformation of the anticodon loop. Tropp and Redfield (1983) found that spermidine stabilized the secondary and tertiary structure of tRNA, perhaps at specific sites, and most recently Nöthig-Laslo *et al.* (1985) reported that the binding of Mn^{2+} to tRNA is enhanced in the presence of spermine; this enhancement of Mn^{2+} binding is lost upon destruction of the specific folding of tRNA and does not occur in single-stranded or double-stranded synthetic polynucleotides. The concept that specificity of ligand binding is a function not only of base sequence but also of specific macromolecular structure is an important one that might also be applicable to polyamine binding to DNA, as will be discussed below.

These results point to theories that might explain the interactions that take place between the polyamines and nucleic acids. Bloomfield and Wilson (1981) applied counterion condensation theory, which models DNA as a linear distribution of negative charge and polyamines as point concentrations of positive charge without a specific site of interaction, and showed that this theory gave a fairly accurate description of polyamine binding to DNA as a function of salt concentration. The values of the association constants were not totally predictable by the theory, however, and the authors felt that some degree of "translational localization" or "restriction of internal degrees of freedom" was possible (Bloomfield and Wilson, 1981). In contrast, site-specific models of interaction have been postulated by Tsuboi (1964) and Liquori *et al.* (1967) in which the tetramethylene chain portion of any of the polyamines spans the minor groove and the positively charged amine residues are in proximity to the negatively charged phosphate residues on opposite DNA strands. Drew and Dickerson (1981) have published X-ray diffraction data for a double-stranded B-DNA dodecamer (CGCGAATTCGCG) that, when crystallized in the presence of spermine, bound one spermine molecule per dodecamer.

A model for spermine binding to DNA in the major groove has been evaluated based on these data (Feuerstein *et al.*, 1986). In this model, spermine is docked into the major groove of B-DNA in such a way that the spermine/DNA complex is stabilized by maximizing interactions between proton acceptors on the DNA and proton donors on spermine. Each amine in spermine hydrogen bonds to two sites on the DNA molecule involving either phosphate oxygen or the N^7 of guanine. This is achieved by bending the major groove over spermine and altering oligomer sugar puckering and intrastrand phosphate distances, which produces a 25° bend in the helical axis. Note the similarity of this model to the crystallographic data for tRNA anticodon loop modification discussed above.

It is most likely that both nonspecific and specific interactions between polyamines and DNA are of importance. Superimposed on a background of nonspecific but regulated cationic function is a series of specific interactions secondary to specific macromolecular structure and perhaps secondary to specific base-pair sequences. Observing the specific interactions is difficult because they must be observed in the presence of a preponderance of nonspecific interaction. We will review briefly information that is available in the area of polyamine–DNA interactions.

The ability of polyamines to condense DNA was first reported by Kaiser *et al.* (1963) who found that in the presence of spermine DNA was converted into a form that was protected from shear. Gosule and Schellman (1976) reported that DNA–polyamine complexes can adopt a toroidal shape; later studies with electron microscopy (Chattoraj *et al.*, 1978; Gosule and Schellman, 1978; Marks and Ruben, 1983a, 1984), enzymatic digestion (Marks and Ruben, 1983b, 1984), and flow linear dichroism (Gosule and Schellman, 1978) support this structure, although variant structures that depend on conditions of fixation and isolation have been described. Wilson and Bloomfield (1979) applied counterion condensation theory to show that DNA collapsed into compact structural forms when 89–90% of the phosphate charge was neutralized regardless of ionic conditions. Putrescine and Mg^{2+}, which do not have condensing activity in aqueous buffer, cannot reach the theoretical 89–90% threshold. This argument for nonspecific ionic interaction is further strengthened by a recent study by Porschke (1984), which indicated that polyamines in condensed DNA were highly mobile. He also found that in the presence of spermine DNA was capable of fast bending motion in the formation of the condensate.

In contrast, results of other studies that employed spermidine, $Co(NH_3)_6^{3+}$ (Thomas and Bloomfield, 1983; Widom and Baldwin, 1980, 1983), and a series of spermidine homologues (Allison *et al.*, 1981) showed that condensation of DNA depends not only on charge but also on the distribution of charge or the structure of the cation. Schellman and Parthasarathy (1984) recently showed that the Bragg spacing and the calculated interhelical spacing for a hexagonal packing model of cation-collapsed DNA vary with the length of the methylene bridge for a series of aliphatic di- and triamines. They suggest "that the arrangement of DNA strands in the complexes is determined by the structure of the polycation, and argues against suggestions that the structure of the collapsed state is maintained by the balance of long-range electrostatic repulsive and attractive forces." In a study of the conformational changes that forestall DNA condensation, Marquet *et al.* (1985) found that in contrast to free DNA, at low spermine concentrations DNA–spermine complexes displayed a permanent dipole component and noted that "at higher spermine concentrations, bending of the DNA molecules occurred so that condensation into toroidal particles became possible." This is in agreement with Porschke's (1984) bending data described above, and the permanent dipole is consistent with site-specific binding and with the crystallographic data of Drew and Dickerson (1981), which showed asymmetric binding of spermine to the dodecamer. The bending of the DNA also is consistent with the bend predicted in the major groove model of Feuerstein *et al.* described above.

Porschke's (1984) finding that polyamines in condensed DNA are highly mobile may not be inconsistent with the results just described; specificity of binding during torus formation may give way to nonspecific counterion effects once the torus is formed, or may be such a small component in the formed torus that it is virtually unmeasurable.

Results from another series of experiments must also be considered with regard to these possibilities, however. Using the Drew–Dickerson dodecamer in solution, Wemmer *et al.* (1985) found that although spermine interacts with the DNA with a binding constant at 10^6 at low ionic strength, the nature of the complex and the lifetime of the ligand on the DNA are such that the mobility of the spermine molecule is effectively independent of that of the DNA molecule. While this evidence seems to conflict with the seeming specificity of binding of spermine to DNA in the crystal (Drew and Dickerson, 1981), recent evidence (D.E. Wemmer, K.S. Srivenugopal and D.R. Morris, unpublished results) indicates that the mobility of spermine in condensed calf thymus DNA is reduced. Thus it appears that specific and nonspecific polyamine–DNA interactions may not only be a function of charge and base-pair sequence, but also of specific macromolecular structure, a situation similar to that found for tRNA.

While toroidal condensation of DNA by polyamines and other cations has important implications for the organization of nucleic acids in viral capsids, in nucleosome formation, and in chromosome condensation, the ability of polyamines to facilitate the B–Z transition has similar implications for DNA functional control. A number of investigators have suggested that the B–Z transition is important for the control of DNA function (Behe and Felsenfeld, 1981; Brahmachari *et al.*, 1985; Klysik *et al.*, 1981; Lafer *et al.*, 1981; Nordheim *et al.*, 1981; Nordheim and Rich, 1983).

The earliest evidence that polyamines may facilitate DNA conformational changes was published by Minyat *et al.* (1978), who showed that the B–A transition of DNA was facilitated by spermidine and spermine. It has been shown that polyamines also facilitate the B–Z transition in the heteropolymers poly(dG-m^5dC) (Behe and Felsenfeld, 1981) and poly(dG-dC) (Ivanov and Minyat, 1981). A spermine to nucleotide ratio of 1 : 40–50 is adequate to effect the transition from B to Z DNA in poly(dG-m^5dC) (Behe and Felsenfeld, 1981; Chen *et al.*, 1984). This fact is not easily explained by nonspecific ionic interactions.

As found for DNA condensation, cation-specific effects were observed in the B–Z transition of poly(dG-m^5dC) treated with a series of divalent cations, spermidine, and $Co(NH_3)_6^{3+}$ (Behe and Felsenfeld, 1981) and by

N^1- and N^8-acetylspermidines (Thomas *et al.*, 1985). In the latter study, the Z-forming ability of N^1- and N^8-acetylspermidine in poly(dG-m^5dC) varied inversely with their ability to stabilize T4 DNA to thermal melting. These data are consistent with the presence of different binding affinities (sites) in the two heteropolymers, but the question of whether these differences are a consequence of different base sequences or different conformations was not addressed. These data also indicate that polyamine structure is important for binding, because isomers with the same charge and only slightly different structures each bind to DNA differently and both bind to DNA differently than the parent spermidine. Thomas and Bloomfield (1985) found that structural effects that are independent of charge are important to determine the effect of a specific polyamine ligand on the thermal stability of calf thymus DNA. Thus, as discussed above, the structural specificity of the ligand, as well as of the nucleic acid, is important.

Few studies have been performed on the effects of polyamines on DNA conformation within a cell. Relatively few of the available physicochemical techniques have been applicable to this most important problem. Obviously the true nature of polyamine–DNA interactions and their relationship to function does not lie in the isolated DNA crystal, in DNA in solution, or in DNA as a toroidal structure; rather, the physiologically important interaction is between polyamines and DNA in its normal protein cloak and with its macromolecular structure. Using viscoelastometry, Hung *et al.* (1983) showed that the conformation of DNA in 9L rat brain tumor cells depleted of polyamines by treatment with α-difluoromethylornithine (DFMO) was significantly altered and that this alteration in conformation could be reversed with exogenous putrescine. Although the exact molecular mechanism for this conformational modification is unclear, the fact that DNA conformation might be altered by manipulation of intracellular polyamine levels has served as the basis for studies of the modification of the cytotoxicity of DNA-directed chemotherapeutic agents following treatment with polyamine biosynthesis inhibitors. This area has been reviewed recently (Marton, 1985).

Thus, there is an increasing amount of information that points to an important and perhaps in some instances specific interaction between polyamines and nucleic acids. This information was obtained both deductively and inductively: models were derived based on experimental evidence, and models were used to design experiments. This most important area of investigation will clearly benefit from the continued input of excellent investigators in molecular biology and physical chemistry.

III. MACROMOLECULAR SYNTHESIS

It is appropriate to begin a discussion of the influence of depletion of intracellular polyamine levels on macromolecular synthesis with a word of caution. Adequate intracellular levels of polyamines are clearly required for optimum growth and replication in all cell types that have been tested. Thus, depletion of polyamines generally leads to decreased growth rate or in some instances to growth arrest or cell death. Six years ago a question central to this problem was raised (Morris and Harada, 1980): Do the effects of polyamine depletion involve the synthesis of protein, RNA, and DNA directly, or are these effects expressed as secondary ramifications of cellular growth control? Experimental results have provided definitive answers in only a limited number of instances. Cell growth depends on many variables including the physical environment (temperature, pH, etc.), the availability of growth factors, and the presence of essential nutrients. Some of these variables are *directly* involved in macromolecular synthesis and some are not; obviously, any factor that affects cell growth will ultimately affect the rates of synthesis of protein, RNA, and DNA. Because polyamines might be considered to be essential nutrients, the effects of polyamine depletion on macromolecular synthesis might be considered to be a general regulatory response of cells to a nutrient deficiency. Many metabolic and regulatory steps may be involved between the primary site of polyamine action and the point at which the influence of polyamines on synthesis of macromolecules is observed.

This caveat must be considered during the interpretation of the results of all experiments designed to examine the influence of polyamine depletion on macromolecular synthesis. Two experimental approaches are immediately obvious. One approach is to examine the relationship between a direct influence of polyamines on the synthesis of macromolecules in cell-free systems and a predicted effect in intact cells. Although this approach is conceptually simple and potentially useful, it has been of limited value because it is difficult to attach biological relevance to the many actions of polyamines that have been described in cell-free systems (discussed in Tabor and Tabor, 1984). A second approach is to compare the effects on cells of polyamine depletion with the effects of other methods of limiting cell growth. A clearly defined difference in the mechanism of inhibition of macromolecular synthesis between the two conditions would strongly suggest that polyamine depletion was not triggering a "standard" response of the particular cell type to nutrient deficiency.

A. Protein Synthesis

1. General Considerations

The first evidence that polyamines were functionally involved in protein synthesis came from studies with systems isolated from both procaryotes and eucaryotes (for references, see Tabor and Tabor, 1964, 1976, 1984). Results of early studies suggested that addition of polyamines to these cell-free systems simply lowered the concentration of Mg^{2+} required for synthesis with no influence on the velocity of the reaction; that is, it was merely a cation for cation substitution. Subsequent work has shown that, compared with the rate of protein synthesis observed at the optimum Mg^{2+} concentration, polyamines may increase the rate of synthesis by several fold. Nevertheless, because of the small amount of stimulation caused by polyamines and the complex relationship with the concentration of Mg^{2+}, interpretation of the physiologic significance of these experiments depends critically on the concentrations that are assumed to exist for intracellular cations.

The first evidence that polyamines influenced protein synthesis in intact cells came from studies of polyamine-deficient mutants of *Escherichia coli*. Growth of these mutants in the absence of polyamines depressed the rate of protein synthesis and the rate of growth to the same extent (Morris and Jorstad, 1973), and addition of putrescine caused an almost immediate enhancement of protein synthesis that preceded changes in the rate of synthesis of both RNA and DNA (Young and Srinivasan, 1972). In addition, Young and Srinivasan (1974) have shown that the synthesis of proteins in the RNA phage f2 is inhibited in host cells grown in the absence of polyamines. Thus, results obtained in bacterial systems seem to suggest that polyamines are involved in protein synthesis; these results may be related to results obtained in cell-free systems described above.

An apparent conflict in the results of studies with animal cells requires some discussion. In exponentially growing HTC cells, inhibition of polyamine synthesis with DFMO clearly blocks protein synthesis, as shown by a decrease in the incorporation of radioactive amino acids and by changes in the patterns of distribution of polysomes (Rudkin *et al.*, 1984). In quiescent lymphocytes stimulated with mitogens, the situation is more complex. These cells show marked elevations of both polyamines and the rate of protein synthesis in the transition from G_0 to G_1 phase. For example, in mitogen-stimulated bovine lymphocytes, cellular levels of putrescine, spermidine, and spermine were increased approximately 6-, 2-, and 1.3-fold, respectively, in the first 24 hr after stimulation (Fillingame and Morris, 1973). This increase in polyamine levels preceded initiation of

DNA replication. The rate of incorporation of [^{3}H]leucine into proteins was stimulated 6-fold over the same period, during which polyamine levels increased (Morris *et al.*, 1977). Extensive studies have shown that inhibition of polyamine synthesis has little (Hölttä and Hovi, 1985), if any (Morris *et al.*, 1977; Seyfried and Morris, 1979), influence on the marked stimulation of translation during the G_0–G_1 transition. On the contrary, at later times when activated lymphocytes have fully entered the cell cycle, protein synthesis becomes sensitive to inhibition of polyamine synthesis (Igarashi and Morris, 1984; Hölttä and Hovi, 1985), as was the case with cycling HTC cells. Thus, it seems that basal levels of polyamines in quiescent animal cells are sufficient to support protein synthesis in the early, prereplicative stages of mitogenic activation, but as cells continue to grow in the presence of inhibitors and dilute out the basal levels, the polyamine requirement for translation becomes apparent.

Thus, results obtained to date suggest that polyamines may be directly involved in protein synthesis in cells. The specific sites in the translational machinery at which polyamines may be acting are discussed in detail in the remainder of this section.

2. *Elongation versus Initiation*

Polyamines may affect either the initiation step or the elongation of nascent polypeptide chains. This variable was first investigated in polyamine-deprived procaryotes using three independent experimental approaches that clearly showed that the rate of elongation is slowed proportionally to the reduction in the rate of cell growth (Morris and Hansen, 1973; Jorstad and Morris, 1974; Morris, 1981). Moreover, the fact that the size distribution of polysomes did not change with polyamine limitation (Jorstad and Morris, 1974) suggests that the rate of initiation is also reduced to a similar extent. This result is unusual because in the normal state, the rate of polypeptide chain elongation is relatively invariant for a given cell type for both bacteria (reviewed in Nierlich, 1978) and animal cells (Kay *et al.*, 1975). This suggests that polyamines affect protein synthesis by direct participation in translation and not merely as a secondary effect on cellular growth control mechanisms.

Two studies have examined the profile of polysomes from intact animal cells treated with DFMO (Rudkin *et al.*, 1984; Hölttä and Hovi, 1985). Both groups interpreted their results in terms of a polyamine effect on initiation. Polysome distributions found in both studies, however, were composed of messages of widely varying abundances, sizes, and possibly initiation frequencies; the resulting heterogeneity perhaps masked changes in pattern. In addition, it is difficult to compare polysome size distributions meaningfully by absorbance profiles, particularly when the ratio of monosomes to polysomes is very different. The approach of choice would involve an examination of the average size of polysomes containing a particular message, an approach that is now feasible using recombinant DNA probes. Thus, in the absence of additional information, the mechanism through which polyamines stimulate translation in animal cells cannot be described with confidence.

Results of studies of elongation and initiation in cell-free systems have produced conflicting conclusions (reviewed in Takemoto *et al.*, 1983); some authors find that polyamines stimulate initiation and some find that polyamines influence elongation. A recent study may have resolved the issue, at least in a eucaryotic system. Takemoto and co-workers (1983) found that spermidine stimulation of a protein synthesis system from wheat germ could be fully accounted for by effects of roughly equal magnitude on *both* initiation and elongation, a result that is certainly consistent with the studies in intact *E. coli* discussed above.

Results of several other studies, all of which might be consistent with a stimulatory function for polyamines in translational initiation, should be mentioned. Igarashi and co-workers (1980) observed that spermidine stimulated the ribosomal binding of fMet-tRNA$_f^{Met}$. The activity of protein kinase 380, which leads to inhibition of eukaryotic initiation factor 2-α, is itself inhibited by polyamines (Kuroda *et al.*, 1982). Another initiation factor, eIF-4D, contains the polyamine metabolite hypusine bound in covalent linkage (Cooper *et al.*, 1983). Although the function of hypusine in eIF-4D is not yet understood, it is intriguing that the conjugate is formed upon stimulation of animal cells by mitogens. It should be noted that fMet-tRNA$_f^{Met}$ deacylase is inhibited by spermidine (Igarashi *et al.*, 1981b). The role of this enzyme in protein biosynthesis and the biological significance of spermidine stimulation have not been defined, but this effect is clearly of potential importance for the interpretation of the influence of polyamines in crude cell-free protein synthesis systems.

3. Translational Fidelity

Equally as important as the influence of polyamines on the rate of protein synthesis is the effect on fidelity of translation: the frequency with which amino acids are correctly incorporated into a polypeptide in response to a given codon. Results of studies both in cell-free and in intact cell systems strongly suggest that polyamines can have a significant effect on translational fidelity.

Studies of the influence of polyamines on fidelity of translation in cell-free systems have been reviewed (Tabor and Tabor, 1984) and will be discussed only briefly here. The specific effect observed, either increased or decreased translational fidelity in the presence of polyamines, seems to depend both on the concentration of Mg^{2+} and on the temperature of incubation (see Tabor and Tabor, 1984, for specific references). A phenomenon that may be related to fidelity is the ability of polyamines in cell-free systems to induce "read-through" (inaccurate translation) of termination codons in viral and mammalian mRNAs (Hirashima *et al.*, 1979; Morch and Benicourt, 1980; Hryniewicz and Vonder Haar, 1983). On the other hand, it should be noted that exogenous polyamines are often included in the medium for experiments with cell-free systems that have been optimized for both rate and fidelity of translation (for an example, see Kurland, 1982).

Studies in intact cell systems have suggested that ribosomes, particularly the ribosomal protein S12, which is the product of gene *rpsL* (or *strA*) in *E. coli,* are involved. This protein is the site of action of streptomycin, and the well-known influence of the antibiotic on translational fidelity is exerted through interaction with this protein. One early hint of the effect of polyamines on translation came from a study of Goldemberg and Algranati (1981), which showed that sensitivity to streptomycin is decreased in polyamine-starved *E. coli*. Moreover, it has been shown that certain *rpsL* mutations, when expressed in polyamine auxotrophs, generated an absolute requirement for polyamines for growth (Tabor *et al.*, 1981; reviewed in Tabor and Tabor, 1985). An observation made by the Tabors (1982) may be related. They found that the growth of wild-type phage T7 is relatively unaffected by starvation of *E. coli* polyamine auxotrophs, but that suppression of three different T7 *amber* mutations in two different *sup* strains of *E. coli* was strongly inhibited by polyamine depletion (Tabor and Tabor, 1982).

Because of the well-known relationship between streptomycin action, translational fidelity, and the small ribosomal subunit, it is tempting to speculate that the observations in whole cells discussed above may be related to interactions between polyamines and the 30 S subunit. Moreover, because codon–anticodon interaction is one determinant of elongation rate, polyamine interaction with the 30 S subunit may affect polypeptide elongation.

4. Ribosome Structure and Function

Two effects of polyamines will be considered. First, there are effects that can best be considered as general cationic interactions. Second, results obtained in intact cells suggest that polyamines may affect the structure and function of ribosomes in specific ways.

Although secondary redistribution is a major problem that affects quantitative interpretation of the association of polyamines with isolated ribosomes (Tabor and Kellogg, 1967), since the pioneering work of Cohen and

Lichtenstein (1960) it has been known that ribosomes extracted from whole cells contain polyamines. Polyamines and Mg^{2+} compete for binding to ribosomes (Choi and Carr, 1967), but the influence on structure and function of the particles is complex and depends on the amount of each cation bound. As first shown by Cohen and Lichtenstein (1960), polyamines may partially replace Mg^{2+} in promoting the association between 30 and 50 S subunits. Polyamines can also replace Mg^{2+} at approximately 80% of the cation binding sites and maintain the native structure of isolated subunits (Weiss and Morris, 1973; Kimes and Morris, 1973). Interestingly, Mg^{2+} is specifically required at the remaining 20% of sites and it has been suggested that this cation is an integral part of the native structures of the ribosomal subunits (Weiss *et al.*, 1973). Replacement of Mg^{2+} with polyamines at these critical sites leads to unfolding of the particles with concomitant inactivation. This possibly reflects a steric factor for cations in maintaining the tertiary structure of the ribosome.

Defects in structure were found in 30 S ribosomal subunits isolated from polyamine-depleted *E. coli* cells. Extracts from these mutants deprived of polyamines have less activity in protein synthesis than preparations from unstarved cells, and the defect apparently resides in the 30 S ribosomal subunit (Echandi and Algranati, 1975a,b). Igarashi and coworkers (1979, 1981a) found two abnormalities in the small subunit isolated from several different mutant strains of *E. coli*. These particles have both a deficiency in protein S1 and an undermethylation of adenine in 16 S rRNA. They have suggested that these structural abnormalities are a result of a defect in the assembly of the 30 S particle that is caused by a lack of polyamines. The molecular mechanisms that cause these alterations in 30 S structure and the changes in translational fidelity in polyamine-deficient cells have yet to be determined.

5. *Structure and Activity of Transfer RNA*

The relationship between the level of charging of tRNA species and translational elongation and fidelity discussed above are obvious; suboptimal loading of even one species could seriously affect both of these processes. The influence of polyamines on the structure and activity of tRNA has been well studied and the older literature has been reviewed by Tabor and Tabor (1976). A number of aminoacyl tRNA synthetases are stimulated by polyamines; the question of whether spermine can or cannot completely replace the requirement of these enzymes for Mg^{2+} is controversial. In a recent study of the transfer of amino acids from aminoacyl adenylates to tRNAs, Kim and Mehler (1981) defined three groups of aminoacyl tRNA synthetases: those that have an absolute requirement for a divalent cation such as Mg^{2+} or Mn^{2+}, those that are activated by a variety of cations including protonated spermine, and those that are fully active in the absence of activating cations. In many instances in which polyamines cause activation, it has not been established whether the functional interaction is with the enzyme or the tRNA.

One of the best-studied cases is that of yeast $tRNA^{Phe}$. In a careful kinetic analysis, Loftfield *et al.* (1981) found evidence consistent with a model in which two spermines are involved in the transition state of the charging reaction; they postulated that these spermine molecules were bound to the tRNA. This intermediate is consistent with crystallographic studies of yeast $tRNA^{Phe}$ mentioned above, in which two molecules of spermine were found bound to the tRNA (Quigley *et al.*, 1978). In solution, this tRNA binds several spermines with high affinity (Schreier and Schimmel, 1975) and interaction with spermine seems to cause changes in the structure of the anticodon loop (Nilsson *et al.*, 1983). In addition to influencing binding of the tRNA to the homologous synthetase, structural transitions induced by spermine also could influence codon–anticodon interaction on the ribosome.

6. Influence of Polyamines on the Synthesis of Specific Proteins

A variety of studies have been published that report the influence of polyamines on the synthesis of specific proteins or classes of proteins. Results of some of these studies are summarized in this section. In cell-free systems from both mammalian (Atkins *et al.*, 1975) and plant (Takemoto *et al.*, 1983) sources, polyamines were found to stimulate the synthesis of high-molecular-weight proteins to a greater extent than stimulation of synthesis of low-molecular-weight proteins. A similar observation has been made in polyamine auxotrophs of *E. coli* (Goldemberg and Algranati, 1981). It is possible to interpret these results in terms of premature termination of polypeptide elongation in the absence of polyamines, which might be the result either of sluggish elongation that allows unscheduled termination or of abnormalities in translational fidelity.

Igarashi and his colleagues have examined in detail the cell-free synthesis of the three proteins encoded by the RNAs of phages Qβ and MS2, and observed profound differential effects (Watanabe *et al.*, 1981). Spermidine seems to affect initiation specifically, possibly by mediating changes in the tertiary structure of the RNA. In two early studies of DNA-directed protein synthesis systems, Fuchs (1976) and Treadwell *et al.* (1976) found that spermidine stimulated synthesis. This has been examined in more detail using T7 and $\lambda rif^{d}18$ DNA-directed systems (Watanabe *et al.*, 1983). Spermidine differentially affected the synthesis of the proteins encoded by both DNAs. Although in some instances the amount of stimulation of specific proteins correlated with molecular weight, the correlation was not absolute. The molecular mechanism of this effect is not understood and analysis of these systems is complicated by interpretation of the effects of coupled transcription and translation.

Using polyamine auxotrophs of *E. coli,* Igarashi and his co-workers have examined the effects of replenishment of polyamines on the synthesis of specific proteins in polyamine-starved cells. They observed differential effects on the synthesis of a protein termed "PI" (Mitsui *et al.*, 1984a) and the β and β' subunits of RNA polymerase (Mitsui *et al.*, 1984b). Because the β and β' subunits have high molecular weights (M_r 160,000), these findings could reflect the general effect on high-molecular-weight proteins discussed above. On the other hand, the effects on PI, which has a M_r of 62,000, probably reflect another mechanism. Because purified PI and spermidine stimulate the synthesis of phage MS2 RNA replicase in a cell-free system, the authors proposed a cascade model in which polyamines act first to stimulate the synthesis of PI, which then acts synergistically with the polyamines to enhance the synthesis of other proteins (Mitsui *et al.*, 1984a). PI is probably related to a factor described by Rosano *et al.* (1983) that enhanced polyamine stimulation of a cell-free translation system from *E. coli*. Preliminary results obtained with mitogen-activated lymphocytes have shown that inhibition of polyamine biosynthesis affects the synthesis of specific proteins (Igarashi and Morris, 1984).

B. DNA Replication

Inhibition of macromolecular synthesis, and of DNA replication in particular, follows naturally from the fact that depletion of polyamines in cells inhibits growth and division. There is good reason to think, however, that the polyamine-related depression of DNA synthesis in both procaryotes and eucaryotes is a unique phenomenon and not simply the result of growth regulation.

1. Studies in Intact Cells

Early evidence suggested that polyamines affected DNA replication in bacteria; infection of spermidine-starved *E. coli* with bacteriophage T4 caused suboptimal rates of DNA accumulation (Dion and Cohen, 1972). Uninfected *E. coli* also depends on polyamines for optimum DNA replication, and the defect in polyamine-starved cells has been defined precisely. Three independent experimental approaches showed that the rate of replication fork movement was slowed in polyamine-starved cells (Geiger and Morris, 1978; 1980a) and no evidence was found for an effect of polyamines on initiation of chromosome replication. This was an unusual response, because the rate of chromosome replication is normally adjusted to the growth rate of the cells by variations in the frequency of initiation (Cooper and Helmstetter, 1968). On the basis of these observations, it was suggested that in bacteria polyamines may act as cofactors for DNA replication (Geiger and Morris, 1978) and not as regulatory molecules.

Studies in polyamine-depleted mammalian cells, induced by treatment with polyamine inhibitors, also have suggested a role for polyamines in DNA replication. Lymphocyte cultures activated with mitogen in the presence of methylglyoxal bis(guanylhydrazone) (MGBG) initiated DNA replication at the same time as in control cultures, but the rates of [^{3}H]thymidine incorporation and of entry into mitosis were slowed substantially (Fillingame *et al.*, 1975). Subsequent studies with other inhibitors of polyamine biosynthesis in which intracellular thymine deoxynucleotide pools were monitored confirmed the initial conclusion that appropriate polyamine levels were necessary for DNA replication, but not for entry into S phase (Morris *et al.*, 1977; Seyfried and Morris, 1979). This relationship between polyamine accumulation and DNA synthesis has now been observed in a variety of systems (for reviews see Morris and Harada, 1980; Heby and Jänne, 1981; Heby, 1981). Of particular note are cell cycle studies with inhibitor-treated cells (Harada and Morris, 1981) and mutant cell lines (Anehus *et al.*, 1984) that showed S phase is lengthened in the polyamine-deficient cells; lengthening of S phase is not a normal mechanism by which eukaryotic cells regulate cycle time (Pardee *et al.*, 1978).

Results of studies with MGBG-treated HeLa cells suggested that the major effect of polyamine deficiency was on replicon initiation, and that polyamine depletion had only a minor effect on DNA chain elongation (Krokan and Eriksen, 1977). This result is quite different than that obtained in *E. coli,* and should be confirmed with other inhibitors, because MGBG may produce effects that are not related to inhibition of polyamine biosynthesis.

2. *Studies with Isolated Systems*

Despite the fact that depletion of polyamines in both prokaryotic and eukaryotic cells clearly causes inhibition of DNA replication, there is no compelling evidence for or against a *direct* functional interaction of polyamines at the site of replication. To answer this question definitively, the influence of polyamines on DNA replication must be studied in defined cell-free systems.

Polyamines influence a variety of cell-free DNA replication systems (reviewed in Tabor and Tabor, 1976), including those from both prokaryotic and eukaryotic organisms using single-stranded and duplex templates. In most instances, however, the biological relevance of these effects has not been assessed (discussed in Morris, 1981). As much as a 20-fold polyamine stimulation has been found for the replication of single-stranded ϕX174 DNA in a cell-free system from *E. coli* (Geiger and Morris, 1980b). This system is considered to be a good model for some aspects of chromosomal DNA replication (Shekman *et al.,* 1974), which also might depend on polyamines in intact cells. However, an examination of the specificity of the spermidine requirement of this reaction led to the inescapable conclusion that the rate-limiting, polyamine-dependent step assayed in this system was not the same as that observed for replication of cellular DNA (Geiger and Morris, 1980b; Morris, 1981).

When nuclei isolated from mammalian cells are incubated with appropriate substrates, replicons that had been initiated previously are elongated (Hershey *et al.,* 1973; Hallick and Namba, 1974). Nuclei isolated from either HeLa cells (Krokan and Eriksen, 1977) or lymphocytes (Knutson and Morris, 1978) treated with polyamine biosynthesis inhibitors were found to have reduced DNA replication activity. Unfortunately, it was not possible to reactivate nuclei from either source by replenishment of polyamines; this apparent irreversibility of the reaction has confounded attempts to extend these studies to provide a more refined analysis of possible direct polyamine function in DNA replication. There are clearly many explanations for this negative result (Knutson and Morris, 1978). The major effect of polyamines may be on replicon initiation (Krokan and Eriksen, 1977); if this is the case, stimulation should not occur *in nuclei,* because these systems only elongate and do not initiate (Hershey *et al.,* 1973; Hallick and Namba, 1974).

Molecules other than DNA polymerases must be considered as possible sites for the interaction with polyamines. Replication proteins such as DNA binding proteins, helicases, primases, or topoisomerases (Kornberg, 1982) could all interact with polyamines and influence replication. The striking stimulation by polyamines of the catalysis of DNA reassociation by the *E. coli* single-stranded DNA binding protein, a reaction certainly relevant to replication, has been reported (Christiansen and Baldwin, 1977). Studies on DNA topoisomerases will be discussed below.

C. RNA Biosynthesis

The stimulation of RNA polymerases by polyamines has been studied extensively, and results have been reviewed in detail by others (Tabor and Tabor, 1964, 1976; Blair, 1985). It has been reported that the rate of β-

galactosidase mRNA elongation in *E. coli* auxotrophs is slowed after depletion of polyamines (Morris and Hansen, 1973); this experiment should be repeated for other mRNAs and for the stable RNAs. The relationship between spermidine stimulation of cellular mRNA polymerization and cell-free stimulation of RNA polymerase is not necessarily straightforward (discussed in Morris, 1981).

In early studies a correlation was found between RNA accumulation and polyamine levels, which led to suggestions that polyamines might be regulating transcriptional events. Although this correlation certainly exists, later studies have shown that a regulatory role for the polyamines in this process is not probably (for reviews, see Tabor and Tabor, 1976; Heby and Jänne, 1981; Heby, 1981). It has also been suggested that ornithine decarboxylase is a regulatory subunit of RNA polymerase I (see Russell and Durie, 1978, for a discussion of this model). The initial studies were performed with highly impure preparations of ornithine decarboxylase, and subsequent work by others (Seeley *et al.*, 1984) performed with highly purified enzyme failed to confirm the hypothesis.

Other reactions of RNA, including methylation and nucleolytic degradation (reviewed in Tabor and Tabor, 1976), are known to be affected by polyamines. It has been shown recently that a specific nucleolytic step in the processing of a tRNA precursor is stimulated by spermidine, probably through an effect on the structure of the RNA molecule (Peebles *et al.*, 1983). The physiological significance of these effects has not yet been assessed.

IV. TOPOISOMERASE ACTIVITY

The topoisomerases are a class of enzymes that regulate the superhelical density of DNA (for reviews, see Wang and Liu, 1979; Cozzarelli, 1980; Gellert, 1981a,b; Liu, 1983; Wang, 1985). Maintenance of the appropriate degree of supercoiling is now recognized to be a critical factor in both expression of bacterial genes and replication of the procaryotic chromosome. In eucaryotes, topoisomerases seem to be major components of chromosome structure (Earnshaw *et al.*, 1985; Earnshaw and Heck, 1985) and also appear to be necessary for chromosome segregation at cell division (Dinardo *et al.*, 1984; Uemura and Yanagida, 1984). These enzymes are thought to be important sites of action of some chemotherapeutic agents (Nelson *et al.*, 1984; Zwelling *et al.*, 1985).

Gellert *et al.* (1976) were the first to show that polyamines influence the activity of a topoisomerase; in their pioneering study of *E. coli* DNA gyrase, they showed that spermidine greatly stimulated the enzyme at physiological concentrations of Mg^{2+}. Since that time, both stimulatory and inhibitory effects of polyamines have been observed for particular topoisomerases (discussed in Srivenugopal and Morris, 1985).

It is useful to divide these enzymes into two classes: topoisomerases that recognize and act upon single-stranded regions in supercoiled DNA and topoisomerases that interact at double-stranded regions. The first class seems to be made up largely of the type 1 prokaryotic topoisomerases, including *E. coli* topoisomerase I, or omega protein (Wang, 1973), and topoisomerase III (Srivenugopal *et al.,* 1984). These enzymes are inhibited by spermidine, apparently through reduction in the single-strandedness of the supercoiled substrate (Srivenugopal and Morris, 1985), which is required for enzyme activity. The second class of topoisomerase is composed of bacterial topoisomerase II (DNA gyrase) and the eukaryotic topoisomerases. In those cases examined, representatives of this second class of enzymes were strongly stimulated by polyamines (reviewed in Srivenugopal and Morris, 1985), but the detailed mechanism of this stimulation is not yet understood.

A physiological significance of the interactions of polyamines and the topoisomerases was first suggested to explain some properties of polyamine auxotrophs of *E. coli* (Morris and Lockshon, 1981). It was pointed out that the net result of spermidine action on the known *E. coli* topoisomerase reactions should be to increase the superhelical density of cellular DNA. This leads to the prediction that polyamine starvation of these auxotrophs would cause a decrease in DNA supercoiling and thereby produce the observed effects on DNA transcription and replication. In an initial test of this hypothesis using a rapid lysis technique, the predicted changes in supercoiling of DNA plasmids upon polyamine starvation were not found (Lockshon and Morris, 1983). It has since become clear, however, that compensatory mutations are strongly selected for in topoisomerase mutants (Dinardo *et al.,* 1982; Pruss *et al.,* 1982). The possibility that such secondary mutations have accumulated has not been controlled for in any of the polyamine auxotrophs currently available.

Polyamines profoundly affect the catenation of duplex rings catalyzed by topoisomerases. Catenation is performed by all topoisomerases and has an absolute requirement for polycations. Krasnow and Cozzarelli (1982) showed that spermidine controlled the formation and resolution of catenanes by influencing the aggregation state of the substrate DNA (see previous discussion). The possibility that polyamines exert physiological control of the passage of DNA strands through one another clearly could be of significance in removal of topological barriers during chromosome replication and segregation.

Interactions of polyamines with mammalian topoisomerases is now of interest in the context of the mechanism of action of certain antineoplastic drugs. The DNA-intercalating agents have been shown to stimulate DNA cleavage by purified topoisomerase II (Nelson *et al.*, 1984) and in isolated nuclei (Pommier *et al.*, 1985). It is thought that polyamines stabilize a normal intermediate in the catalytic breaking–rejoining cycle of the enzyme. Recently, it was found that the frequency of DNA scissions induced by 4′-(9-acridinylamino)methanesulfon-*m*-aniside (*m*-AMSA) is increased in L1210 cells depleted of polyamines by treatment with DFMO (Zwelling *et al.*, 1985). At this point it is not clear whether polyamines directly affect the activity of topoisomerase II or the DNA substrate.

V. MEMBRANE STRUCTURE AND FUNCTION

The finding that small amounts of spermine could stabilize osmotically fragile bacteria was the earliest evidence that the polyamines might interact with either the cell wall or the cell membrane (for references see Bachrach, 1973; Tabor and Tabor, 1964). The same effect was found for the protection of bacterial protoplasts by C.W. Tabor (1960, 1962) and Mager (1962), which showed that interaction of polyamines with the cell membrane was involved in the stabilization. Spermine did not interfere with the action of lysozyme used in the preparation of protoplasts, but it did prevent the leakage of intracellular constituents (Grossowicz and Ariel, 1963). Although other organic cations produced the same effect as spermine, Mg^{2+} did not protect protoplasts from lysis (Grossowicz and Ariel, 1963).

As Tabor and Tabor (1964) pointed out, compared with the concentration of sucrose usually employed (0.5 *M*), the low concentration of spermine (1 m*M*) necessary to prevent protoplast lysis indicates that spermine does not act through an osmotic mechanism that has been proposed for the action of sucrose. They postulated that spermine forms complexes with acidic groups in the cell membrane, which reduces repulsive forces without diminishing cohesive forces and thereby stabilizes the membrane.

Other early studies indicated that mammalian membranes are also stabilized by polyamines. The delayed hemolysis of reticulocytes that occurs when they are suspended in saline is partially inhibited by spermine (Little, 1962), and diamines were shown to prevent the hemolysis of sensitized sheep red blood cells caused by hemolytic complement (Wirtz and Becker, 1961). It has been shown more recently that red blood cells are an excellent model of polyamine–membrane interactions. Chun *et al.* (1976) found that polyamine levels in blood are elevated in patients with sickle cell disease. They also found that the normal electrokinetic properties of the red blood cell membrane can be altered by polyamines (Chun *et al.*, 1976, 1977). Natta *et al.* (1980) have reported that the erythrocyte stroma fraction obtained from sickle cells contains an abnormally large amount of spermine. They noted that putrescine and spermidine binding to erythrocyte stroma from sickle cells appears to be independent of spermine binding and postulated that specificity of binding for these compounds or differential accessibility were responsible for these effects.

Schindler *et al.* (1980) further characterized the interactions between polyamines and red blood cell membranes. They found that the lateral mobility of fluorescein-labeled membrane glycoproteins was decreased by spermine. Ballas *et al.* (1983) showed that the polyamines decreased erythrocyte deformability and stabilized the membrane skeleton, making it more resistant to fragmentation. In control experiments performed in the latter study, it was shown that Ca^{2+} or Mg^{2+} could not replace the polyamines, which suggests that charge per se was not responsible for the observed effects. Furthermore, this study showed that the observed effects on the erythrocyte membrane were not the result of the known reaction in which polyamines are cross-linked to erythrocyte membrane proteins in the presence of transglutaminase and Ca^{2+} (Folk and Finlayson, 1977; Schrode and Folk, 1978). Because a variety of diseases such as sickle cell anemia, uremia, and Duchenne muscular dystrophy that are characterized by increased erythrocyte polyamine levels are also characterized by rheologic abnormalities of the erythrocyte, with decreased deformability and decreased survival, the role of the polyamines in the pathobiology of these diseases deserves further study.

Significant levels of polyamines have also been found in platelets (Cooper *et al.*, 1976) and exogenous polyamines have been shown to inhibit drug-induced platelet aggregation (Rennert *et al.*, 1976). Platelets from patients with psoriasis, who have high blood polyamine levels, display decreased ADP-induced platelet aggregation without the addition of exogenous polyamines (Rennert *et al.*, 1976).

A study of the aggregation and fusion of unilamellar vesicles into liposomes showed that polyamines promoted aggregation of the vesicles and that spermine significantly enhanced both Ca^{2+}- and Mg^{2+}-mediated fusion (Hong *et al.*, 1983). The authors suggest that polyamines could be important for the modulation of Ca^{2+}-mediated membrane fusion intracellularly. It is informative to quote these authors at some length. They state that "intracellular membrane flow, i.e. endocytic vesicle–lysosome fusion and Golgi vesicle–plasma membrane fusion, is required for cell growth," referring to a study by Harada *et al.* (1981) that showed inhibition of ornithine decarboxylase caused the accumulation of intracellular vacuoles and a decline in growth rate, both of which are reversed with exogenous putrescine, and conclude that "By controlling membrane fusion rates, polyamines may regulate intracellular membrane flow, i.e., that limitation of polyamine synthesis may cause a buildup of intracellular membrane due to the lack of fusogenic activity necessary for normal membrane flow." The results of a study performed by Kirschbaum (1984) lend further support to the involvement of spermine in membrane aggregation. He showed that the aggregation of isolated kidney brush border membranes induced by gentamicin could be blocked by low concentrations (20 μM) of spermine and that at higher concentrations spermine itself proved to be an effective aggregator.

The swelling of isolated mitochondria by treatment with a number of agents such as oleate and succinate or by suspension in hypotonic solutions is markedly reduced by putrescine, cadaverine, spermidine, and spermine (Herbst and Witherspoon, 1960; Tabor, 1960). Polyamines have also been shown to interfere with small cation transfer in mitochondria (Åkerman, 1977) and to neutralize the negative charge of the mitochondrial membrane (Huunan- Seppälä, 1971). Byczkowski and Porter (1983) have shown that the decreased respiration and membrane aggregation caused by treatment of intact mitochondrial membranes with either MGBG or 4,4′-diacetyldiphenylurea bis(guanylhydrazone) were prevented or reversed by polyamines. They postulated that competition between these drugs and the polyamines for low-affinity binding sites occurs at the outer surface of the inner mitochondrial membrane. Results of another recent study showed that the size of mitochondria was increased significantly in 9L rat brain tumor cells in which ornithine decarboxylase had been inhibited by treatment with DFMO (Oredsson *et al.,* 1983). The increase in mitochondrial size was reversed by adding putrescine to the culture medium 24 hr after treatment with DFMO. It should be noted that other cell lines treated with ornithine decarboxylase inhibitors did not show a similar change in mitochondrial structure (Harada *et al.,* 1981; Pleshkewych *et al.,* 1980).

In a thought-provoking study, Koenig *et al.* (1983) reported that testosterone-induced endocytosis, hexose transport, and amino acid transport in the proximal tubules of mouse kidney depended on very rapid intracellular changes in ornithine decarboxylase activity and polyamine levels. The Ca^{2+} fluxes and intracellular calcium mobilization (increased cytosolic Ca^{2+} and decreased mitochondrial Ca^{2+}) associated with the receptor-mediated response to testosterone were blocked by treatment with DFMO, which suppressed the rapid increase in ornithine decarboxylase activity and the increased levels of spermidine and spermine, but not of putrescine. The authors postulated that the receptor-mediated activity was the result of an intimate relationship between the polyamines and membrane structure and function.

ACKNOWLEDGMENTS

Supported in part by NIH Program Project Grant CA-13525 and National Cooperative Drug Discovery Grant CA-37606 (to L.J.M.) and National Science Foundation Grant PCM-8301985 (to D.R.M.). We express our gratitude to Drs. K.S. Srivenugopal and B.G. Feuerstein for their suggestions based on their scholarly knowledge of the polyamine literature, and we thank Jeanne B. Elliott for preparing the manuscript and Neil Buckley for editing it.

REFERENCES

Åkerman, K.E.O. (1977). *J. Bioenerg. Biomembr.* **9,** 65–72.

Allison, S.A., Herr, J.C., and Schurr, J.M. (1981). *Biopolymers* **20,** 469–488.

Anehus, S., Pohjanpelto, P., Baldetorp, B., Långström, E., and Heby, O. (1984). *Mol. Cell. Biol.* **4,** 915–922.

Atkins, J.F., Lewis, J.B., Anderson, C.W., and Gestleland, R.F. (1975). *J. Biol. Chem.* **250,** 5688–5695.

Bachrach, U. (1973). "Function of Naturally Occurring Polyamines." Academic Press, New York.

Ballas, S.K., Mohandas, N., Marton, L.J., and Shohet, S.B. (1983). *Proc. Natl. Acad. Sci. U.S.A.* **80,** 1942–1946.

Behe, M., and Felsenfeld, G. (1981). *Proc. Natl. Acad. Sci. U.S.A.* **78,** 1619–1623.

Blair, D.G.R. (1985). *Int. J. Biochem.* **17,** 23–30.

Bloomfield, V.A., and Wilson, R.W. (1981). *In* "Polyamines in Biology and Medicine" (D.R. Morris and L.J. Marton, eds.), pp. 183–206. Dekker, New York.

Brahmachari, S.K., Ramesh, N., Das, M.R., and Parnaik, V.K. (1985). *Biochem. Int.* **11,** 281–290.

Byczkowski, J.Z., and Porter, C.W. (1983). *Gen. Pharmacol.* **14,** 615–621.

Chattoraj, D.K., Gosule, L.C., and Schellman, J.A. (1978). *J. Mol. Biol.* **121,** 327–329.

Chen, H.H., Behe, M.J., and Rau, D.C. (1984). *Nucleic Acids Res.* **12,** 2381–2389.

Choi, Y.S., and Carr, C.W. (1967). *J. Mol. Biol.* **25,** 331–339.

Christiansen, C., and Baldwin, R.L. (1977). *J. Mol. Biol.* **115,** 441–454.

Chun, P.W., Rennert, O.M., Saffern, E.E., and Taylor, W.J. (1976). *Biochem. Biophys. Res. Commun.* **69,** 1095–1101.

Chun, P.W., Saffen, E.E., Ditore, R.J., Rennert, O.M., and Weinstein, N.H. (1977). *Biophys. Chem.* **6,** 321–335.

Cohen, S.S. (1971). "Introduction to the Polyamines." Prentice-Hall, New York.

Cohen, S.S., and Lichtenstein, J. (1960). *J. Biol. Chem.* **235,** 211–216.

Cooper, H.L., Park, M.H., Folk, J.E., Safer, B., and Braverman, R. (1983). *Proc. Natl. Acad. Sci. U.S.A.* **80,** 1854–1857.

Cooper, K.D., Shukla, J.B., and Rennert, O.M. (1976). *Clin. Chim. Acta* **73,** 71–88.

Cooper, S., and Helmstetter, C.E. (1968). *J. Mol. Biol.* **31,** 519–540.

Cozzarelli, N.R. (1980). *Science* **207,** 953–960.

Dinardo, S., Voelkel, K.A., Sternglanz, R., Reynolds, A.E., and Wright, A. (1982). *Cell* **31,** 45–51.

Dinardo, S., Voelkel, K., and Sternglanz, R. (1984). *Proc. Natl. Acad. Sci. U.S.A.* **81,** 2616–2620.

Dion, A.S., and Cohen, S.S. (1972). *Proc. Natl. Acad. Sci. U.S.A.* **69,** 213–217.

Drew, H.R., and Dickerson, R.E. (1981). *J. Mol. Biol.* **151,** 535–536.

Earnshaw, W.C., and Heck, M.M.S. (1985). *J. Cell Biol.* **100,** 1716–1725.

Earnshaw, W.C., Halligan, B., Cooke, C.A., Heck, M.M.S., and Liu, L.F. (1985). *J. Cell Biol.* **100,** 1706–1715.

Echandi, G., and Algranati, I.D. (1975a). *Biochem. Biophys. Res. Commun.* **62,** 313–319.

Echandi, G., and Algranati, I.D. (1975b). *Biochem. Biophys. Res. Commun.* **67,** 1185–1191.

Feuerstein, B.G., Pattabiraman, N., and Marton, L.J. (1986). *Proc. Natl. Acad. Sci. U.S.A.* **83,** 5948–5952.

Fillingame, R.H., and Morris, D.R. (1973). *Biochemistry* **12,** 4479–4487.

Fillingame, R.H., Jorstad, C.M., and Morris, D.R. (1975). *Proc. Natl. Acad. Sci. U.S.A.* **72,** 4042–4045.

Folk, J.E., and Finlayson, J.S. (1977). *Adv. Protein Chem.* **31,** 1–133.

Fuchs, E. (1976). *Eur. J. Biochem.* **63,** 15–22.

Geiger, L.E., and Morris, D.R. (1978). *Nature (London)* **272,** 730–732.

Geiger, L.E., and Morris, D.R. (1980a). *J. Bacteriol.* **141,** 1192–1198.

Geiger, L.E., and Morris, D.R. (1980b). *Biochim. Biophys. Acta* **609,** 264–281.

Gellert, M. (1981a). *Annu. Rev. Biochem.* **50,** 879–910.

Gellert, M. (1981b). *In* "The Enzymes" (P.D. Boyer, ed.), Vol. 14, pp. 345–366. Academic Press, New York.

Gellert, M., Mizuuchi, K., O'Dea, M.H., and Nash, H.A. (1976). *Proc. Natl. Acad. Sci. U.S.A.* **73,** 3872–3876.

Goldemberg, S.H., and Algranati, I.D. (1981). *Eur. J. Biochem.* **117,** 251–255.

Gosule, L.C., and Schellman, J.A. (1976). *Nature (London)* **259,** 333–335.

Gosule, L.C., and Schellman, J.A. (1978). *J. Mol. Biol.* **121,** 311–326.

Grossowicz, N., and Ariel, M. (1963). *J. Bacteriol.* **85,** 293–300.

Hallick, L.M., and Namba, M. (1974). *Biochemistry* **13,** 3152–3158.

Harada, J.J., and Morris, D.R. (1981). *Mol. Cell. Biol.* **1,** 594–599.

Harada, J.J., Porter, C.W., and Morris, D.R. (1981). *J. Cell. Physiol.* **107,** 413–426.

Heby, O. (1981). *Differentiation* **19,** 1–20.

Heby, O., and Jänne, J. (1981). *In* "Polyamines in Biology and Medicine" (D.R. Morris and L.J. Marton, eds.), pp. 243–310. Dekker, New York.

Herbst, E.J., and Witherspoon, B.H. (1960). *Fed. Proc., Fed. Am. Soc. Exp. Biol.* **19,** 138.

Hershey, H.V., Stieber, J.F., and Mueller, G.C. (1973). *Eur. J. Biochem.* **34,** 383–394.

Hirashima, A., Harigai, H., and Watanabe, I. (1979). *Biochem. Biophys. Res. Commun.* **88,** 1046–1051.

Hölttä, E., and Hovi, T. (1985). *Eur. J. Biochem.* **152,** 229–237.

Hong, K., Schuber, F., and Papahadjopoulos, D. (1983). *Biochim. Biophys. Acta* **732,** 469–472.

Hryniewicz, M.M., and Vonder Haar, R.A. (1983). *Mol. Gen. Genet.* **190,** 336–343.

Hung, D.T., Marton, L.J., Deen, D.F., and Shafer, R.H. (1983). *Science* **221,** 268–270.

Huunan-Seppälä, A. (1971). *Bioenergetics* **2,** 197–207.

Igarashi, K., and Morris, D.R. (1984). *Cancer Res.* **44,** 5332–5337.

Igarashi, K., Kashiwagi, K., Kishida, K., Watanabe, Y., Kogo, A., and Hirose, S. (1979). *Eur. J. Biochem.* **93,** 345–353.

Igarashi, K., Kojima, M., Watanabe, Y., Maeda, K., and Hirose, S. (1980). *Biochem. Biophys. Res. Commun.* **97,** 480–486.

Igarashi, K., Kashiwagi, K., Kishida, K., Kakegawa, T., and Hirose, S. (1981a). *Eur. J. Biochem.* **114,** 127–131.

Igarashi, K., Matsunaka, M., Kashiwagi, K., Mitsui, K., and Hirose, S. (1981b). *Biochim. Biophys. Acta* **656,** 240–245.

Ivanov, V.I., and Minyat, E.E. (1981). *Nucleic Acids Res.* **9,** 4682–4700.

Jorstad, C.J., and Morris, D.R. (1974). *J. Bacteriol.* **119,** 857–860.

Kaiser, D., Tabor, H., and Tabor, C.W. (1963). *J. Mol. Biol.* **6,** 141–147.
Kay, J.E., Ahern, T., Lindsay, V.J., and Sampson, J. (1975). *Biochim. Biophys. Acta* **378,** 241–250.
Kim, J.J.P., and Mehler, A.H. (1981). *Arch. Biochem. Biophys.* **209,** 465–470.
Kimes, B.W., and Morris, D.R. (1973). *Biochemistry* **12,** 442–449.
Kirschbaum, B.B. (1984). *J. Pharmacol. Exp. Ther.* **229,** 409–416.
Klysik, J., Stirdivant, S.M., Larson, J.E., Hart, P.A., and Wells, R.D. (1981). *Nature (London)* **290,** 672–677.
Knutson, J.C., and Morris, D.R. (1978). *Biochim. Biophys. Acta* **520,** 291–301.
Koenig, H., Goldstone, A., and Lu, C.Y. (1983). *Nature (London)* **305,** 530–534.
Kornberg, A. (1982). "Supplement to DNA Replication." Freeman, San Francisco.
Krasnow, M.A., and Cozzarelli, N.R. (1982). *J. Biol. Chem.* **257,** 2687–2693.
Krokan, H., and Eriksen, A. (1977). *Eur. J. Biochem.* **72,** 501–508.
Kurland, C.G. (1982). *Cell* **28,** 201–202.
Kuroda, Y., Merrick, W.C., and Sharma, R.K. (1982). *Science* **215,** 415–416.
Lafer, E.M., Moller, A., Nordheim, A., Stollar, B.D., and Rich, A. (1981). *Proc. Natl. Acad. Sci. U.S.A.* **78,** 3546–3550.
Liquori, A.M., Constantino, L., Crescenzi, V., Elia, V., Giglio, E., Puliti, R., DeSanti-Savino, M., and Vitagliano, V. (1967). *J. Mol. Biol.* **24,** 113–122.
Little, J.R. (1962). *Proc. Soc. Exp. Biol.* **109,** 910–912.
Liu, L.F. (1983). *CRC Crit. Rev. Biochem.* **15,** 1–24.
Lockshon, D., and Morris, D.R. (1983). *Nucleic Acids Res.* **11,** 2999–3017.
Loftfield, R.B., Eigner, E.A., and Pastuszyn, A. (1981). *J. Biol. Chem.* **256,** 6729–6735.
McMahon, M.E., and Erdmann, V.A. (1982). *Biochemistry* **21,** 5280–5288.
Mager, J. (1962). *Biochim. Biophys. Acta* **62,** 202–204.
Marks, K.A., and Ruben, G.C. (1983a). *Nucleic Acids Res.* **11,** 1839–1854.
Marks, L.A., and Ruben, G.C. (1983b). *Biochim. Biophys. Acta* **741,** 279–287.
Marks, L.A., and Ruben, G.C. (1984). *J. Biomol. Struct. Dyn.* **1,** 1109–1132.
Marquet, R., Houssier, C., and Fredericq, E. (1985). *Biochim. Biophys. Acta* **825,** 365–374.
Marton, L.J. (1985). *In* "Polyamines in Clinical Disorders" (D.F. Tierney, moderator), *West. J. Med.* **142,** 63–73.
Minyat, E.E., Ivanov, V.I., Kritzyn, A.M., Minchenkova, L.E., and Schyolkina, A.K. (1978). *J. Mol. Biol.* **128,** 397–409.
Mitsui, K., Igarashi, K., Kakegawa, T., and Hirose, S. (1984a). *Biochemistry* **23,** 2679–2683.
Mitsui, K., Ohnishi, R., Hirose, S., and Igarashi, K. (1984b). *Biochem. Biophys. Res. Commun.* **123,** 528–534.
Morch, M.D., and Benicourt, C. (1980). *Eur. J. Biochem.* **105,** 445–451.
Morris, D.R. (1981). *In* "Polyamines in Biology and Medicine" (D.R. Morris and L.J. Marton, eds.), pp. 223–242. Dekker, New York.
Morris, D.R., and Hansen, M.T. (1973). *J. Bacteriol.* **116,** 588–592.
Morris, D.R., and Harada, J.J. (1980). *In* "Polyamines in Biomedical Research" (J.M. Gaugus, ed.), pp. 1–16. Wiley, London.
Morris, D.R., and Jorstad, C.M. (1973). *J. Bacteriol.* **113,** 271–277.
Morris, D.R., and Lockshon, D. (1981). *In* "Advances in Polyamine Research" (C.M. Caldarera, V. Zappia, and U. Bachrach, eds.), Vol. 3, pp. 299–307. Raven, New York.
Morris, D.R., Jorstad, C.M., and Seyfried, C.E. (1977). *Cancer Res.* **37,** 3169–3172.
Natta, C.L., Motyczka, A.A., and Kremzner, L.T. (1980). *Biochem. Med.* **23,** 144–149.
Nelson, E.M., Tewey, K.M., and Liu, L.F. (1984). *Proc. Natl. Acad. Sci. U.S.A.* **81,** 1361–1365.

Nierlich, D.P. (1978). *Annu. Rev. Microbiol.* **32,** 393–432.
Nilsson, L., Rigler, R., and Wintermeyer, W. (1983). *Biochim. Biophys. Acta* **740,** 460–465.
Nordheim, A., and Rich, A. (1983). *Nature (London)* **303,** 674–678.
Nordheim, A., Pardue, M.L., Lafer, E.M., Mooler, A., Stollar, B.D., and Rich, A. (1981). *Nature (London)* **294,** 417–422.
Nöthig-Laslo, V., Weygand-Durasevic, I., Zivkovic, T., and Kucan, Z. (1981). *Eur. J. Biochem.* **117,** 263–267.
Nöthig-Laslo, V., Weygand-Durasavic, I., and Kucan, Z. (1985). *J. Biomol. Struct. Dyn.* **5,** 941–951.
Oredsson, S.M., Friend, D.S., and Marton, L.J. (1983). *Proc. Natl. Acad. Sci. U.S.A.* **80,** 780–784.
Pardee, A.B., Dubrown, R., Hamlin, J.L., and Kletzien, R.F. (1978). *Annu. Rev. Biochem.* **47,** 715–750.
Peebles, C.L., Gegenheimer, P., and Abelson, J. (1983). *Cell* **32,** 525–536.
Pleshkewych, A., Kramer, D.L., Kelly E., and Porter, C.W. (1980). *Cancer Res.* **40,** 4533–4540.
Pochon, F., and Cohen, S.S. (1972). *Biochem. Biophys. Res. Commun.* **47,** 720–726.
Pommier, Y., Minford, J.K., Schwartz, R.E., Zwelling, L.A., and Kohn, K.W. (1985). *Biochemistry* **24,** 6410–6416.
Porschke, D. (1984). *Biochemistry* **23,** 4821–4828.
Pruss, G., Manes, S.H., and Drlica, K. (1982). *Cell* **31,** 35–42.
Quigley, G.J., Teeter, M.M., and Rich, A. (1978). *Proc. Natl. Acad. Sci. U.S.A.* **75,** 64–68.
Rennert, O., Buehler, B., Miale, T., and Lawson, D. (1976). *Life Sci.* **19,** 257–264.
Rosano, C.L., Bunce, S.C., and Hurwitz, C. (1983). *J. Bacteriol.* **153,** 326–334.
Rudkin, B.B., Mamont, P.S., and Seiler, N. (1984). *Biochem. J.* **217,** 731–741.
Russell, D.H., and Durie, B.G.M. (1978). "Polyamines as Biochemical Markers of Normal and Malignant Growth," pp. 89–105. Raven, New York.
Sakai, T.T., and Cohen, S.S. (1973). *In* "Polyamines in Normal and Neoplastic Growth" (D.H. Russell, ed.), pp. 41–54. Raven, New York.
Schellman, J.A., and Parthasarathy, N. (1984). *J. Mol. Biol.* **175,** 313–329.
Schindler, M., Koppel, D.E., and Sheetz, M.P. (1980). *Proc. Natl. Acad. Sci. U.S.A.* **77,** 1457–1461.
Schreier, A.A., and Schimmel, P.R. (1975). *J. Mol. Biol.* **93,** 323–329.
Schrode, J., and Folk, J.E. (1978). *J. Biol. Chem.* **253,** 4837–4840.
Seely, J.E., Stetler, D.A., Jacob, S.T., and Pegg, A.E. (1984). *Biochem. Biophys. Res. Commun.* **120,** 219–225.
Seyfried, C.E., and Morris, D.R. (1979). *Cancer Res.* **39,** 4861–4867.
Shekman R., Weiner, A., and Kornberg, A. (1974). *Science* **186,** 987–993.
Srivenugopal, K.S., and Morris, D.R. (1985). *Biochemistry* **24,** 4766–4771.
Srivenugopal, K.S., Lockshon, D., and Morris, D.R. (1984). *Biochemistry* **23,** 1899–1906.
Stevens, L. (1970). *Biol. Rev.* **45,** 1–27.
Tabor, C.W. (1960). *Biochem. Biophys. Res. Commun.* **2,** 117–120.
Tabor, C.W. (1962). *J. Bacteriol.* **83,** 1101–1111.
Tabor, C.W., and Kellogg, P.D. (1967). *J. Biol. Chem.* **242,** 1044–1052.
Tabor, C.W., and Tabor, H. (1976). *Annu. Rev. Biochem.* **45,** 285–306.
Tabor, C.W., and Tabor, H. (1984). *Annu. Rev. Biochem.* **53,** 749–790.
Tabor, C.W., and Tabor, H. (1985). *Microbiol. Rev.* **49,** 81–99.
Tabor, H., and Tabor, C.W. (1964). *Pharmacol. Rev.* **16,** 245–300.
Tabor, H., and Tabor, C.W. (1972). *Adv. Enzymol.* **36,** 203–268.
Tabor, H., and Tabor, C.W. (1982). *Proc. Natl. Acad. Sci. U.S.A.* **79,** 7087–7091.

Tabor, H., Tabor, C.W., Cohn, M.S., and Hafner, E.W. (1981). *J. Bacteriol.* **147,** 702–704.
Takemoto, T., Nagamatsu, Y., and Oka, T. (1983). *Biochim. Biophys. Acta* **740,** 73–79.
Thomas, T.J., and Bloomfield, V.A. (1983). *Biopolymers* **22,** 1097–1106.
Thomas, T.J., and Bloomfield, V.A. (1985). *Biopolymers* **23,** 1295–1306.
Thomas, T.J., Bloomfield, V.A., and Canallakis, Z.N. (1985). *Biopolymers* **24,** 725–729.
Treadwell, B.V., Kung, H.F., and Weissbach, H. (1976). *Arch. Biochem. Biophys.* **176,** 799–801.
Tropp, J.S., and Redfield, A.G. (1983). *Nucleic Acids Res.* **11,** 2121–2134.
Tsuboi, M. (1964). *Bull. Chem. Soc. Jpn.* **37,** 1514–1522.
Uemura, T., and Yanagida, T. (1984). *EMBO J.* **3,** 1737–1744.
Wang, J.C. (1973). *In* "DNA Synthesis *In Vitro*" (R.B. Inman and R.D. Wells, eds.), pp. 163–174. Univ. Park Press, Baltimore.
Wang, J.C. (1985). *Annu. Rev. Biochem.* **54,** 665–697.
Wang, J.C., and Liu, L.F. (1979). *In* "Molecular Genetics" (J.H. Taylor, ed.), Part III, pp. 65–88. Academic Press, New York.
Watanabe, Y., Igarashi, K., and Hirose, S. (1981). *Biochim. Biophys. Acta* **656,** 134–139.
Watanabe, Y., Igarashi, K., Mitsui, K., and Hirose, S. (1983). *Biochim. Biophys. Acta* **740,** 362–368.
Weiss, R.L., and Morris, D.R. (1973). *Biochemistry* **12,** 435–441.
Weiss, R.L., Kimes, B.W., and Morris, D.R. (1973). *Biochemistry* **12,** 450–456.
Wemmer, D.E., Srivenugopal, K.S., Reid, B.R., and Morris, D.R. (1985). *J. Mol. Biol.* **185,** 457–459.
Widom, J., and Baldwin, R.L. (1980). *J. Mol. Biol.* **144,** 431–453.
Widom, J., and Baldwin, R.L. (1983). *Biopolymers* **22,** 1595–1620.
Wilson, R.W., and Bloomfield, V.A. (1979). *Biochemistry* **18,** 2192–2196.
Wirtz, G.H., and Becker, E.L. (1961). *Immunology* **4,** 473–485.
Young, D.V., and Srinivasan, P.R. (1972). *J. Bacteriol.* **112,** 30–39.
Young, D.V., and Srinivasan, P.R. (1974). *J. Bacteriol.* **117,** 1280–1288.
Zwelling, L.A., Kerrigan, D., and Marton, L.J. (1985). *Cancer Res.* **45,** 1122–1125.

5

The Use of Inhibitors to Study the Biochemistry and Molecular Biology of Polyamine Biosynthesis and Uptake

ANTHONY E. PEGG

Department of Physiology and Cancer Research Center
Milton S. Hershey Medical Center
Pennsylvania State University College of Medicine
Hershey, Pennsylvania 17033

I. INTRODUCTION

The use of inhibitors of polyamine biosynthesis to study the physiological function of polyamines and the potential of the polyamine pathway as a target for chemotherapeutics are discussed at length in the other chapters of this book. In addition to these important areas the development of these inhibitors has provided extremely valuable research tools to study the biochemistry underlying the regulation of cellular polyamine levels. This work is outlined here. In three separate areas the availability of these inhibitors has led to significant increase in the understanding of this regulation:

1. The study of ornithine decarboxylase (ODC)[1] where the availability of DFMO has facilitated cloning of the gene and permitted the specific labeling of the enzyme by covalent binding of the radioactive form of the

[1] Abbreviations: ODC, ornithine decarboxylase; AdoMetDC, *S*-adenosylmethionine decarboxylase; DFMO, DL-α-difluoromethylornithine; MGBG, methylglyoxal bis-guanylhydrazone); $AdoS^+(CH_3)_2$, *S*-methyl-5′-methylthioadenosine.

drug. Such labeling has been used to aid in the detection and purification of the protein; to localize the enzyme by autoradiography and to study its degradation, to provide a convenient labeled antigen for radioimmunoassay, and to determine the active site of ODC.

2. The isolation and cloning of *S*-adenosylmethionine decarboxylase (AdoMetDC). The AdoMetDC inhibitor, MGBG, can be used to increase the amount of the enzyme and to prepare a powerful affinity chromatography adsorbent for its purification. The ODC inhibitors such as DFMO can be used to increase the abundance of the mRNA for AdoMetDC by depletion of polyamines which act as repressors of its synthesis.

3. The study of the polyamine transport system. Polyamine depletion by ODC inhibitors greatly enhances the activity of this system and MGBG can be used as a model, nonmetabolized substrate to study this process.

II. STUDIES OF ORNITHINE DECARBOXYLASE

A. Purification and Structure of ODC Protein

DFMO was conceived as an enzyme-activated irreversible inhibitor which would act by virtue of its ability to serve as a substrate for ODC. It was thought that after decarboxylation a reactive intermediate would be generated which would inactivate the enzyme by forming a covalent bond with a residue at the active site (Metcalf *et al.*, 1978; Bey, 1978). Subsequent experiments with radioactive DFMO labeled with ^{14}C at the 5-position and at the 1-position have confirmed this mechanism (Pritchard *et al.*, 1981; Seely *et al.*, 1982a–c; Pegg, 1986). The partition ratio of decarboxylation to inactivation is about 3.

The binding of DFMO to ODC is highly specific even in crude tissue extracts (Seely *et al.*, 1982a–c; Erwin *et al.*, 1983) and the reaction of such extracts with [^{3}H]DFMO can be used to titrate the number of molecules of active ODC protein present (Seely *et al.*, 1982a,c; Erwin *et al.*, 1983; Kyriakidis *et al.*, 1984). Similarly, the titration of partially purified ODC preparations with radioactive DFMO can be used to evaluate their purity (Pritchard *et al.*, 1981; Seely *et al.*, 1982b; Kameji and Hayashi, 1982). Such titrations have shown that ODC is a very minor component of the total soluble protein amounting to less than 1 part per million of the soluble protein in most cells (Seely *et al.*, 1982b,c; Pegg *et al.*, 1982; Kameji *et al.*, 1982). Although a very large extent of induction can be brought about in the rat liver by a variety of agents, ODC protein is only 1 part in 350,000 after maximal induction (Kameji *et al.*, 1982). The mouse kidney after maximal stimulation with androgens contains much higher amounts of ODC protein than other mammalian tissues and ODC can amount to up to 1 part in 8000 (Persson, 1982; Seely *et al.*, 1982a–c; Isomaa *et al.*, 1983).

The reaction with DFMO can be used to identify the protein corresponding to the subunit of ODC in two ways. If radioactive DFMO is used, the protein can be labeled and its chromatographic or electrophoretic mobility can then be followed (Pritchard *et al.*, 1981; Isomaa *et al.*, 1983; Kameji and Hayashi, 1982; Seely *et al.*, 1982b, 1985; Persson *et al.*, 1984). In this way it was determined that the subunit M_r of rodent ODC is about 53,000 and that there are two forms of the protein differing slightly in charge and having isoelectric points of about 5.3 (Persson *et al.*, 1984; Seely *et al.*, 1985).

Second, it appears that the binding to ODC of the reactive species derived from DFMO slightly alters the charge on the molecule and thus changes its mobility on isoelectric focusing gels. Although the change in mobility is small, it is possible to identify the ODC protein by comparing gels of proteins before and after reaction with DFMO and looking for the spots which assume a different position (McConlogue and Coffino, 1983b; Persson *et al.*, 1984; Seely *et al.*, 1985).

The reaction of ODC protein with [^{3}H]DFMO of high specific activity has also been used to generate a labeled ligand for RIA (Seely and Pegg, 1983a,b; Alhonen-Hongisto *et al.*, 1985a–c) and for the screening of hybridoma cells producing monoclonal antibodies to this protein (Pegg *et al.*, 1984). Since the interaction of DFMO with ODC is so specific, it is not necessary to have a completely homogeneous ODC preparation in order to produce this labeled ligand. In this respect, the use of [^{3}H]DFMO rather than a nonspecific iodination technique is a considerable advantage. Studies using immunochemical methods to quantitate ODC protein have shown that the changes in ODC activity produced by a wide variety of stimuli are due to parallel alterations in the amount of ODC protein (reviewed by Pegg, 1986).

The exact nature of the linkage between DFMO and ODC has not yet been determined but it is possible to carry out specific chemical and/or proteolytic cleavage of the [^{14}C]DFMO-labeled protein from mouse kidney. Analysis of the size of these fragments by gel electrophoresis and comparison of these fragments with the amino acid sequence derived from the published cDNA sequence indicate that the binding appears to occur on lysine at residue 298 (Pegg, 1986). This presumably forms part of the active site or is very close to it. Similar analysis of the DFMO-labeled peptides generated from ODC from other species would be of considerable interest.

B. Investigation of the Degradation of ODC Protein

DFMO has also been used to study the turnover of ODC. Studies by Seely *et al.* (1982a) of the fate of the ODC protein labeled by administration of [^{14}C]DFMO to mice *in vivo* confirmed directly the rapid turnover of this protein and indicated that androgen administration increases its half-life. No labeled peptide fragments were detected in these experiments, indicating that the rate-limiting step in the degradation is likely to be the initial cleavage and that further degradation is even more rapid.

The exact mechanism of the degradation of ODC is still a mystery but it appears that it involves the binding to ODC of the inhibitory protein called antizyme (Fujita *et al.*, 1982; Seely and Pegg, 1983b; Kitani and Fujisawa, 1984; Murakami *et al.*, 1985; Murakami and Hayashi, 1985). This protein was first discovered by Canellakis and his colleagues on the basis of its formation of an inactive complex with ODC (reviewed by Canellakis *et al.*, 1979), but there appears to be only a limited accumulation of antizyme–ODC complexes *in vivo* because of their rapid degradation (Seely and Pegg, 1983b; Murakami *et al.*, 1985; Murakami and Hayashi, 1985). The reversible nature of the complex formed between antizyme and ODC is indicated by (1) the fact that the complexed ODC does react with DFMO, albeit slowly (Kyriakidis *et al.*, 1984) and (2) ODC which has been inactivated with DFMO can be used to displace the active ODC enzyme from antizyme–ODC complexes (Murakami *et al.*, 1985). This displacement forms the basis of a convenient assay for these inactive complexes (Murakami *et al.*, 1985; Murakami and Hayashi, 1985).

Although gene amplification leading to the overproduction of ODC protein appears to be a common mechanism for the development of cells resistant to ODC inhibitors such as DFMO (see Section II,D below), another pathway to resistance involves an increased stability of the enzyme. Several cell lines have been described in which this change has occurred and the increase in half-life produces a higher amount of the enzyme protein at the steady state (McCann *et al.*, 1979; Pritchard *et al.*, 1982; Pösö *et al.*, 1984; Murakami *et al.*, 1985). Comparisons of these cell lines with their normal equivalents provide an important tool for the investigation of the mechanism of degradation of ODC. The use of cloned ODC genes in mammalian expression vectors and the introduction of labeled ODC into cells by cell fusion or microinjection techniques also provide exciting new approaches to this question.

C. Autoradiographic Localization of ODC

The highly specific nature of the covalent binding of labeled DFMO to ODC permits it to be used to study the localization of the enzyme by autoradiography (Pegg *et al.*, 1982; Emmanuelson and Heby, 1982, 1983; Zagon *et al.*, 1983, 1984). These results have indicated that a fraction of the ODC is present in the nucleus both in mouse kidney cells (Zagon *et al.*, 1984) and in the cells of a polychaete, *Ophryotrocha labronica* (Emmanuelson and Heby, 1983). In the male mouse kidney it was found that the majority of the ODC was present in the cells of the proximal tubules with considerably lower amounts in the distal tubules and in the collecting tubules (Pegg *et al.*, 1982; Zagon *et al.*, 1984).

D. Gene Amplification, cDNA Production, and Analysis of mRNA Levels

Several groups have demonstrated that prolonged exposure of cultured cells to DFMO leads to the development of resistance to the antiproliferative effects of this drug and that this resistance is mediated via the overproduction of the ODC protein (McConlogue and Coffino, 1983a,b; Choi and Scheffler, 1983; Kahana and Nathans, 1984). In most cases, this overproduction is due to a striking amplification of the ODC gene (McConlogue *et al.*, 1984; Kahana and Nathans, 1984; Alhonen-Hongisto *et al.*, 1985a,b) and the mRNA for the enzyme is correspondingly elevated, thus providing a suitable material for facile cloning of the cDNA (McConlogue *et al.*, 1984; Kahana and Nathans, 1984). These cDNA probes have been used to deduce the complete sequence of the ODC protein, to study the regulation of ODC synthesis, and to study the ODC gene.

The nucleotide sequences corresponding to cDNAs of the ODC mRNA have been published by two groups (Gupta and Coffino, 1985; Kahana and Nathans, 1985a). Their results for the region encompassing the coding portion and hence the predicted amino acid sequence are in full agreement with each other although they differ substantially in the 5′ leader sequence. The cDNA sequences indicate that the enzyme subunit contains 461 amino acids and has a predicted isoelectric point of 5.1 and an M_r of about 51,000, which agrees well with values obtained for the purified mouse kidney protein (Seely *et al.*, 1982b). The reason for the disagreement of the 5′ sequences is unclear and may relate to cloning artifacts, but the very long 5′ noncoding sequence may be of regulatory importance. According to Kahana and Nathans (1985a) it contains a number of termination codons which follow possible initiation sites and may be the cause of the very poor translation of ODC mRNA in the reticulocyte lysate system. Removal of the sequence facilitates the translation of the mRNA in reticulocyte lysates (Kahana and Nathans, 1985b).

Analysis by the Northern blotting technique using cDNA probes for ODC has shown that there are two forms of ODC mRNA with sizes of 2.2 and 2.7 kb (Berger *et al.*, 1984; Kontula *et al.*, 1984; McConlogue *et al.*,

1984). These two forms reflect the presence of two possible polyadenylation sites in the 3′ noncoding segment (Hickok *et al.*, 1986). There is also some evidence from Southern blot analysis that there is a family of ODC genes (Berger *et al.*, 1984; Alhonen-Hongisto *et al.*, 1985a,b; Hickok *et al.*, 1986). It appears that more than one of these may be active (Berger *et al.*, 1984; Alhonen-Hongisto *et al.*, 1985a,b) but some may represent pseudogenes and further investigations involving sequencing of genomic clones are needed. It should be noted that although there is complete agreement between the published sequences of the coding regions for ODC from mouse X63-Ag8.653 myeloma cells (Kahana and Nathans, 1985a) and mouse S49 lymphoma cells (Gupta and Coffino, 1985), the sequences obtained from cDNA clones corresponding to mouse kidney ODC differ slightly (Hickok *et al.*, 1986). This may also indicate the presence of multiple ODC genes and the studies with probes obtained from cells in which gene amplification has occurred may not reflect the complete spectrum of genes which are normally expressed.

The majority of the DFMO-resistant cell lines which have been examined so far have a greatly increased content of an ODC gene and it appears that this amplification occurs quite readily when polyamine synthesis and growth are blocked by this inhibitor. However, the increased expression of ODC may not always be due to gene amplification. It has been reported by Alhonen-Hongisto *et al.* (1985c) that L1210 cells exposed to DFMO and cadaverine overproduce ODC mRNA and protein without substantial gene amplification.

The remarkable increases in ODC protein which occur in response to a variety of stimuli are due in part to an increased rate of synthesis. In some cases this increased rate of synthesis is clearly paralleled by an increased accumulation of the mRNA (Berger *et al.*, 1984; Kahana and Nathans, 1984; Kontula *et al.*, 1984, 1985; Alhonen-Hongisto *et al.*, 1985c; Pohjanpelto *et al.*, 1985; Feinstein *et al.*, 1985; Gilmour *et al.*, 1985), but it is also clear that regulation of the translation of the ODC mRNA can also occur (Berger *et al.*, 1984; Kontula *et al.*, 1984, 1985; Kameji *et al.*, 1984; Pohjanpelto *et al.*, 1985; Persson *et al.*, 1985; Gilmour *et al.*, 1985; Kahana and Nathans, 1985b; Dirks *et al.*, 1986; Sertich and Pegg, 1986).

The rapid decline of ODC activity in response to the exogenous application of polyamines appears not to involve a decline in the level of the mRNA (Kahana and Nathans, 1985b; Dirks *et al.*, 1986; Sertich and Pegg, 1986). These experiments provide clear evidence for translational control under some conditions, but the situation is complex since a substantial part of the loss of ODC protein appears to be mediated via an increased rate of enzyme degradation rather than by a reduction in the rate of synthesis (Persson *et al.*, 1984; Murakami *et al.*, 1985; Dirks *et al.*, 1986).

III. STUDIES OF *S*-ADENOSYLMETHIONINE DECARBOXYLASE (AdoMetDC)

A. Purification and Structure of AdoMetDC

Administration of the inhibitor MGBG leads to a large, apparently paradoxical increase in the amount of AdoMetDC (see reviews by Williams-Ashman and Pegg, 1981; Pegg, 1984; Tabor and Tabor, 1984). This increase appears to be due to a great increase in the half-life of the enzyme protein (Pegg, 1979, 1984; Williams-Ashman and Pegg, 1981) and no change in the mRNA level was seen in response to MGBG (Shirahata and Pegg, 1986).

Since MGBG is a reversible inhibitor and can readily be removed from the enzyme by dialysis, pretreatment with MGBG can be used to enrich the starting material for enzyme purification by 20- to 30-fold. Furthermore, an excellent matrix for purification of AdoMetDC by affinity chromotography can be made by linking MGBG to Sepharose (Pegg, 1974, 1984; Pösö and Pegg, 1982; Tabor and Tabor, 1984). Two forms of AdoMetDC with differing isoelectric points were found to be present in rat tissues, with one form predominating in skeletal muscle (psoas) and the other in liver (Pösö and Pegg, 1982). These forms could be distinguished on the basis of their sensitivity to MGBG and could be separated by chromatography on MGBG–Sepharose. The psoas form was more strongly inhibited by MGBG and bound more tightly to the affinity column (Pösö and Pegg, 1982).

Although no irreversible inhibitor comparable to DFMO is available for AdoMetDC, a procedure for titration and labeling of this enzyme similar to the one described above using DFMO for ODC has been developed (Shirahata *et al.*, 1985). Sodium cyanoborohydride was used to reduce the Schiff base between ^{35}S-decarboxylated AdoMet and the pyruvate prosthetic group on the enzyme. It was shown that this procedure was highly specific and that even in crude extracts the only protein labeled was the subunit of AdoMetDC. This subunit of M_r 32,000 was labeled stoichiometrically by this procedure. The method was used to assess the purity of AdoMetDC preparations and to provide a labeled ligand for radioimmunoassay of the protein (Shirahata *et al.*, 1985; Shirahata and Pegg, 1985). The analysis by radioimmunoassay indicated that the changes in AdoMetDC activity brought about by a variety of stimuli, including androgen treatment, polyamine depletion, and excess MGBG, were due to changes in the amount of the enzyme protein (Pegg, 1984; Shirahata *et al.*, 1985; Shirahata and Pegg, 1985). Immunochemical methods have also been used to demonstrate that both the rate of synthesis and the half-life of the protein are increased by mitogenic stimulation of bovine lymphocytes (Seyfried *et al.*, 1982).

B. Cloning of cDNA and *in Vitro* Synthesis of AdoMetDC Protein

AdoMetDC is closely regulated by the polyamine content. Both spermidine and spermine can bring about this regulation and when the content of these polyamines is increased AdoMetDC content decreases. Conversely, when inhibitors of polyamine synthesis are used, the amount of AdoMetDC protein is increased (Mamont *et al.*, 1981, 1982; Pegg, 1984; Tabor and Tabor, 1984). It appears that spermine is the more important polyamine in this regulation since AdoMetDC increases by about 8-fold in cells depleted of spermidine by DFMO (Alhonen-Hongisto, 1980; Mamont *et al.*, 1981; Pegg, 1984; Shirahata *et al.*, 1985) and by up to 25-fold in cells depleted of spermine by $AdoS^{+}(CH_3)_2$ (Pegg *et al.*, 1987). The increased activity of AdoMetDC in cells or tissues depleted of spermidine by DFMO treatment was due both to a stabilization of the active protein and to an increased rate of synthesis (Shirahata and Pegg, 1985; Pegg and Shirahata, 1986). Administration of DFMO in the drinking water for 3 days increased the content of AdoMetDC mRNA in rat prostate by about 9-fold. Although treatment with MGBG produced a larger increase in the amount of AdoMetDC protein than DFMO, MGBG did not increase the mRNA for the enzyme at all under these conditions (Pegg and Shirahata, 1986).

A highly purified preparation of AdoMetDC mRNA was obtained from prostates of rats treated with DFMO in this way. The polysomes synthesizing AdoMetDC were purified by immunoprecipitation with a monospecific antiserum to AdoMetDC. The mRNA isolated from these polysomes appeared to be about 30% pure mRNA for AdoMetDC when translated in the reticulocyte lysate system (Pegg and Shirahata, 1986). This mRNA has been used for cloning of cDNA to AdoMetDC and for the study of the synthesis of the AdoMetDC protein. It was found that a precursor of M_r 37,000 was produced in the reticulocyte lysate system and that this was converted to the active enzyme subunit of M_r 32,000 during continued incubation (Pegg and Shirahata, 1986). This conversion presumably generates the pyruvate prosthetic group at the amino terminus of the subunit but it is not yet clear whether the conversion is spontaneous or requires the presence of processing proteins in the lysate.

A cDNA clone to bovine AdoMetDC has been obtained by Mach *et al.* (1985). They isolated the enzyme from bovine liver using affinity chromatography on MGBG–Sepharose (Seyfried *et al.*, 1982) and obtained the amino sequence of some of the peptides generated by a lysine-specific protease, Lys-C. An oligonucleotide probe corresponding to a sequence Tyr-Trp-Thr-Ile-His-Ile, which was present in one of these peptides, was synthesized and used to screen a bovine cDNA library in λgt11. A resulting clone was confirmed by sequencing and was used to demonstrate that a rapid fourfold increase in AdoMetDC mRNA occurs within 6 hr of stimulation of bovine lymphocytes with concanavalin (Mach *et al.*, 1985).

IV. STUDIES OF POLYAMINE TRANSPORT

A. Enhancement of Uptake by Polyamine Depletion

It is well established that a wide range of mammalian cells contain a transport system for the uptake of polyamines (see Rinehart and Chen, 1984, and references therein), but little is known of the underlying biochemistry and function of this system. Since most, if not all, of the cells possessing this transport system are capable of *de novo* synthesis of polyamines, it is not clear what the purpose of the system actually is. However, it appears to be regulated by cellular polyamine content since an increased activity of the uptake system occurs in response to polyamine depletion brought about by ODC inhibitors (Alhonen-Hongisto *et al.*, 1980; Jänne *et al.*, 1983). The mechanism by which this increase is brought about is not known but the ability to enhance the uptake in this way should aid in the isolation of the carrier protein.

The uptake system is not totally specific for the naturally occurring polyamines and a variety of derivatives and analogs are also taken up by it, including MGBG (Mandel and Flintoff, 1978; Seppänen *et al.*, 1981; Porter *et al.*, 1982, 1984, 1985; Williams-Ashman and Seidenfeld, 1986). It has been suggested that priming with DFMO may be used to enhance the uptake of MGBG into tumors and thus improve its therapeutic potential (Seppänen *et al.*, 1980, 1981; Jänne *et al.*, 1983), but this procedure has not yet shown any significant advantage in minimizing the toxicity of MGBG. Kramer *et al.* (1985) were unable to confirm earlier reports (Kallio *et al.*, 1983; Jänne *et al.*, 1983) that the pretreatment with DFMO selectively increased uptake into tumor cells. It appears that the increased uptake is a general response of all cells when intracellular polyamine levels become depleted.

B. Use of MGBG as a Substrate for Polyamine Transport

Since MGBG is an extremely cytotoxic substance, its application to cells following exposure to a mutagenic agent provides a powerful method for the selection of those mutations which have rendered the transport system to be inoperative. Such mutants in CHO and L6 cells were obtained in this way by Mandel and Flintoff (1978). Their availability provides another tool for the study of the biochemistry and function of the polyamine transport system. The importance of this system in permitting the uptake of polyamines and putrescine into cells is clearly demonstrated by the fact that the adverse effects of DFMO on growth cannot be reversed by the application of polyamines to these mutant CHO cells (Byers *et al.,* 1987).

MGBG is a useful model substrate with which to study the transport system since it is not metabolized to any substantial extent (Porter *et al.,* 1981; Williams-Ashman and Seidenfeld, 1986). The uptake of the extremely poisonous herbicide, paraquat, into mammalian cells appears to involve its transport by the same system which transports polyamines or MGBG. The uptake of paraquat is competitively inhibited by these polyamines (Gordon-Smith *et al.,* 1983; Smith, 1985; Rannels *et al.,* 1985) and the mutant CHO cells which are resistant to MGBG are also resistant to paraquat (Byers *et al.,* 1986).

More detailed characterization of the polyamine transport system is obviously desirable and it appears that suitable experimental systems for this are now available. MGBG can be used as a convenient nonmetabolized substrate for the system. The opportunities to increase activity by treatment with DFMO, and to compare cell types which have large amounts of the activity with mutants which lack the ability to transport polyamines, should facilitate this work.

REFERENCES

Alhonen-Hongisto, L. (1980). *Biochem. J.* **190,** 747–754.

Alhonen-Hongisto, L., Seppänen, P., and Jänne, J. (1980). *Biochem. J.* **192,** 941–945.

Alhonen-Hongisto, L., Kallio, A., Sinervirta, R., Seppänen, P., Kontula, K.K., Jänne, O.A., and Jänne, J. (1985a). *Biochem. Biophys. Res. Commun.* **126,** 734–740.

Alhonen-Hongisto, L., Kallio, A., Sinervirta, R., Jänne, O.A., Gahmberg, C.G., and Jänne, J. (1985b). *Biochem. J.* **229,** 711–715.

Alhonen-Hongisto, L., Sinervirta, R., Jänne, O.A., and Jänne, J. (1985c). *Biochem. J.* **232,** 605–607.

Berger, F.G., Szymanski, P., Read, E., and Watson, G. (1984). *J. Biol. Chem.* **259,** 7941–7946.

Bey, P. (1978). *In* "Enzyme-Activated Irreversible Inhibitors" (N. Seiler, M.J. Jung, and J. Koch-Weser, eds.), pp. 27–42. Elsevier, Amsterdam.
Byers, T., Kameji, R., Rannels, D.E., and Pegg, A.E. (1987). *Am. J. Physiol.*, in press.
Canellakis, E.S., Viceps-Madore, D., Kyriakidis, D.A., and Heller, J.S. (1979). *Curr. Top. Cell. Regul.* **15,** 155–202.
Choi, J.H., and Scheffler, I.E. (1983). *J. Biol. Chem.* **258,** 12601–12608.
Dirks, L., Grens, A., Slezynger, T.C., and Scheffler, I.E. (1986). *J. Cell. Physiol.* **126,** 371–378.
Emmanuelson, H., and Heby, O. (1982). *Cell Biol. Int. Rep.* **6,** 951–954.
Emmanuelson, H., and Heby, O. (1983). *Eur. J. Cell Biol.* **30,** 288–294.
Erwin, B.G., Seely, J.E., and Pegg, A.E. (1983). *Biochemistry* **22,** 3027–3032.
Feinstein, S.C., Dana, S.L., McConlogue, L., Shooter, E.M., and Coffino, P. (1985). *Proc. Natl. Acad. Sci. U.S.A.* **82,** 5761–5765.
Fujita, K., Murakami, Y., and Hayashi, S. (1982). *Biochem. J.* **204,** 647–652.
Gilmour, S.K., Avdalovic, N., Madara, T., and O'Brien, T.G. (1985). *J. Biol. Chem.* **260,** 16439–16444.
Gordon-Smith, R.H., Brooke-Taylor, S., Smith, L.L., and Cohen, G.M. (1983). *Biochem. Pharmacol.* **32,** 3701–3709.
Gupta, M., and Coffino, P. (1985). *J. Biol. Chem.* **260,** 2941–2944.
Hickok, N.J., Seppänen, P., Kontula, K.K., Jänne, J.A., Bardin, C.W., and Jänne, O.A. (1986). *Proc. Natl. Acad. Sci. U.S.A.* **83,** 594–598.
Isomaa, V.V., Pajunen, A.E.I., Bardin, C.W., and Jänne, O.A. (1983). *J. Biol. Chem.* **258,** 6735–6740.
Jänne, J., Hölttä, E., Kallio, A., and Käpyaho, K. (1983). *Spec. Top. Endocrinol. Metab.* **5,** 227–293.
Kahana, C., and Nathans, D. (1984). *Proc. Natl. Acad. Sci. U.S.A.* **81,** 3645–3649.
Kahana, C., and Nathans, D. (1985a). *Proc. Natl. Acad. Sci. U.S.A.* **82,** 1673–1677.
Kahana, C., and Nathans, D. (1985b). *J. Biol. Chem.* **260,** 15390–15393.
Kallio, A., Sëppanen, P., Alhonen-Hongisto, L., and Jänne, J. (1983). *Cancer Res.* **43,** 324–327.
Kallio, A., Nikula, P., and Jänne, J. (1984). *Biochem. J.* **218,** 641–644.
Kameji, T., and Hayashi, S. (1982). *Biochim. Biophys. Acta* **705,** 405–407.
Kameji, T., Murakami, Y., Fujita, K., and Hayshi, S. (1982). *Biochim. Biophys. Acta* **717,** 111–117.
Kameji, T., Fujita, K., Noguchi, T., Takiguchi, M., Mori, M., Tatibana, M., and Hayashi, H. (1984). *Eur. J. Biochem.* 35–39.
Kitani, T., and Fujisawa, H. (1984). *J. Biol. Chem.* **259,** 12307–12310.
Kontula, K.K., Torkkeli, T.K., Bardin, C.W., and Jänne, O.A. (1984). *Proc. Natl. Acad. Sci. U.S.A.* **81,** 731–735.
Kontula, K.K., Seppänen, P.J., Van Duyne, P., Bardin, C.W., and Jänne, O.A. (1985). *Endocrinology* **116,** 226–233.
Kramer, D.L., Paul, B., and Porter, C.W. (1985). *Cancer Res.* **45,** 2512–2515.
Kyriakidis, K.A., Flamigni, F., Pawlak, J.W., and Canellakis, E.S. (1984). *Biochem. Pharmacol.* **33,** 1575–1578.
McCann, P.P., Tardif, C., Hornsperger, J.-M., and Böhlen, P. (1979). *J. Cell. Physiol.* **99,** 183–190.
McConlogue, L., and Coffino, P. (1983a). *J. Biol. Chem.* **258,** 12083–12086.
McConlogue, L., and Coffino, P. (1983b). *J. Biol. Chem.* **258,** 8384–8388.
McConlogue, L., Gupta, M., Wu, L., and Coffino, P. (1984). *Proc. Natl. Acad. Sci. U.S.A.* **81,** 540–544.

Mach, M., Neubauer, M.G., Degen, J.L., Seyfried, C.E., and Morris, D.R. (1985). *In* "Polyamines: Basic and Clinical Aspects" (K. Imahori, F. Suzuki, O. Suzuki, and U. Bachrach, eds.), pp. 155–164. VNU Science Press, Utrecht.

Mamont, P.S., Joder-Ohlenbusch, A.-M., Nussli, M., and Grove, J. (1981). *Biochem. J.* **196,** 411–422.

Mamont, P.S., Danzin, C., Wagner, J., Siat, M., Joder-Ohlenbusch, A.-M., and Claverie, N. (1982). *Eur. J. Biochem.* **123,** 499–504.

Mandel, J.-L., and Flintoff, W.F. (1978). *J. Cell. Physiol.* **97,** 335–344.

Metcalf, B.W., Bey, P., Danzin, C., Jung, M.J., Casara, P., and Vevert, J.P. (1978). *J. Am. Chem. Soc.* **100,** 2551–2553.

Murakami, Y., and Hayashi, S. (1985). *Biochem. J.* **226,** 893–896.

Murakami, Y., Fujita, K., Kameji, T., and Hayashi, S. (1985). *Biochem. J.* **225,** 669–697.

Pegg, A.E. (1974). *Biochem. J.* **141,** 581–583.

Pegg, A.E. (1979). *J. Biol. Chem.* **254,** 3249–3253.

Pegg, A.E. (1984). *Cell Biochem. Function* **2,** 11–15.

Pegg, A.E. (1986). *Biochem. J.* **234,** 249–262.

Pegg, A.E., and Shirahata, A. (1986). *Fed. Proc., Fed. Am. Soc. Exp. Biol.* **45,** 1459.

Pegg, A.E., Seely, J.E., and Zagon, I.S. (1982). *Science* **217,** 68–70.

Pegg, A.E., Seely, J.E., Persson, L., Herlyn, M., Ponsell, K., and O'Brien, T.G. (1984). *Biochem. J.* **217,** 123–128.

Pegg, A.E., Wechter, R., and Pajunen, A. (1987). *Biochem. J.,* in press.

Persson, L. (1982). *Acta Chem. Scand. Ser. B* **36,** 685–688.

Persson, L., Seely, J.E., and Pegg, A.E. (1984). *Biochemistry* **23,** 3777–3783.

Persson, L., Oredsson, S.M., Anehus, S., and Heby, O. (1985). *Biochem. Biophys. Res. Commun.* **131,** 239–245.

Pohjanpelto, P., Hölttä, E., Jänne, O.A., Knuutila, S., and Alitalo, K. (1985). *J. Biol. Chem.* **260,** 8532–8537.

Porter, C.W., Dave, C., and Mihich, E. (1981). *In* "Polyamines in Biology and Medicine" (D.R. Morris and L.J. Marton, eds.), pp. 407–436. Dekker, New York.

Porter, C.W., Bergeron, R.J., and Stolowich, N.J. (1982). *Cancer Res.* **42,** 4072–4078.

Porter, C.W., Miller, J., and Bergeron, R.J. (1984). *Cancer Res.* **44,** 126–128.

Porter, C.W., Cavanaugh, P.F., Stolowich, N., Ganis, B., Kelly, E., and Bergeron, R.J. (1985). *Cancer Res.* **45,** 2050–2057.

Pösö, H., and Pegg, A.E. (1982). *Biochemistry* **21,** 3116–3222.

Pösö, H., Karvonen, E., Suomalainen, H., and Andersson, L.C. (1984). *J. Biol. Chem.* **259,** 12307–12310.

Pritchard, M.L., Seely, J.E., Pösö, H., Jefferson, L.S., and Pegg, A.E. (1981). *Biochem. Biophys. Res. Commun.* **100,** 1597–1603.

Pritchard, M.L., Pegg, A.E., and Jefferson, L.S. (1982). *J. Biol. Chem.* **257,** 5892–5899.

Rannels, D.E., Pegg, A.E., Clark, R.S., and Addison, J.L. (1985). *Am. J. Physiol.* **249,** E506–E513.

Rinehart, C.A., and Chen, K.Y. (1984). *J. Biol. Chem.* **259,** 4750–4756.

Seely, J.E., and Pegg, A.E. (1983a). *J. Biol. Chem.* **258,** 2496–2500.

Seely, J.E., and Pegg, A.E. (1983b). *Biochem. J.* **216,** 701–717.

Seely, J.E., Pösö, H., and Pegg, A.E. (1982a). *J. Biol. Chem.* **257,** 7549–7553.

Seely, J.E., Pösö, H., and Pegg, A.E. (1982b). *Biochemistry* **21,** 3394–3399.

Seely, J.E., Pösö, H., and Pegg, A.E. (1982c). *Biochem. J.* **206,** 311–318.

Seely, J.E., Persson, L., Sertich, G., and Pegg, A.E. (1985). *Biochem. J.* **226,** 577–586.

Seppänen, P., Alhonen-Hongisto, L., Siimes, M., and Jänne, J. (1980). *Int. J. Cancer* **26,** 571–576.

Seppänen, P., Alhonen-Hongisto, L., and Jänne, J. (1981). *Biochim. Biophys. Acta* **674,** 169–177.
Seppänen, P., Ruohola, H., and Jänne. J. (1984). *Biochim. Biophys. Acta* **803,** 331–337.
Sertich, G., and Pegg, A.E. (1986). *Fed. Proc., Fed. Am. Soc. Exp. Biol.* **45,** 388.
Seyfried, C.E., Oleinik, O.E., Degen, J.L., Resing, K., and Morris, D.R. (1982). *Biochim. Biophys. Acta* **716,** 169–177.
Shirahata, A., and Pegg, A.E. (1985). *J. Biol. Chem.* **260,** 9583–9588.
Shirahata, A., and Pegg, A.E. (1986). *J. Biol. Chem.* **261,** 13833–13837.
Shirahata, A., Christman, K., and Pegg, A.E. (1985). *Biochemistry* **24,** 4417–4423.
Smith, L.L. (1985). *Biochem. Soc. Trans.* **13,** 332–334.
Tabor, C.W., and Tabor, H. (1984). *Adv. Enzymol. Rel. Areas Mol. Biol.* **56,** 251–282.
Williams-Ashman, H.G., and Pegg, A.E. (1981). *In* "Polyamines in Biology and Medicine" (D.R. Morris and L.J. Marton, eds.), pp. 43–73. Dekker, New York.
Williams-Ashman, H.G., and Seidenfeld, J. (1986). *Biochem. Pharmacol.* **35,** 1217–1225.
Zagon, I.S., Seely, J.E., and Pegg, A.E. (1983). *In* "Methods in Enzymology" (H. Tabor and C. W. Tabor, eds.), Vol. 94 pp. 169–176. Academic Press, New York.
Zagon, I.S., McLaughlin, P.J., Seely, J.E., Hoeksema, G.W., and Pegg, A.E. (1984). *Cell Tissue Res.* **235,** 371–377.

6

Inhibitors of Polyamine Biosynthesis: Cellular and *in Vivo* Effects on Tumor Proliferation

PRASAD S. SUNKARA,* STEPHEN B. BAYLIN,† AND GORDON D. LUK† [1]

* *Merrell Dow Research Institute*
2110 East Galbraith Road
Cincinnati, Ohio 45215

† *The Oncology Center and Department of Medicine*
The Johns Hopkins University School of Medicine
Baltimore, Maryland 21205

I. INTRODUCTION

The naturally occurring polyamines, putrescine, spermidine, and spermine, are ubiquitous in eukaryotic organisms. Although the discovery of spermine as precipitating crystals by Leeuwenhoek in human semen dates back to 1677 (Mann, 1964), a critical role for these amines has been recognized only since specific inhibitors of polyamine-biosynthetic enzymes became available in the last decade (Abdel-Monem *et al.*, 1974; Metcalf *et al.*, 1978). A number of studies have suggested that polyamines play an important role in cell proliferation and differentiation (Jänne *et al.*, 1978; Luk *et al.*, 1981; Heby, 1981; Pegg and McCann, 1982; Sunkara and Prakash, 1984; see also Chapter 8).

[1] Present address: Department of Medicine, Wayne State University School of Medicine, Detroit, Michigan 48201.

In this chapter we will examine the role of polyamines in rapid cell growth and transformation. In addition, the biochemical and pharmacological consequences of inhibiting polyamine biosynthesis by DL-α-difluoromethylornithine (DFMO), a specific inhibitor of ornithine decarboxylase (ODC), and the potential therapeutic use of this inhibitor as an antitumor and antimetastatic agent are also described in some detail.

II. POLYAMINE BIOSYNTHESIS IN NORMAL AND TRANSFORMED CELLS

A. Polyamine Synthesis during Viral-Induced Transformation

Bachrach and co-workers (Don *et al.*, 1975) have employed the virus-induced transformation of primary cultures of chick fibroblasts as a model to study the role of polyamines in neoplastic growth. Infection of chick embryos with oncogenic viruses resulted in an enhanced accumulation of putrescine and spermidine, while infection with a variety of nononcogenic viruses did not show any effect. A temperature-sensitive mutant of Rous sarcoma virus caused an increase in the cellular polyamine content at 37°C, but not at the nonpermissive temperature of 42°C. This observation suggests that the enhanced polyamine synthesis is a result of virus-induced transformation but not of viral multiplication, since the virus multiplies equally well at both temperatures but induces transformation only at the permissive temperature (Don and Bachrach, 1975). An early stimulation of ODC was seen in chick embryo fibroblasts infected with Rous sarcoma virus and in 3T3 cells infected with murine sarcoma virus (Don *et al.*, 1975; Gazdar *et al.*, 1976). The increase in the putrescine synthesis may be due to either stabilization of ODC (Bachrach, 1976) or a decrease in the activity of the catabolic enzyme, diamine oxidase. It is of interest to determine whether the changes in the enzyme activities are the primary effect of transformation, the result of increased proliferation, or secondary to some general inhibition of proteolysis resulting from transformation. The studies just described clearly demonstrate a specific association between the increase in polyamine synthesis and some primary and/or secondary events involved in viral-induced cell transformation.

B. Polyamine Synthesis during Chemical Carcinogenesis

The most widely used system for studying the role of polyamines in chemical carcinogensis is the mouse skin tumor model. The induction of skin tumors in mice involves two phases, initiation and promotion. Initiation is achieved by a single subcarcinogenic dose of a variety of chemicals, followed by promotion with the repeated application of certain phorbol esters (Boutwell, 1977). A single application of promoting agents, but not their nonpromoting analogs, resulted in a transient stimulation of ODC and *S*-adenosyl-L-methionine decarboxylase (AdoMetDC) activities (O'Brien *et al.*, 1975; Yuspa *et al.*, 1976). Nonpromoting but hyperplastic agents had little or no effect on epidermal ODC activity, ruling out the possibility that the increase is due to general enhancement of protein synthesis (O'Brien *et al.*, 1975). Furthermore, the inhibition of skin tumor promotion by a number of retinoids closely correlated with their ability to inhibit the induction of ODC (Verma and Boutwell, 1977). The induction of ODC and increase in putrescine concentration were also seen in mouse epidermis exposed to UV radiation (Seiler and Knödgen, 1979).

In the model system of diaminoazobenzene-induced liver carcinogenesis, Scalabrino *et al.* (1978) observed that liver ODC and AdoMetDC activities increased in a bimodal fashion. The first peak of activity, observed 1 month after initiation of the carcinogenic diet, was followed by a second peak 3 months later. Thereafter, the activity remained elevated until the tumors began to appear. The levels of putrescine also remained high during hepatocarcinogenesis. These changes in the enzymes appear to be specific to the liver, since the hepatocarcinogenic diet produced a marked drop in the activity of ODC in the kidneys. Administration of a single dose of liver carcinogen, diethyl nitrosamine, to rats resulted in a rapid increase in ODC activity within 3 hr after carcinogen administration. ODC remained elevated for 7 days (Olsen and Russell, 1979). Similarly, administration of colonic carcinoma-inducing carcinogens to rats resulted in a rapid initial increase in ODC activity in colonic mucosa (Luk *et al.*, 1982a, 1986b; Rozhin *et al.*, 1984). With continued carcinogen administration, this ODC increase became sustained prior to the appearance of colonic tumors (Luk *et al.*, 1982a, 1986b; Rozhin *et al.*, 1984). The incidence of these carcinogen-induced colonic tumors was reduced by the inhibition of ODC activity (Kingsnorth *et al.*, 1983a). The foregoing studies clearly demonstrated an increase in polyamine biosynthesis during chemical carcinogenesis.

C. Polyamines in Experimental Animal Tumors

The earliest studies of polyamine synthesis in animal tumors were performed with Ehrlich ascites carcinoma cells grown in the peritoneal cavities of mice (Siimes and Jänne, 1967; Bachrach *et al.*, 1967). The concentration of spermidine was found to be rather high compared with that found in regenerating rat liver (Siimes and Jänne, 1967; Russell and Snyder, 1968). The changes in tumor polyamine levels, especially those of cellular putrescine and ODC, showed a close correlation with tumor cell multiplication (Anderson and Heby, 1972).

Williams-Ashman and co-workers (1972, 1973) performed a systematic study to correlate the concentration of polyamines with the growth rate of various transplantable hepatomas. Their studies with several hepatomas having widely different growth rates revealed that, in general, the concentration of putrescine and the ODC activity of the tumor cells were substantially higher in tumor cells than in normal rat liver. Further, putrescine was approximately 10-fold higher in the most rapidly growing hepatomas. The concentration of spermidine was also found to be elevated. In contrast, very little change was observed in the concentration of spermine in tumor cells.

Marton and Heby (1974) observed high levels of putrescine, spermidine, and their biosynthetic enzymes in the peripheral part of the brain when a chemically induced brain tumor was transplanted into the brain. High concentrations of putrescine were observed in nitrosoethylurea-induced rat gliomas (Seiler *et al.*, 1975). Kremzner and co-workers (1970) had found that the concentration of putrescine in human brain tumors was several fold higher than in normal human brain. A sixfold increase in ODC activity over that in normal mouse liver was observed in transplanted mouse L1210 leukemia cells 4 days after implantation (Russell and Levy, 1971).

D. Polyamines in Clinical Cancer

The presence of polyamines in high concentrations in neoplastic tissues suggests that increased amounts of these compounds may be found in extracellular body fluids. Russell and co-workers (1971) reported that urinary excretion of polyamines was considerably greater in leukemia, lymphoma, and solid tumor patients than in healthy persons with nonmalignant diseases. Since then, several laboratories have been engaged in the measurement of polyamine levels from different body fluids and red blood cells for use as biochemical markers for human malignancies (Marton *et al.*, 1976; Rennert *et al.*, 1977; Nishioka and Romsdahl, 1974; Takami *et al.*, 1979). This topic was extensively reviewed by Savory and Shipe (1975), Bachrach (1976), and Jänne *et al.* (1978). In general, it appears that polyamine levels in body fluids are of little or no value in the diagnosis of cancer (Waalkes *et al.*, 1975; Raina and Jänne, 1975). However, in certain cases, such as suspected brain tumors, polyamine determinations may be of value when applied in conjunction with other diagnostic techniques (Marton, 1981). In contrast, there is ample evidence to suggest that polyamine determinations could be useful in assessing tumor response to chemotherapy (Russell, 1977) and the prognosis of certain tumors (Cohen, 1977).

III. INHIBITORS OF POLYAMINE BIOSYNTHESIS

The ubiquitous occurrence of polyamines, their association with nucleic acids, and accumulation of these polycations in rapidly dividing cells stimulated considerable research into the design of specific inhibitors of polyamine-biosynthetic enzymes. It is not altogether surprising that early efforts were focused on inhibitors of ODC, the rate-limiting enzyme. Several substrate analogs were synthesized as potential competitive inhibitors of the enzyme (see Chapter 1 for details), and a considerable body of information about the enzyme and the role of polyamines was accumulated by employing these inhibitors. However, the rebound of the enzyme activity when the inhibitor was removed from the system rendered these competitive inhibitors unsatisfactory for the pharmacological manipulation of polyamines *in vivo*. The availability of DL-α-difluoromethylornithine, a "suicide inhibitor" of ODC, made it possible to obtain a sustained ODC inhibition and a decrease in putrescine and spermidine concentrations *in vivo,* by either oral or systemic administration to animals. Since then a number of other substrate and product analogs with specific and irreversible inhibitory activity have been made (Sjoerdsma and Schechter, 1983), with the aim of obtaining a more potent and longer acting compound than DFMO (Chapter 1).

A. DL-α-Difluoromethylornithine

DL-α-Difluoromethylornithine is a potent and specific irreversible inhibitor of ornithine decarboxylase (Bey, 1978; Metcalf *et al.,* 1978). DFMO was shown to inhibit ODC, to decrease intracellular putrescine and spermidine concentrations, and to inhibit cell growth in a number of cells in culture (Mamont *et al.,* 1978; Sunkara *et al.,* 1980; Seidenfeld *et al.,* 1981; Luk *et al.,* 1981) and in animal tumors *in vivo* (Prakash *et al.,* 1978, 1980; Marton *et al.,* 1981; Sunkara *et al.,* 1982, 1983b).

DFMO, given short term, has been generally well tolerated by experimental animals. Single oral doses of 5 g/kg had no apparent adverse effects on mice. Administration of a total daily dose of 4 g/kg for 4 weeks to mice and rats did not slow their normal weight gain and no organ toxicity was found. The only side effect seen was diarrhea in rats, which was easily reversible when the administration of the drug was stopped. The administration of DFMO for longer periods of time or in greater doses to animals and humans resulted in more toxic side effects. Sustained administration of DFMO to athymic nude mice bearing human lung tumor xenografts for 6 weeks or more resulted in weight loss, diarrhea, and debilitation (Luk *et al.*, 1983a, 1986a). Rats given DFMO developed significant thrombocytopenia by 3 weeks with lesser decreases in peripheral red and white blood cell counts (Luk *et al.*, 1983b; Ota *et al.*, 1986). Dogs given DFMO developed diarrhea and showed villous atrophy and atrophy of microvilli in the absorptive surface of the intestinal mucosa (Yarrington *et al.*, 1983). Importantly, when administration of DFMO in combination with other drugs is considered, it must be remembered that DFMO can severely retard the recovery of intestinal mucosa and bone marrow after ara-C (1-β-D-arabinofuranosylcytosine) injury in rats (Luk *et al.*, 1981, 1983b).

The long-term administration of DFMO to man resulted in some of the same toxicities seen in animals, with a similar time course for thrombocytopenia, especially in patients given prior chemotherapy (Abeloff *et al.*, 1984). Diarrhea was also seen. An unusual toxicity of the drug was a reversible loss of hearing acuity which has been found in up to 54% of patients taking oral DFMO (Abeloff *et al.*, 1984, 1986).

B. Other Inhibitors of Polyamine Biosynthesis

Numerous inhibitors of the polyamine biosynthetic enzymes have been synthesized and tested (see Chapter 2). In particular, analogs of ornithine and putrescine have been synthesized and tested for activity as ODC inhibitors (Sjoerdsma and Schechter, 1984; see also Chapter 1).

A new suicide inhibitor of ODC deserves a brief mention here. (2*R*,5*R*)-6-Heptyne-2,5-diamine (δ-methylacetylenic putrescine, MAP) is a more potent inhibitor than DFMO, *in vitro* and *in vivo*, with similar antitumor effects, albeit at lower doses (Bartholeyns *et al.*, 1984; Mamont *et al.*, 1984). Since DFMO seems to be a highly specific inhibitor and most studies have been performed with this drug, the rest of the chapter is mainly devoted to the antiproliferative and antitumor activities of DFMO.

IV. CELLULAR EFFECTS OF POLYAMINE INHIBITORS

A. Polyamine Inhibitors: Inhibition of Cell Growth in Culture

The initial studies on the correlation of polyamine levels and rapid cell proliferation indicated a requirement for polyamines in DNA synthesis and cell proliferation. A search for specific inhibitors of polyamine biosynthesis was initiated to clarify the exact role of polyamines in cell growth. Initially a number of ornithine analogs such as α-hydrazinoornithine (Johansson and Skinner, 1973; Harik and Snyder, 1973) and α-methylornithine (Abdel-Monem *et al.,* 1974) were made. These compounds are reversible competitive inhibitors of ODC. Although they were effective in inhibiting the accumulation of polyamines (Inoue *et al.,* 1975; Mamont *et al.,* 1976; Heby *et al.,* 1978; Sunkara *et al.,* 1979a,b), their effects were easily reversible *in vivo.*

The availability of DL-α-difluoromethylornithine, an irreversible catalytic inhibitor of ODC, made it possible to confirm the role of polyamines in rapid cell proliferation in a number of systems. The initial studies indicated antiproliferative activity of DFMO against HTC cells, L1210 leukemia cells, and MA-160 human prostatic adenoma cells grown in cell culture (Mamont *et al.,* 1978). These inhibitory effects were completely reversible by the addition of exogenous putrescine and spermidine. Antiproliferative effects of DFMO were confirmed in a number of other cells such as human embryo fibroblasts (Hölttä *et al.,* 1979), concanavalin-A-stimulated lymphocytes (Seyfried and Morris, 1979), HeLa cells (Sunkara *et al.,* 1980), Ehrlich ascites carcinoma cells (Oredsson *et al.,* 1980), and 9L rat gliosarcoma cells (Seidenfeld *et al.,* 1980).

Each of these earlier studies indicated that DFMO was a cytostatic agent. However, initial observations with human small-cell carcinoma cells indicated that some tumor cell types can not survive during chronic inhibition of ODC (Luk *et al.,* 1981, 1982b). These results have since been extrapolated to HL-60 human promyelocytic leukemia cells (Luk *et al.,* 1982c), HeLa cells (Sunkara *et al.,* 1983a), and mouse B_{16} melanoma (Sunkara *et al.,* 1983b). In the case of small-cell lung carcinoma cells, polyamine depletion induced by DFMO was followed by pronounced cell loss and rapid decrease in viability. In the cases of HeLa and B_{16} melanoma cells, a decrease in plating efficiency was found. Further, B_{16} melanoma seemed to be quite sensitive, and an increase in the cytotoxicity (as determined by the plating efficiency), reaching about 80% of the cell survival over the control at the end of 7 days, was observed in the DFMO-treated cultures (Sunkara *et al.,* 1985a). These results indicate that DFMO is a cytostatic or cytotoxic agent depending on the cell line employed, and some of the potential mechanisms involved are commented upon below.

B. Polyamine Inhibitors: Effect on Cell Cycle Traverse of Normal and Transformed Cells

The cell cycle-specific fluctuations in polyamine levels, especially a peak in late G_1, and the requirement for polyamines for optimal rates of DNA replication, indicate that these physiological cations may play an important role in the traverse of cells through G_1 into S phase (Otani *et al.*, 1974; Filligame *et al.*, 1975). Seyfried and Morris (1979) found that DFMO inhibited DNA replication in activated lymphocytes without altering the number of cells in S phase. Harada and Morris (1981) employed α-methylornithine in CHO cells and found that the longer doubling times in inhibitor-treated cells was due to increase in the length of G_1 and S phases. Heby and Anderson (1980) found an accumulation of Ehrlich ascites cells in S and G_2 phases of the cell cycle when the cells were treated with α-methylornithine or methylglyoxal bis(guanylhydrazone) (MGBG). In fact, several studies indicate that inhibition of polyamine biosynthesis by MGBG in a number of normal cells such as WI-38, rat embryo, mouse 3T3, and human skin fibroblasts (PA_2) causes an arrest of these cells in G_1 phase (Boynton *et al.*, 1976; Rupniak and Paul, 1980; Sunkara *et al.*, 1979a). This arrest is easily reversed with spermidine or spermine. However, treatment of transformed cells (HeLa, SV-3T3, and CHO) with MGBG or DFMO results in the arrest of a majority of cells in S phase (Rupniak and Paul, 1978; Sunkara *et al.*, 1981). We (P.S.) treated exponentially growing normal (human skin fibroblast line PA_2), mouse 3T3 (WI-38), and transformed (HeLa, SV-3T3, 2RA) cells with MGBG (40 μM) or DFMO (2.5 mM) for 48 or 96 hr, respectively. At the end of this period, cell cycle analysis was performed according to the premature chromosome condensation (PCC) method (Rao *et al.*, 1977). The results indicated that inhibition of putrescine and spermidine synthesis resulted in a G_1 block in normal cells, whereas a majority of transformed cells were arrested in S phase. Similar G_0 arrest of 3T3 cells after DFMO treatment was also reported by Tucker *et al.* (1985). These data suggested that a specific polyamine-requiring process exists in G_1 and was necessary for entry of cells into S period. Normal cells recognized this control signal and stopped in G_1. However, transformed cells did not recognize the polyamine limitation or restriction, and continued to progress through the cell cycle; a majority of them entered S phase but failed to complete DNA synthesis and were thus arrested in S phase. This differential effect of inhibition of polyamine biosynthesis on the cell cycle traverse of normal and transformed cells was exploited to kill transformed cells preferentially in combination with other S-phase-specific anticancer drugs.

Rupniak and Paul (1980) showed that a combination MGBG and hydroxyurea can preferentially kill transformed SV-3T3 cells over normal 3T3 cells. By pretreating proliferating cultures of 3T3 and SV-3T3 cells with MGBG, and subsequently treating the cultures with hydroxyurea, an S-phase-specific cytotoxic drug, they obtained highly selective killing of the still-cycling, transformed SV-3T3 cells. In contrast, normal 3T3 cells were protected from the cytotoxic effects of hydroxyurea by virtue of their MGBG-induced growth arrest in the insensitive G_1 phase. Other studies also indicated that the use of an S-phase-specific drug such as ara-C following DFMO treatment resulted in a synergistic killing of tumor cells (HeLa and 2RA) (Sunkara *et al.*, 1980, 1981). However, the normal human diploid fibroblasts (WI-38) did not suffer the cytotoxic effects of ara-C, since they were arrested in G_1 phase as a result of the DFMO treatment. The effect of DFMO on the cell cycle traverse seems to depend also on the cell line studied. Oredsson *et al.* (1983) observed that DFMO did not block 9L brain tumor cells in any particular phase of the cell cycle, and, in fact, DFMO pretreatment seemed to reduce ara-C toxicity. Further, four of the human carcinoma cell lines showed a marked increase in the G_1 fraction and decrease in S-phase population after DFMO induced inhibition of cell proliferation (Seidenfeld and Komar, 1985). These observations indicated that the effect of the inhibitors of polyamine synthesis might vary from cell to cell and also with the duration of treatment with the inhibitor.

The mechanisms underlying the differences in response to DFMO of various cell types will be important to elucidate. Recently, several potential factors have become apparent. The sensitivity of different types of tumor cells to DFMO is linked to factors which modulate their growth behavior *in vitro*. Human small-cell lung cancer, as previously noted, exhibits a cytotoxic response to DFMO *in vitro* (Luk *et al.*, 1981, 1982b).

Established cell lines of this tumor invariably grow as non-anchorage-dependent cell aggregates. In contrast, established lines of non-small-cell lung cancer, which grow as anchorage-dependent monolayers, show only a cytostatic response to DFMO (Luk *et al.*, 1982b). If the small-cell lung cancer cells are made to grow as monolayers and the non-small-cell cells as suspended aggregates, then their respective response patterns become reversed, i.e., small-cell lung cancer cells now exhibit cytostasis while the non-small-cell lung cancer cells are killed by DFMO (Luk and Baylin, 1986).

The precise mechanisms governed by the growth behavior of the lung cancer cells, which determine DFMO sensitivity, are not known. However, recent data suggest that differences among the cell types in the capacity for polyamine biosynthesis is important. Comparison of a sensitive small-cell lung cancer cell line with a more resistant non-small-cell lung cancer cell line has revealed a marked decrease in polyamine biosynthesis in the sensitive cell line. The determinant of this low rate of polyamine biosynthesis appears to be extremely low levels of steady-state ODC mRNA (Casero *et al.*, 1986). Studies in other cell types have shown that induction and/or selection of DFMO-resistant cells are associated with amplification of the ODC gene (McConlogue *et al.*, 1984; Kahana and Nathans, 1984; Berger *et al.*, 1984; Pohjanpelto *et al.*, 1985). Future studies should determine whether these are universal findings underlying cellular responses to polyamine biosynthesis inhibition and whether differences in ODC mRNA levels and/or ODC gene copy number correlate with different types of tumor cell growth *in vitro* and *in vivo*.

C. Polyamine Inhibitors: Effects on Growth of Cells in Culture in Combination with other Antitumor Agents

Several studies have indicated the usefulness of DFMO in potentiating the cytotoxicity of other antitumor agents in a number of cell systems grown in culture (Sunkara *et al.*, 1980, 1981, 1983a,b; Marton, 1981; Kingsnorth *et al.*, 1983b). As discussed earlier, we exploited the differential cell cycle response of normal and tumor cells to DFMO and showed that we could preferentially kill the tumor cells both *in vitro* (Sunkara *et al.*, 1980, 1981) and *in vivo* (Prakash and Sunkara, 1983) by a combination of DFMO and the S-phase-specific drug ara-C. Kingsnorth *et al.* (1983b) reported the potentiation of the antiproliferative effect of 5-fluorouracil against human adenocarcinoma cells by DFMO. Marton and his co-workers (1981) have found that DFMO enhanced the cytotoxicity of 1,3-bis(2-chloroethyl)-1-nitrosourea (BCNU) in rat 9L gliosarcoma cells in culture. Other results from their laboratory also suggest that polyamine depletion by DFMO modifies the DNA structure and conformation and might sensitize the 9L cells to nitrosoureas (Hung *et al.*, 1983). Oredsson *et al.* (1982) observed that pretreatment of 9L cells with DFMO reduced the cytotoxicity of *cis*-platinum. Seidenfeld *et al.* (1981) reported that pretreatment of 9L cells with DFMO did not potentiate the X-ray-induced cell kill. These results clearly suggest that the effects of DFMO seem to vary with the companion cytotoxic agent used.

Jänne and his colleagues (Alhonen-Hongisto *et al.*, 1980; Seppänen *et al.*, 1981) observed that uptake of MGBG, a cytotoxic polyamine antimetabolite, was enhanced after pretreatment of Ehrlich ascites cells. This enhanced uptake was also associated with increased cytotoxicity. Sunkara *et al.* (1983a) also observed a three- to fourfold increase in MGBG uptake and corresponding cytotoxicity of MGBG to HeLa cells following DFMO pretreatment. In another set of studies we have observed that mouse type I interferon potentiated the cytotoxic activity of DFMO to B_{16} melanoma cells grown in culture (Sunkara *et al.*, 1983b).

The foregoing studies in cell cultures laid the foundation for the investigations to determine the usefulness of DFMO as an antitumor agent, either by itself or in combination with other cytotoxic agents in experimental tumors, and, eventually, in the treatment of human cancers.

V. POLYAMINE INHIBITORS: *IN VIVO* EFFECTS AGAINST EXPERIMENTAL TUMORS

A. Monotherapy of Experimental Tumors with DFMO

1. Nonhuman

Early studies with L1210 leukemia in mice (Prakash *et al.,* 1978) showed antitumor activity when DFMO was administered systemically on a frequent schedule. The increase in survival time observed was 23% compared to untreated controls. Subsequent work with the same tumor model, but with DFMO administered in the drinking water, indicated that only a marginal increase in survival time was attainable with larger oral doses, despite significant decreases in intracellular tumor putrescine and spermidine concentrations (Prakash and Sunkara, 1983; Bartholeyns and Koch-Weser, 1981). However, when administered in the drinking water, the compound inhibited EMT_6 solid tumor growth in mice by 80% by day 35 compared to untreated controls (Prakash *et al.,* 1980). Extensive work with a variety of other experimental tumor models has indicated that, in general, the compound as a single agent is moderately effective in bringing about substantial reduction in tumor burden (Fozard and Koch-Weser, 1982; Sjoerdsma and Schechter, 1984), even when the therapy is initiated during the early stages of tumor growth.

Among the several experimental tumors studied to date, the B_{16} melanotic melanoma in mice was found to be most sensitive to the antiproliferative activity of DFMO. Greater than 80% inhibition of tumor growth was observed when the drug was given orally (Sunkara *et al.,* 1983a). Inhibition of growth and depletion of polyamines in this tumor by DFMO were accompanied by the induction of the enzyme tyrosinase and also accumulation of melanin (Sunkara *et al.*, 1985a), indicating that at least in this tumor model, inhibition of polyamine biosynthesis brings about other biochemical changes characteristic of differentiated function. Induction of differentiated biochemical functions by DFMO has also been reported in Cloudman S91 melanoma cells in culture (Käpyaho and Jänne, 1983). Although melamona in humans constitutes only 1–2% of the total cancer incidence, the virulence and the poor chemotherapeutic response observed render this disease an extremely difficult one to manage clinically. The fairly nontoxic nature of DFMO and its effectiveness as a single agent against B_{16} melanoma and the unique biochemical changes observed offer a strong impetus for clinical studies with the compound against human melanomas.

2. Human

A study on the viability of human small-cell lung carcinoma cells during polyamine depletion by DFMO indicates that these cells are unable to maintain structural integrity during prolonged polyamine starvation and eventually die under sustained polyamine deprivation (Luk *et al.*, 1981, 1982b). This study appears to be the first to report clear *in vitro* cytotoxicity for DFMO toward a mammalian cell. Established human small-cell variant lung carcinoma implants in athymic nude mice were later reported to respond successfully to oral therapy with DFMO (Luk *et al.*, 1983a, 1986a). These experimental results raise the possibility that this drug could also provide beneficial therapeutic effects against this malignancy in humans.

Human small-cell lung carcinoma is an unusual solid tumor because, early in the disease, it often responds well to combination chemotherapy. Thus, many patients with small-cell lung cancer achieve complete or partial remission following initial therapy. However, the disease almost always returns, and patients die within an average of 13 months (Greco and Oldham, 1979). In the context of the sensitivity of the tumor to DFMO and its clinical behavior, "consolidation" treatment with DFMO might be useful. Patients could be maintained with DFMO following initial chemotherapy in an effort to prevent tumor regrowth. Recent results on a study of DFMO dosage schedules in athymic nude mice bearing human small-cell lung cancer xenografts (Luk *et al.*, 1986a) suggest that long-term tumor treatment with minimal toxicity may be possible. A trial of consolidation therapy in patients with small-cell lung cancer is now under way (Abeloff *et al.*, 1984, 1986) and the results should ascertain the potential for use of DFMO in this manner.

B. Combination Chemotherapy with DFMO and Other Antitumor Agents

Extensive *in vitro* and *in vivo* studies and also available pharmacokinetic and toxicological data on DFMO (Fozard and Koch-Weser, 1982; Sjoerdsma and Schechter, 1984) suggest that the chemotherapeutic potential of the compound can best be exploited by combining it with appropriate cytotoxic agents. Early studies with DFMO and cyclophosphamide against EMT_6 solid tumors in mice showed only an additive antitumor effect (Prakash *et al.*, 1980). Subsequently this combination was shown to be effective in suppressing the growth of established DMBA-induced tumors in rats even though the tumors were not very responsive to monotherapy with either of the two agents alone (Fozard and Prakash,

1982). This combination was also reported to provide synergistic antitumor activity in a rat prostate tumor model (Heston *et al.*, 1982). Synergistic antitumor activity was reported with DFMO in combination with adriamycin or vindesine against L1210 leukemia, hepatoma solid tumor in rats, and EMT_6 solid tumor in mice (Bartholeyns and Koch-Weser, 1981). In these studies, the selection of the cytotoxic drugs was arbitrary and was primarily based on the premise that DFMO is likely to provide at least an additive therapeutic effect by delaying the recovery of the tumor cells from the cytotoxic drug-induced injury.

Studies with two other DNA-reactive drugs and DFMO demonstrated that synergistic or even antagonistic therapeutic effects could be expected depending on the cytotixic agent selected. Thus, synergistic therapeutic effects were observed with BCNU and DFMO against 9L rat brain tumor (Marton *et al.*, 1981), whereas mutual antagonism was reported with DFMO and *cis*-platinum against 9L tumor cells (Oredsson *et al.*, 1982). Clearly, it would be highly desirable to be able to select combinations, based on biochemical and pharmacological findings, which are likely to provide synergistic or at least additive antitumor effects. The usefulness of the sister chromatid exchange assay was reported in predicting the synergistic or antagonistic effects of DNA-reactive cytotoxic agents in combination with DFMO (Tofilon *et al.*, 1982). Extensive further work is needed to test the reliability of this procedure in ascertaining the suitability of cytotoxic agents for combination protocols with DFMO.

The differential effect of polyamine depletion on cell cycle traverse of normal and transformed cells (Rupniak and Paul, 1980; Sunkara *et al.*, 1981; Sunkara and Rao, 1981), as described earlier, has been the basis for the rational design of therapeutic protocols involving DFMO and phase-specific cytotoxic drugs in a murine leukemia model (Prakash and Sunkara, 1983). Preferential accumulation of L1210 tumor cells in S phase compared to normal bone marrow cells was also observed when DFMO was administered to mice. Therapeutic synergism with a combination of DFMO and ara-C could be demonstrated in this tumor model when ara-C administration was timed to coincide with the maximum differential accumulation of tumor cells in S phase (Prakash and Sunkara, 1983). Ideally, in human neoplasia with relatively high growth rates, namely, leukemias and lymphomas, polyamine limitation might lead to tumor cell selective accumulation in S phase, priming them for a synergistic cell kill by S-phase-specific cytotoxic drugs such as ara-C and hydroxyurea.

Sequential therapy with DFMO and MGBG is another example of a combination regimen arrived at primarily on the basis of biochemical and pharmacological studies with the two compounds. The structural analogy with spermidine and the observation that *in vivo* antineoplastic activity could partially be reversed by exogenous spermidine (Mihich, 1963) indicate some degree of polyamine antagonism for MGBG. Polyamines and MGBG were also reported to share a common cellular transport system (Dave and Caballeri, 1973). These experimental observations led Jänne and colleagues to pretreat tumor cells with DFMO to deplete intracellular polyamines, and subsequently to expose these cells to MGBG, an experimental manipulation that led to a three- to fourfold increase in the uptake of MGBG into the tumor cells (Alhonen-Hongisto *et al.*, 1980). DFMO pretreatment was later reported to potentiate the antitumor activity of MGBG against L1210 leukemia (Burchenal *et al.*, 1981). Work in our laboratory (Sunkara *et al.*, 1983b) employing L1210 leukemia in mice showed that polyamine depletion in L1210 cells by DFMO resulted in a two- to threefold increase in the accumulation of MGBG in tumor cells. The extent of MGBG uptake correlated inversely with intracellular spermidine concentration. Spermidine was also found to be a more potent inhibitor of MGBG uptake than was putrescine in this system. A distinct synergistic increase in survival time was also observed with a combination of DFMO and MGBG in this tumor model.

Finally, the report on the efficacy of DFMO in combination with interferon in suppressing the growth of subcutaneous B_{16} melanomas in mice (Sunkara *et al.*, 1983a) is indicative of the potential for effective therapeutic protocols involving combinations of polyamine-biosynthetic inhibitors and immunomodulators. Although the dose of interferon employed in the study was insufficient to provide any significant antitumor effect, the combination of DFMO and interferon resulted not only in pronounced synergistic antitumor effect but also tumor eradication in some animals. Work in our laboratory showed that the DFMO–interferon combination was equally effective against Lewis lung (3LL) carcinoma, a tumor with poor therapeutic response to either DFMO or interferon therapy alone. In addition, other immunomodulatory agents such as tuftsin and *Corynebacterium parvum* were also found to be very effective in suppressing B_{16} melanoma growth in mice (Sunkara *et al.*, 1984; Bowlin *et al.*, 1985).

C. Polyamine Inhibitors: Effect on Tumor Metastases

Although a vast amount of the literature presented in earlier sections of this chapter indicates an important role for polyamines in tumor cell growth, there is no information available on the role of these cations in the process of tumor metastases. The effect of DFMO on the growth and pulmonary metastasis of 3LL tumor in mice has been examined (Sunkara *et al.*, 1982; Bartholeyns, 1983). The Lewis lung tumor of C57BL mice provides an excellent experimental model for studying tumor metastases.

The primary tumor grows rapidly at the inoculation site and metastasizes to lungs in about 3 weeks after tumor transplantation, and these pulmonary foci can easily be counted.

The results of a recent study showed that administration of DFMO to animals bearing tumors resulted in a 69 and 75% decrease in the intracellular levels of putrescine and spermidine, respectively, in the primary tumor, without any change in the spermine concentrations. The decreased polyamine levels were also associated with a 43% inhibition of tumor growth. Inhibition of polyamine biosynthesis as a result of DFMO administration not only inhibited growth of the primary tumor but also dramatically decreased the secondary pulmonary metastases by 79%, with 25% of the animals showing no visible metastases. Further, this inhibition of tumor growth and metastases could be reversed by simultaneous administration of putrescine to the animals. Administration of putrescine to animals receiving DFMO did not result in any significant elevation of tumor putrescine concentrations, although spermidine concentrations were substantially elevated compared to tumors from animals that received DFMO alone. The lack of any effect on tumor putrescine concentration following exogenous putrescine administration to DFMO-treated animals is most likely due to its rapid conversion to spermidine. Since polyamine depletion in tumor cells leads to an increase of AdoMetDC, the rate-limiting enzyme involved in the conversion of putrescine to spermidine, DFMO treatment is likely to facilitate the conversion of exogenous putrescine to spermidine in the solid tumor employed in the study. Essentially similar results were obtained by Bartholeyns (1983). Our recent studies with B_{16} melanoma, a spontaneously metastasizing tumor, indicate that DFMO inhibits the treatment of tumor-bearing animals and results in a dose-dependent inhibition of primary tumor growth and pulmonary metastases. The antimetastatic activity of DFMO does not seem to be an indirect effect due to the inhibition of primary tumor growth. At 0.5% in drinking water, DFMO did not have any effect on the primary tumor but still showed about 55% inhibition of pulmonary metastases (Sunkara *et al.*, 1985b).

The novel findings of these studies are that inhibition of polyamine biosynthesis resulted in a decrease in the spread of the tumor cells to lungs. The fact that the inhibition of metastases could be reversed by simultaneous administration of putrescine suggests that polyamines may play a role not only in tumor growth (Prakash *et al.*, 1980; Marton *et al.*, 1981; Sunkara *et al.*, 1983a,b) but also in the process of tumor metastases. To understand the mechanism of tumor metastasis by DFMO, one can envisage that the depletion of cellular polyamines in the primary tumor could have affected (1) the invasion of the tumor cells into the lymphatics or blood vessels, (2) their transport into distant organs, and (3) the establishment of a microenvironment unfavorable to the growth of pulmonary metastases (Poste and Fidler, 1980).

To understand which phase in the metastatic cascade is inhibited by DFMO, we have employed the model of experimental pulmonary metastases of B_{16} melanoma (Sunkara *et al.*, 1985a; Sunkara and Rosenberger, 1987). The tumor cells are directly introduced into the bloodstream intravenously via the tail vein and the spread and metastases of the cells into lungs were determined. Our results indicated that DFMO did not have any significant effect on the experimental pulmonary metastases. Further, cells obtained from DFMO-pretreated tumors showed about 50% reduction in their ability to metastasize to lungs when introduced into bloodstream by intravenous injection into the tail vein. Based on these data, one can hypothesize that DFMO might interfere with the initial spread of the metastatic cells from the primary tumor and their transport through the bloodstream into the target organ. Once the metastatic cells reach the target organ, further treatment with DFMO does not seem to have an effect on the growth of these cells to form metastatic foci. These results indicate that DFMO and other specific inhibitors of polyamine synthesis merit consideration as potential therapeutic agents in the clinical management of metastasis.

VI. SUMMARY

Based on the available experimental evidence, it appears that polyamines play an important role in the proliferation of normal and transformed cells. Although the requirement for polyamines for DNA replication is well established, the molecular events in which the polyamines are essential are yet to be determined. DL-α-Difluoromethylornithine, a specific irreversible inhibitor of ornithine decarboxylase, has proved to be important in understanding the role of polyamines in cell proliferation and differentiation. DFMO has shown interesting antitumor and antimetastatic activities in a number of experimental tumors, either administered alone or in combination with other cytotoxic agents. Currently DFMO is in phase II clinical trials in combination with leukocyte interferon against melanoma in the United States. Consolidation therapy using DFMO alone is also being tried as discussed above. Hopefully, this exciting compound will prove useful as an adjunct in the treatment against cancer and other proliferative diseases.

REFERENCES

Abdel-Monem, M.M., Newton, N.E., and Weeks, C.E. (1974). *J. Med. Chem.* **17,** 447–451.

Abeloff, M.D., Slavik, M., Luk, G.D., Griffin, C.A., Hermann, J., Blanc, O., Sjoerdsma, A., and Baylin, S.B. (1984). *J. Clin. Oncol.* **2,** 124–130.

Abeloff, M.D., Rosen, S.T., Luk, G.D., Baylin, S.B., Zeltzman, M., and Sjoerdsma, A. (1986). *Cancer Treat. Rep.* **70,** 843–845.

Alhonen-Hongisto, L., Seppänen, P., and Jänne, J. (1980). *Biochem J.* **192,** 941–945.

Anderson, G., and Heby, O. (1972). *J. Natl. Cancer Inst.* **48,** 165–172.

Bachrach, U. (1976). *Ital. J. Biochem.* **25,** 77–93.

Bachrach, U., Bekierkunst, A., and Abzug, S. (1967). *Isr. J. Med. Sci.* **3,** 474–477.

Bachrach, U., Son, S., and Wiener, H. (1974). *Cancer Res.* **34,** 1577–1580.

Bartholeyns, J. (1983). *Eur. J. Cancer Clin. Oncol.* **19,** 567–572.

Bartholeyns, J., and Koch-Weser, J. (1981). *Cancer Res.* **41,** 5158–5161.

Bartholeyns, J., Mamont, P., and Casara, P. (1984). *Cancer Res.* **44,** 4972–4977.

Berger, F. G., Szymanski, P., Read, E., and Watson, G. (1984). *J. Biol. Chem.* **259,** 7941–7946.

Bey, P. (1978). *In* "Enzyme-Activated Irreversible Inhibitors" (N. Seiler, M.J. Jung, and J. Koch-Weser, eds.), pp. 27–41. Elsevier, Amsterdam.

Boutwell, R.K. (1977). *In* "Origins of Human Cancer" (H.H. Hiatt, J.D. Watson, and J.A. Winsten, eds.), pp 773–783. Cold Spring Harbor Lab., Cold Spring Harbor, New York.

Bowlin, T.L., Rosenberger, A.L., and Sunkara, P.S. (1985). *Cancer Immunol. Immunother.* **20,** 214–218.

Boynton, A.L., Whitield, J.F., and Isaacs, R.J. (1976). *J. Cell. Physiol.* **89,** 481–488.

Burchenal, J.H., Lokys, L., Smith, R., Cartmell, S. and Warrel, R. (1981). *Proc. Am. Assoc. Cancer Res.* **22,** 230.

Casero, R.A., Baylin, S.B., Nelkin, B.D., and Luk, G.D. (1986). *Biochem. Biophys. Res. Commun.* **134,** 572–579.

Cohen, S.S. (1977). *Cancer Res.* **37,** 939–942.

Dave, C., and Caballeri, L. (1973). *Fed. Proc., Fed. Am. Soc. Exp. Biol.* **32,** 763.

Don, S., and Bachrach, U. (1975). *Cancer Res.* **35,** 3618–3622.

Don, S., Wiener, H., and Bachrach, U. (1975). *Cancer Res.* **35,** 194–198.

Fillingame, R.H., Jorstad, C.M., and Morris, D.R. (1975). *Proc. Natl. Acad. Sci. U.S.A.* **72,** 4042–4045.

Fozard, J.R., and Koch-Weser, J. (1982). *Trends Pharmacol. Sci.* **3,** 107–110.

Fozard, J.R., and Prakash, N.J. (1982). *Nauyn-Schmiedebergs Arch. Pharmakol.* **320,** 72–77.

Gazdar, A.F., Stull, H.B., Kilton, L.J., and Bachrach, U. (1976). *Nature (London)* **262,** 696–698.

Greco, F.A., and Oldham, R.K. (1979). *N. Engl. J. Med.* **301,** 355–358.

Harada, J.J., and Morris, D.R. (1981). *Mol. Cell. Biol.* **1,** 594–599.

Harik, S.I., and Snyder, S.H. (1973). *Biochim. Biophys. Acta* **327,** 501–509.

Heby, O. (1981). *Differentiation* **19,** 1–20.

Heby, O., and Andersson, G. (1980). *In* "Polyamines in Biomedical Research" (J.M. Gaugas, ed.), pp. 17–34. Wiley, New York.

Heby, O., Andersson, G., and Gray, J. W. (1978). *Exp. Cell Res.* **111,** 461–464.

Heston, W.D.W., Fair, W.R., and Kadmon, D. (1982). *Cancer Lett. (Shannon, Irel.)* **16,** 71–79.

Höltta, E., Jänne, J., and Hovi, T. (1979). *Biochem. J.* **178,** 109–117.
Hung, D.T., Marton, L.J., Deen, D.F., and Shafer, R.H. (1983). *Science* **221,** 368–370.
Inoue, H., Kato, Y., Takigawa, M., Adachi, K., and Takeda, Y. (1975). *J. Biochem.* (*Tokyo*) **77,** 879–893.
Jänne, J., Pösö, H., and Raina, A. (1978). *Biochim. Biophys. Acta* **473,** 241–293.
Johansson, J., and Skinner, W.A. (1973). U.S. Patent No. 3,754.027. *Chem. Abstr.* **79,** 137504.
Kahana, C., and Nathans, D. (1984). *Proc. Natl. Acad. Sci. U.S.A.* **81,** 3645–3649.
Käpyaho, K., and Jänne, J. (1983). *Biochem. Biophys. Res. Commun.* **113,** 8–23.
Kingsnorth, A.N., King, W.W., McCann, P.P., Diekema, K.A., Ross, J.S., and Malt, R.A. (1983a). *Cancer Res.* **43,** 2545–2549.
Kingsnorth, A.N., Russell, W.E., McCann, P.P., Diekema, K., and Malt, R.A. (1983b). *Cancer Res.* **43,** 4035–4038.
Kremzner, L.T., Barrett, R.E., and Terrano, J.J. (1970). *Ann. N.Y. Acad. Sci.* **171,** 735–748.
Luk, G.D., and Baylin, S.B. (1986). *Cancer Res.* **46,** 1844–1848.
Luk, G.D., Goodwin, G., Marton, L.J., and Baylin, S.G. (1981). *Proc. Natl. Acad. Sci. U.S.A.* **78,** 2355–2358.
Luk, G.D., Hamilton, S.R., O'Ceallaigh, D., Smith, J., Natuzzi, E., and Baylin, S.B. (1982a). *Gastroenterology* **82,** 1121.
Luk, G.D., Goodwin, G., Gazdar, A.F., and Baylin, S.B. (1982b). *Cancer Res.* **42,** 3070–3073.
Luk, G.D., Civin, C.I., Weissman, R.M., and Baylin, S.B. (1982c). *Science* **216,** 75–77.
Luk, G.D., Abeloff, M.D., Griffin, C.A., and Baylin, S.B. (1983a). *Cancer Res.* **43,** 4239–4243.
Luk, G.D., Sharkis, S.J., Abeloff, M.D., McCann, P.P., Sjoerdsma, A., and Baylin, S.B. (1983b). *Proc. Natl. Acad. Sci. U.S.A.* **80,** 5090–5093.
Luk, G.D., Abeloff, M.D., McCann, P.P., Sjoerdsma, A., and Baylin, S.B. (1986). *Cancer Res.*, in press.
Luk, G.D., Abeloff, M.D., McCann, P.P., Sjoerdsma, A., and Baylin, S.B. (1986a). *Cancer Res.*, **46,** 1849–1853.
Luk, G.D., Hamilton, S.R., Yang, P., Smith, J.A., O'Ceallaigh, D., McAvinchey, D., and Hyland, J. (1986b). *Cancer Res.* **46,** 4449–4452.
McConlogue, L., Gupta, M., Wu, L., and Coffino, P. (1984). *Proc. Natl. Acad. Sci. U.S.A.* **81,** 540–544.
Mamont, P.S., Bohlen, P., McCann, P.P., Bey, P., Schuber, F., and Tardif, C. (1976). *Proc. Natl. Acad. Sci. U.S.A.* **73,** 1626–1630.
Mamont, P.S., Duchesne, M.C., Grove, J., and Bey, P. (1978). *Biochem. Biophys. Res. Commun.* **81,** 58–66.
Mamont, P.S., Siat, M., Joder-Ohlenbusch, A.M., Bernhardt, A., and Casara, P. (1984). *Eur. J. Biochem.* **142,** 457–463.
Mann, T. (1964). "The Biochemistry of Semen and of the Male Reproductive Tract." Methuen, London.
Marton, L.J. (1981). *Adv. Polyamine Res.* **3,** 425–430.
Marton, L.J., and Heby, O. (1974). *Int. J. Cancer* **13,** 619–628.
Marton, L.J., Heby, O., Levin, V.A., Lubich, W.P., Crafts, D.C., and Wilson, C.B. (1976). *Cancer Res.* **36,** 973–977.
Marton, L.J., Levin, V.A., Hervatin, S.J., Koch-Weser, J., McCann, P.P., and Sjoerdsma, A. (1981). *Cancer Res.* **41,** 4436–4431.
Metcalf, B.W., Bey, P., Danzin, C., Jung, M.J., Casara, P., and Vevert, J.P. (1978). *J. Am. Chem. Soc.* **100,** 2551–2553.

Mihich, E. (1963). *Pharmacologist* **5,** 270.

Nishioka, K., and Romsdahl, M.M. (1974). *Clin. Chim. Acta* **57,** 155–161.

O'Brien, T.G., Simsiman, R.C., and Boutwell, R.K. (1975). *Cancer Res.* **35,** 2426–2433.

Olson, J.W., and Russell, D.H. (1979). *Cancer Res.* **39,** 3074–3079.

Oredsson, S., Anehus, S., and Heby, O. (1980). *Biochem. Biophys. Res. Commun.* **94,** 151–158.

Oredsson, S.M., Deen, D.F., and Marton, L.J. (1982). *Cancer Res.* **42,** 1296–1299.

Oredsson, S.M., Gray, J.W., Deen, D.F., and Marton, L.J. (1983). *Cancer Res.* **43,** 2541–2644.

Ota, D.M., Nishioka, K., Grossie, B., and Dixon, D. (1986). *Int. J. Cancer,* in press.

Otani, S., Mizoguchi, Y., Matsui, I., and Morisawa, S. (1974). *Mol. Biol. Rep.* **1,** 431–436.

Pegg, A.E., and McCann, P.P. (1982). *Am. J. Physiol.* **243,** C212–C221.

Pohjanpelto, P., Hölttä, E., Janne, O.A., Knuutila, S., and Alitalo, K. (1985). *J. Biol. Chem.* **260,** 8532–8537.

Poste, G., and Fidler, J. (1980). *Nature (London)* **283,** 139–146.

Prakash, N.J., and Sunkara, P.S. (1983). *Cancer Res.* **43,** 3192–3196.

Prakash, N.J., Schechter, P.J., Grove, J., and Koch-Weser, J. (1978). *Cancer Res.* **38,** 3059–3062.

Prakash, N.J., Schechter, P.J., Mamont, P.S., Grove, J., Koch-Weser, J., and Sjoerdsma, A. (1980). *Life Sci.* **26,** 181–194.

Raina, A., and Jänne, J. (1975). *Med. Biol.* **53,** 121–147.

Rao, P.N., Wilson, B., and Puck, T.T. (1977). *J. Cell. Physiol.* **91,** 131–141.

Rennert, O.M., Lawson, D.L., Shukla, J.B., and Miale, T.D. (1977). *Clin. Chim. Acta* **75,** 365–369.

Rozhin, J., Wilson, P.S., Bull, A.W., and Nigro, N.D. (1984). *Cancer Res.* **44,** 3226–3230.

Rupniak, H.T., and Paul, D. (1978). *J. Cell. Physiol.* **94,** 161–170.

Rupniak, H.T., and Paul, D. (1980). *Cancer Res.* **40,** 293–297.

Russell, D.H. (1977). *Clin. Chem.* **23,** 22–27.

Russell, D.H., and Levy, C.C. (1971). *Cancer Res.* **32,** 248–251.

Russell, D.H., and Snyder, S.H. (1968). *Proc. Natl. Acad. Sci. U.S.A.* **60,** 1420–1427.

Russell, D.H., Levy, C.C., Schimpff, S.C., and Hawk, I.A. (1971). *Cancer Res.* **31,** 1555–1558.

Savory, J., and Shipe, J.R. (1975). *Ann. Clin. Lab. Sci.* **5,** 110–114.

Scalabrino, G., Pösö, H., Hölttä, E., Hannonen, P., Kallio, A., and Jänne, J. (1978). *Int. J. Cancer* **21,** 239–245.

Seidenfeld, J., and Komar, A.A. (1985). *Cancer Res.* **45,** 2132–2136.

Seidenfeld, J., Deen, D. F., and Marton, L. J. (1980). *Int. J. Radiat. Biol.* **38,** 223–229.

Seidenfeld, J., Gray, J. W., and Marton, L. J. (1981). *Exp. Cell Res.* **131,** 209–216.

Seiler, N., and Knödgen, B. (1979). *Biochem. Med.* **21,** 168–181.

Seiler, N.A., Lamberty, U., and Al-Therib, J.J. (1975). *J. Neurochem.* **24,** 797–800.

Seppänen, P., Alhonen-Hongisto, L., and Jänne, J. (1981). *Biochim. Biophys. Acta* **674,** 169–177.

Seyfried, C.E., and Morris, D.R. (1979). *Cancer Res.* **39,** 4861–4867.

Siimes, M., and Jänne, J. (1967). *Acta Chem. Scand.* **21,** 815–817.

Sjoerdsma, A. (1981). *Clin. Pharmacol. Ther.* **30,** 3–22.

Sjoerdsma, A., and Schechter, P.J. (1984). *Clin. Pharmacol. Ther.* **35,** 287–300.

Sunkara, P.S., and Prakash, N.J. (1984). *In* "Novel Approaches to Cancer Chemotherapy" (P.S. Sunkara, ed.), pp. 94–117. Academic Press, New York.

Sunkara, P.S., and Rao, P.N. (1981). *Adv. Polyamine Res.* **3,** 347–356.

Sunkara, P.S., Rao, P.N., and Nishioka, K. (1977). *Biochem. Biophys. Res. Commun.* **74,** 1125–1133.

Sunkara, P.S., and Rosenberger, A.L. (1987). *Cancer Res.* **47,** 933–935.
Sunkara, P.S., Pargac, M.B., Nishioka, K., and Rao, P.N. (1979a). *J. Cell. Physiol.* **98,** 451–458.
Sunkara, P.S., Rao, P.N., Nishioka, K., and Brinkley, B.R. (1979b). *Exp. Cell Res.* **119,** 63–68.
Sunkara, P.S., Fowler, S.K., Nishioka, K., and Rao, P.N. (1980). *Biochem. Biophys. Res. Commun.* **95,** 423–430.
Sunkara, P.S., Fowler, S.K., and Nishioka, K. (1981). *Cell Biol. Int. Rep.* **5,** 991–997.
Sunkara, P.S., Prakash, N.J., and Rosenberger, A.L. (1982). *FEBS Lett.* **150,** 397–399.
Sunkara, P.S., Prakash, N.J., Mayer, G.D., and Sjoerdsma, A. (1983a). *Science* **219,** 851–853.
Sunkara, P.S., Prakash, N.J., Chang, C.C., and Sjoerdsma, A. (1983b). *JNCI, J. Natl. Cancer Inst.* **40,** 505–509.
Sunkara, P.S., Prakash, N.J., and Nishioka, K. (1984). *Ann. N.Y. Acad. Sci.* **419,** 268–272.
Sunkara, P.S., Chang, C.C., and Prakash, N.J. (1985a). *Cancer Res.* **45,** 4867–4870.
Sunkara, P.S., Rosenberger, A.L., and Davis, G.F. (1985b). *Proc. Am. Assoc. Cancer Res.* **26,** 59.
Takami, H., Romsdahl, M.M., and Nishioka, K. (1979). *Lancet* **2,** 912.
Tofilon, P.J., Ordesson, S.M., Dean, D.F., and Marton, L.J. (1982). *Science* **217,** 1046–1048.
Tucker, R.W., Meade-Cobun, K., Theiss, W., and Luk, G. (1985). *J. Cell Biol.* **101,** 245a.
Verma, A.K., and Boutwell, R.K. (1977). *Cancer Res.* **37,** 2196–2201.
Waalkes, T.P., Gehrke, C.W., Toomey, D.C., Zumwalt, R.W., Hueser, J.N., Kuo, K.C., Lakings, D.B., Ahmann, D.L., and Moertel, C.G. (1975). *Cancer Chemother. Rep.* **59,** 85–92.
Williams-Ashman, H.G., Coppoc, G.L., and Weber, G. (1972). *Cancer Res.* **32,** 1924–1932.
Williams-Ashman, H.G., Coppoc, G.L., Schenone, A., and Weber, G. (1973). *In* "Polyamines in Normal and Neoplastic Growth" (D.H. Russell, ed.), pp. 181–197. Raven, New York.
Yarrington, J.T., Sprinkle, D.J., Loudy, D.E., Diekema, K.A., McCann, P.P., and Gibson, J.P. (1983). *Exp. Mol. Pathol.* **39,** 300–316.
Yuspa, S.H., Lichti, U., Ben, T., Patterson, E., Hennings, H., Slaga, T.J., Colburn, N., and Kelsey, W. (1976). *Nature* (*London*) **262,** 402–404.

7

Polyamine Inhibition *in Vivo* and in Organ Growth and Repair

CHARLES DANZIN AND PIERRE S. MAMONT

Merrell Dow Research Institute
Strasbourg Centre
16, rue d'Ankara
F 67084 Strasbourg Cedex, France

I. INTRODUCTION

Considerable experimental evidence accumulated during the last 20 years leads to the concept that the polyamines putrescine, spermidine, and spermine play essential functions in cellular growth and differentiation (for reviews, see Jänne *et al.*, 1978; Williams-Ashman and Canellakis, 1979; Heby, 1981; Pegg and McCann, 1982; Tabor and Tabor, 1984). This concept has been mainly substantiated by the studies of the biochemical and pharmacological consequences of polyamine deficiency in prokaryotes and eukaryotes. Indeed, this deficiency is achieved in polyamine-auxotroph cells cultured *in vitro* obtained either by mutation (Tabor *et al.*, 1980; Morris, 1981; Paulus *et al.*, 1982; Tabor *et al,*, 1983) or by selection (Knuutila and Pohjanpelto, 1983; Steglich *et al.*, 1985). *In vivo,* the only available approach for study is the use of very potent inhibitors of the biosynthetic enzymes. This explains the intense research effort aimed at the design and synthesis of such inhibitors of polyamine biosynthesis (see Chapters 1 and 2).

In this chapter we review the effects of the various available inhibitors on their target enzymes, on the concentrations of polyamines and their biosynthetic byproducts, and on various biological parameters of organ growth and repair in different animal models.

II. *IN VIVO* BIOCHEMISTRY OF POLYAMINE BIOSYNTHESIS INHIBITORS

A. ODC Inhibitors

1. Putrescine Analogs

Since the pioneer works of Kay and Lindsay (1973), a number of studies have shown that polyamines and various diamines, including putrescine, suppress ODC activity (for reviews, see McCann, 1980; Heby and Jänne, 1981). This reduction of ODC activity is partially explained by the polyamine-elicited induction of a protein called "antizyme" (Heller *et al.*, 1976; Canellakis *et al.*, 1979; Seely and Pegg, 1983) which specifically interacts with the ODC protein to form an inactive enzyme–antizyme complex. Regardless of the actual mechanism of the suppression of ODC activity (Persson *et al.*, 1984; Davis *et al.*, 1985; Murakami and Hayashi, 1985; Kahana and Nathans, 1985), this approach has been used as a means of producing polyamine deficiency *in vivo*. In fact, 1,3-diaminopropane, 1,3-diamino-2-propanol, 1,5-diaminopentane, 1,6-diaminohexane, and 1,7-diaminoheptane decrease ODC activity (Pösö *et al.*, 1977; Kallio, 1978; Pegg *et al.*, 1978; Piik *et al.*, 1978; Seely and Pegg, 1983) and polyamine levels (Kato *et al.*, 1978; Piik *et al.*, 1978; Pegg and Hibasami, 1980) in animal tissues. However, rapid metabolism together with lack of specificity of action and significant toxicity make the use of these diamines unreliable to show the effects of polyamine deprivation in biological models. Moreover, these amines may fulfill the requirement for endogenous polyamines and their experimental use has raised doubts regarding the actual significance of their biological effects (Hölttä *et al.*, 1981). A more classical approach has consisted in inhibiting ODC directly by means of specific inhibitors of enzyme catalysis.

2. Competitive Inhibitors

Since the first publications on studies dealing with ODC inhibitors (Skinner and Johansson, 1972; Russell, 1973), about 30 derivatives of ornithine and putrescine have been synthesized as potential reversible inhibitors of ODC (reviewed in Williams-Ashman *et al.*, 1976; Mamont *et al.*, 1980; Stevens and Stevens, 1980; Heby, 1981; Heby and Jänne, 1981). Indeed, DL-α-hydrazino-δ-aminovaleric acid (DL-HAVA) (Inoue *et al.*, 1975) and diaminobutene (Relyea and Rando, 1975; Kameji *et al.*, 1979), the most potent competitive inhibitors of ODC, block putrescine formation for several hours and significantly decrease putrescine and spermidine concentrations in animal tissues (Inoue *et al.*, 1975; Kato *et al.*,

1978). However, besides the fact that diaminobutene is also a potent inhibitor of diamine oxidase (Relyea and Rando, 1975), this compound is metabolized to the unsaturated spermidine analog and may fulfill the role of the natural polyamine (Kameji *et al.,* 1979). DL-HAVA is also not specific since it inhibits other pyridoxal-phosphate-dependent enzymes (Inoue *et al.,* 1975). Moreover, due to its hydrazino part, the compound may deplete cells of pyridoxal phosphate by formation of a hydrazone with the carbonyl group of this important enzyme cofactor. Similar comments apply to 1-aminoxy-3-aminopropane, a new potent competitive inhibitor of ODC *in vitro* (Khomutov *et al.,* 1985).

More generally, even if sufficiently specific to be devoid of untoward biological effects, competitive inhibitors share the major disadvantage of stabilizing ODC (Harik *et al.,* 1974; McCann *et al.,* 1977; Jänne *et al.,* 1981) and thus increase the amount of enzyme protein (Persson *et al.,* 1985). Consequently, when the concentration of inhibitor in the body falls due to metabolic clearance, reactivation of the enzyme results in supernormal putrescine production. This problem is largely avoided by the use of irreversible inhibitors of ODC.

3. *Enzyme-Activated Irreversible Inhibitors*

The design of specific irreversible inhibitors of ODC has been closely associated with the development of the concept of substrate-induced irreversible inhibition of enzymes (Rando, 1974, 1984; Abeles and Maycock, 1976; Walsh, 1977, 1982, 1984; Seiler *et al.,* 1978b).

The chemistry and mechanism of action of the enzyme-activated irreversible inhibitors of ODC are discussed in detail in the first chapter of this book. In this section, we will focus on the effects produced by the administration to animals of the most potent of these inhibitors of ODC activity and on some related biochemical parameters.

a. Ornithine Analogs. The ornithine analog α-difluoromethylornithine (DFMO; see Fig. 1) is one of the first described and most studied irreversible inhibitors of ODC. Since its discovery (Bey, 1978; Metcalf *et al.,* 1978), there have been more than 500 published reports on its biochemical effects on cultured eucaryotic cells and microorganisms and on plants, animal tissues, and human diseases (for previous reviews, see Mamont *et al.,* 1978; Seiler *et al.,* 1978a; Koch-Weser *et al.,* 1981; Sjoerdsma, 1981; Sjoerdsma and Schechter, 1984). In rats, DFMO produces a rapid and dose-dependent decrease of ODC activity in several tissues (Danzin *et al.,* 1979a). It is noteworthy that the inhibition of ODC appears to be organ selective, the enzyme of ventral prostate being the most affected by the compound. In this organ, a single intraperitoneal injection of 100 mg/

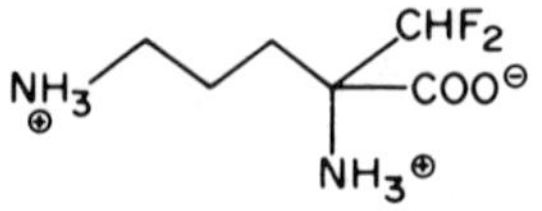

α-difluoromethylornithine
(DFMO, MDL 71782)

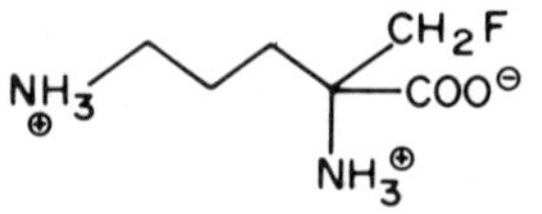

α-monofluoromethylornithine
(MFMO, MDL 71919)

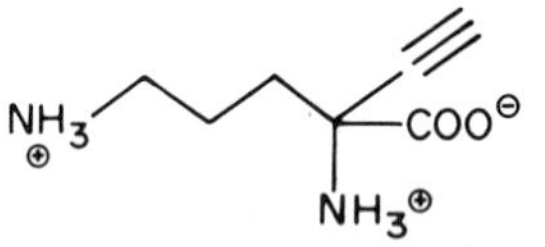

α-ethynylornithine
(MDL 72081)

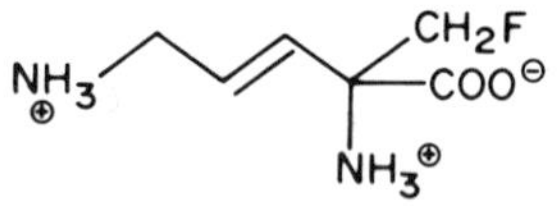

(E)-2-(fluoromethyl)dehydro-
ornithine (MDL 72246)

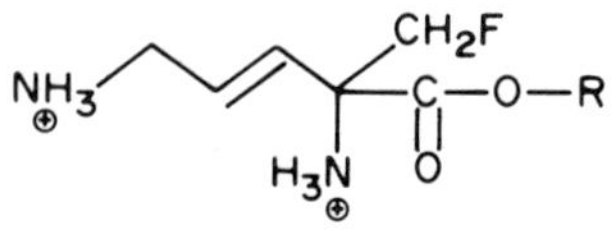

esters of (E)-2-(fluoromethyl)dehydroornithine

R = $-CH_3$	(MDL 72403)
R = $-CH_2-CH_3$	(MDL 72430)
R = $-(CH_2)_7-CH_3$	(MDL 72493)
R = $-CH_2-C_6H_5$	(MDL 72504)

Fig. 1. Formulas of the various enzyme-activated irreversible inhibitors of ODC effective *in vivo:* ornithine analogs.

kg body weight reduces ODC activity to less than 10% of control values within 4 hr. However, 24 hr after drug administration at this dose, prostatic ODC activity is restored to half normal values. Even repeated administration of this dose at intervals of 12 hr allows some recovery of enzyme activity between doses (Danzin *et al.*, 1979a). This is explained by the high turnover of ODC (Pegg and McCann, 1982) and by the pharmacokinetics of the drug, which vary as a function of the species (see Table I). Methods to measure DFMO in body fluids and tissues by automatic amino acid analysis have been published (Danzin *et al.*, 1979b; Grove *et al.*, 1981).

TABLE I

Plasma Half-Life of DFMO in Various Species

Species	Oral dose (mg/kg)	Half-life (min)
Sprague–Dawley rat	200	83[a]
New Zealand white rabbit	10	121[a]
Beagle dog	10	108–132[a]
Cynomolgus monkey	10	353[a]
Healthy man	10	199[b]

[a] Data from Grove *et al.* (1981).
[b] Data from Haegele *et al.* (1981).

Although DFMO is slightly more effective at reducing ODC activity when injected intraperitoneally, it can be administered orally either by gavage or in solution in drinking water of experimental animals (Seiler *et al.*, 1978a). As a result of its inhibitory effect on ODC, repeated intraperitoneal injections of DFMO (200 mg/kg body weight every 6 hr, or 1 g/kg body weight every 12 hr), or administration of the compound in solution in water (20 g/liter) as the sole drinking fluid of animals, cause marked decreases of putrescine and, to a lesser extent, of spermidine concentrations in several rat tissues, but barely affect spermine concentrations (Danzin *et al.*, 1979a,b; Wagner *et al.*, 1982). Moreover, an enhanced activity of *S*-adenosyl-L-methionine decarboxylase (EC 4.1.1.50; AdoMetDC) is observed in tissues of DFMO-treated animals in which concentrations of polyamines, and more particularly of spermidine, have been depleted (Prakash *et al.*, 1978; Danzin *et al.*, 1982a; Mamont *et al.*, 1982a; Pegg *et al.*, 1982a). This finding is in agreement with the hypothesis of a negative regulation of ADoMetDC by spermidine, originally proposed by Mamont *et al.* (1978) and developed further in the works of Hopkins and Manchester (1980), Mamont and Danzin (1981), Mamont *et al.* (1981), Pösö and Pegg (1981), Shirahata and Pegg (1985), Shirahata *et al.* (1985). Furthermore, the decreases of tissue concentrations of putrescine and spermidine are accompanied by increases of decarboxylated *S*-adenosylmethionine, the cosubstrate of spermidine and spermine synthases (Danzin *et al.*, 1982a; Mamont *et al.*, 1982a,b; Pegg *et al.*, 1982a,c; Wagner *et al.*, 1982, 1985a), and, to a lesser extent, of *N*-acetyl decarboxylated *S*-adenosylmethionine, a recently identified metabolite of decarboxylated *S*-adenosylmethionine (Wagner *et al.*, 1985b; Pegg *et al.*, 1986). The increase of decarboxylated *S*-adenosylmethionine is explained both by the increase of AdoMetDC activity and by the limitation of putrescine and spermidine for the aminopropyltransferase reactions.

The low toxicity of DFMO is worth reporting. The LD_{50} of this drug is higher than 5 g/kg body weight in mice (Prakash *et al.*, 1978). Treatment of rats with a daily dose of 2 g/kg body weight for 1 month was associated with only a 7% reduction in weight gain (Seiler *et al.*, 1978a). A similarly low but significant reduction in weight gain of rats has been noticed more recently by Dowling *et al.* (1985). These authors attribute the reduced weight gain to the 20% reduction of food intake of animals given 20 g/liter DFMO in their drinking water. The low toxicity of DFMO thus allows its use as a powerful tool to investigate the pharmacological consequences of polyamine deficiency *in vivo*.

Other ornithine analogs (Fig. 1) such as α-monofluoromethylornithine (Bey, 1978; Kollonitsch *et al.*, 1978) and α-ethynylornithine (Danzin *et al.*, 1981) decrease ODC activity in rat tissues. However, they are somewhat less effective than DFMO in this respect (C. Danzin, unpublished results; Danzin *et al.*, 1981). Among a series of ornithine analogs recently synthesized at Merrell Dow Research Institute (Bey *et al.*, 1983), (*E*)-2-(fluoromethyl)dehydroornithine (Fig. 1) has been found to be an enzyme-activated irreversible inhibitor of ODC 10 times more potent than DFMO *in vitro*. Unfortunately, this compound does not demonstrate better efficiency than DFMO at decreasing ODC activity in rat tissues (Fig. 2) (Mamont *et al.*, 1986), suggesting a poor cellular absorption of the inhibitor. In order to improve the effectiveness of (*E*)-2-(fluoromethyl)dehydroornithine, esters of the compound have been synthesized as potential prodrugs (Fig. 1). The use of ester derivatives is based on the ubiquitous distribution of nonspecific esterases (Krisch, 1971). As expected from their increased cellular concentrations and effective intracellular hydrolysis, the methyl and the ethyl esters used at 10 times lower concentrations are as effective as their parent amino acid or as DFMO at decreasing ODC activity *in situ* (Fig. 2) (Mamont *et al.*, 1986). Moreover, the methyl ester (Fig. 3) and the ethyl ester (not shown) produce a particularly long duration of ODC inhibition in the ventral prostate and in the testes. A single intraperitoneal injection of 25 mg/kg body weight of the methyl ester reduces by 80–90% ODC activity in these two organs. This inhibition remains nearly constant between 3 and 24 hr. The interesting long duration of action of the ester is confirmed by a constant 80% inhibition of ODC observed over a 24-hr period following the eighth daily administration of the same dose. This 8-day treatment of rats results in an effective reduction of putrescine and spermidine concentrations in the ventral prostate and, to a lesser extent, in the thymus and the testes. In the ventral prostate, the marked decrease of polyamine concentrations is accompanied by a striking elevation of decarboxylated *S*-adenosylmethionine (Mamont *et al.*, 1986).

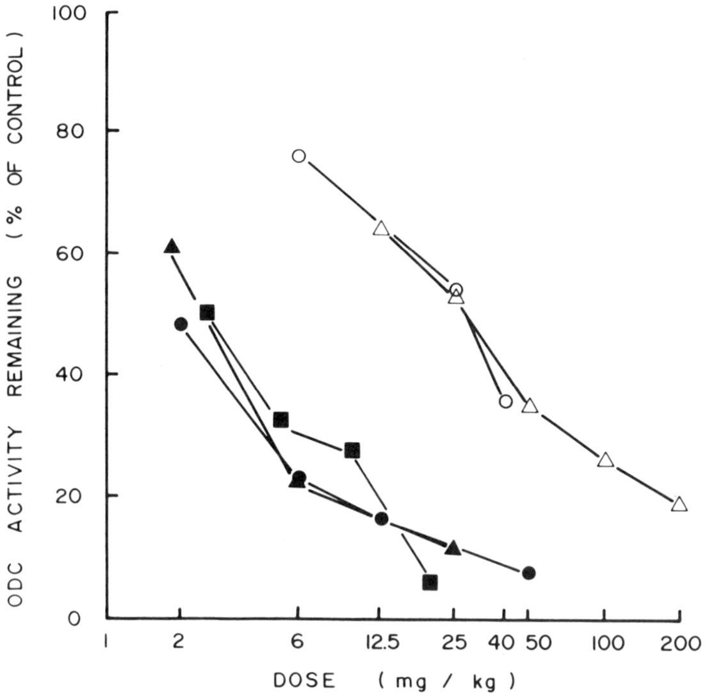

Fig. 2. Comparison of the effects of a single oral administration of DFMO (△), (*E*)-2-(fluoromethyl)dehydroornithine (○), MAP (■), (*E*)-2-(flouromethyl)dehydroornithine methyl ester (●), and (*E*)-2-(fluoromethyl)dehydroornithine ethyl ester (▲) on ODC activity in ventral prostate. ODC activity was measured 6 hr after administration of the compounds to rats.

b. Putrescine Analogs. 5-Hexyne-1,4-diamine, or acetylenic putrescine (see Fig. 4), another potent enzyme-activated irreversible inhibitor of mammalian ODC, was discovered at the same time as DFMO (Metcalf *et al.,* 1978; Danzin *et al.,* 1979c). It is disappointing that this putrescine analog inhibits 4-aminobutyrate aminotransferase (EC 2.6.1.19; GABA-T) *in vivo* and increases the levels of the neurotransmitter 4-aminobutyric acid (GABA) in brain, thus producing secondary pharmacological effects (Danzin *et al.,* 1979d). The effects are due to the oxidation of 5-hexyne-1,4-diamine to 4-aminohex-5-ynoic acid, a potent irreversible inhibitor of GABA-T (Jung *et al.,* 1977). The first step of this process is catalyzed by mitochondrial monoamine oxidase (EC 1.4.3.4.; MAO) (Danzin *et al.,* 1979d, 1983a). Taking advantage of the fact that α-branched primary amines are essentially not substrates of MAO (Blaschko, 1952; Silver-

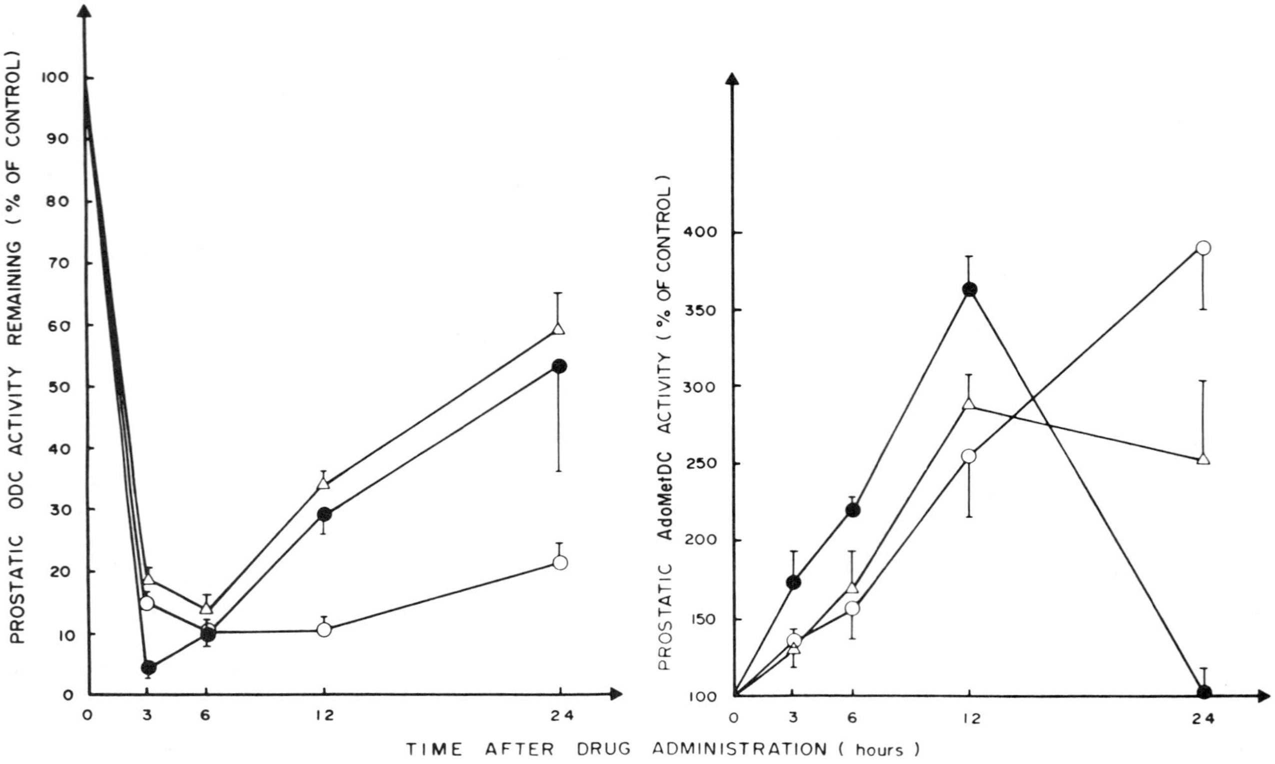

Fig. 3. Comparison of the effects of a single oral administration of 200 mg/kg DFMO (△), 25 mg/kg MAP (●), and 25 mg/kg (*E*)-2-(fluoromethyl)dehydroornithine methyl ester (○) on ODC and AdoMetDC activities in ventral prostate. Rats were given compounds by gavage at time zero. At given intervals, animals were sacrificed and enzyme activities were immediately measured.

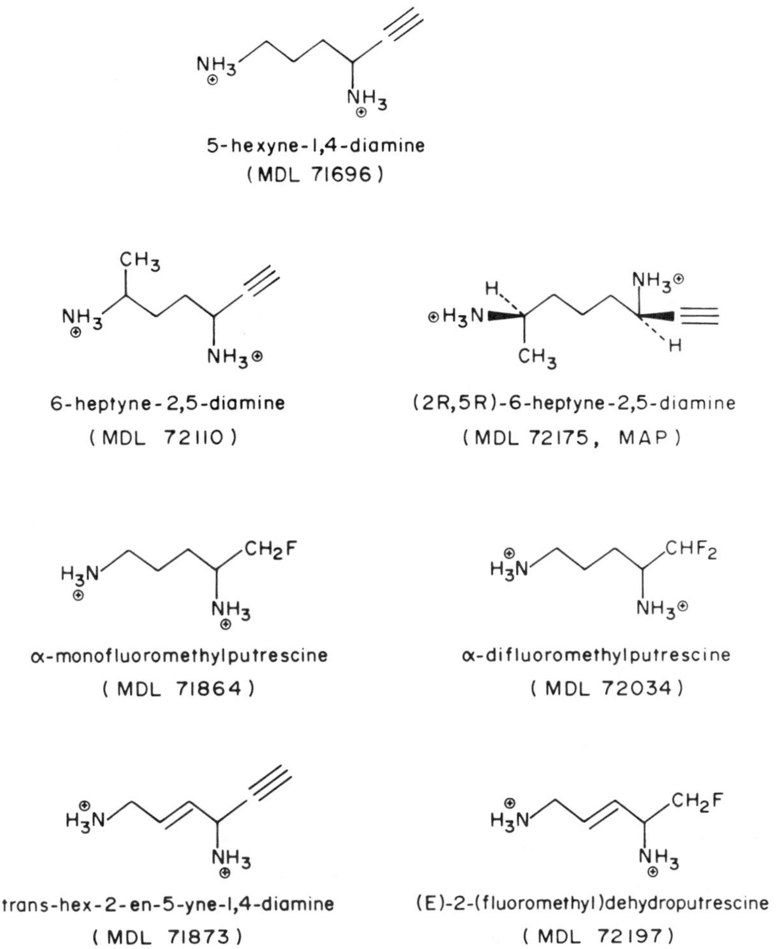

Fig. 4. Formulas of the various enzyme-activated irreversible inhibitors of ODC effective *in vivo:* putrescine analogs.

man, 1984), 6-heptyne-2,5-diamine has been synthesized as the racemic mixture and as its four stereoisomers (Casara *et al.*, 1985). *In vitro* studies have demonstrated that most of the ODC inhibitory activity lies in the (2*R*,5*R*) stereoisomer (Danzin *et al.*, 1983b; Casara *et al.*, 1985). *In vivo,* (2*R*,5*R*)-6-heptyne-2,5-diamine (also called MAP for methyl acetylenic putrescine; see Fig. 4) is 10 times more potent than DFMO and as effective as the esters of (*E*)-2-(fluoromethyl)dehydroornithine at inhibiting ODC (Fig. 2). The organ selectivity of this inhibition is similar to that of the ornithine analogs. However, duration of action of MAP is not longer than that of DFMO (Fig. 3). Like other inhibitors of ODC, MAP is slightly more effective when injected intraperitoneally, but this compound is still active when given orally either by gavage or in solution (usual dosage: 2–3 g/liter) in the drinking water of animals. Because of its high potency and its chemical stability, another convenient method to study the effects of chronic administration in rats or mice consists of continuous infusion of MAP via an osmotic minipump (Bartholeyns *et al.*, 1984). Like DFMO, MAP exerts marginal toxicity in mice and rats at effective doses. The 50% lethal dose values are higher than 1 g/kg body weight for both intravenous and *per os* routes of administration. Animals treated with MAP given as 2 g/liter of drinking water solution (drug intake: 300 mg/kg body weight per day) for time periods from 2 weeks to 6 months manifest no toxic signs except a slight decrease in body weight gain compared to control animals (Bartholeyns *et al.*, 1984; C. Danzin and N. Claverie, unpublished results).

MAP concentrations in physiological samples can be determined by high-performance liquid chromatography (Seiler and Knödgen, 1980; Mamont *et al.*, 1984).

Other putrescine analogs that irreversibly inhibit ODC *in vivo* are α-monofluoromethylputrescine and α-difluoromethylputrescine (see Fig. 4), but they are substantially less potent than MAP. Furthermore, like 5-hexyne-1,4-diamine, they are oxidized *in vivo* to their corresponding acids and produce GABA-T inhibition (Danzin *et al.*, 1982b). The dehydroputrescine analogs *trans*-hex-2-en-5-yne-1,4-diamine (Metcalf *et al.*, 1978) and (*E*)-2-(fluoromethyl)dehydroputrescine (Bey *et al.*, 1983) (Fig. 4) are very effective ODC inhibitors *in vivo;* however, because of their significant toxicity, effects of these compounds have not been explored further in animals (C. Danzin, unpublished results).

B. Inhibitors of other Enzymes Involved in the Metabolism of Polyamines

1. AdoMetDC Inhibitors

Since the discovery by Williams-Ashman and Schenone (1972) that the antitumor agent methylglyoxal bis(guanylhydrazone) (MGBG) (Mihich, 1963) is a very potent competitive inhibitor of AdoMetDC *in vitro,* MGBG has been extensively used to inhibit AdoMetDC activity *in vivo* (for reviews, see Heby and Jänne, 1981; Porter *et al.*, 1981; Pegg, 1984). Although MGBG effectively blocks the formation of spermidine and spermine in various cells grown in culture, it seems to be remarkably difficult to achieve any depletion of spermidine and spermine concentrations in normal animal tissues at sublethal doses (Hölttä *et al.,* 1973; Pegg, 1973). Moreover, a number of side effects of MGBG have been discovered during the last decade and have impaired further development of this drug. Particularly, MGBG causes extensive ultrastructural damages to mitochondria (Pine and DiPaolo, 1966; Pathak *et al.,* 1977; Mikles-Robertson *et al.,* 1979) unrelated to its inhibitory effect on AdoMetDC (Pleshkewych *et al.,* 1980). In addition, MGBG is a potent inhibitor of the diamine oxidase enzyme (EC 1.4.3.6; DAO) *in vitro* and *in vivo* (Hölttä *et al.,* 1973; Pegg and McGill, 1978). Because of this effect and of the inhibition of the biosynthesis of decarboxylated *S*-adenosylmethionine, the aminopropyl donor in the biosynthesis of spermidine and spermine, putrescine accumulates in tissues of animals treated with MGBG (Hölttä *et al.,* 1973; Pegg, 1973). Putrescine accumulation is also observed after injection of animals with the MGBG analog 1,1′-[(methylethanediylidene)dinitrilo]-bis(3-aminoguanidine) (MBAG), an irreversible inhibitor of both AdoMetDC and DAO (Pegg and Conover, 1976; Pegg, 1978; Pegg and McGill, 1978). Very recently, Berenil, another structural analog of MGBG, has been shown to inhibit irreversibly AdoMetDC both *in vitro* and *in vivo* but the effect of this compound on polyamine concentrations or on DAO inhibition has not been yet documented (Karvonen *et al.,* 1985).

Like competitive inhibitors of ODC that stabilize their target enzyme (vide supra), administration of MGBG to animals results in a paradoxical enhancement of AdoMetDC activity and increased protein in rat tissues (Pegg *et al.,* 1973; Shirahata and Pegg, 1985). This finding and the other problems encountered with MGBG should stimulate the search for more potent and more selective inhibitors of AdoMetDC. Ideally, an enzyme-activated irreversible inhibitor of this enzyme would be highly desirable.

2. Spermidine and Spermine Synthase Inhibitors

In vivo inhibition of spermidine synthase can be achieved with cyclohexylamine, a competitive inhibitor of the enzyme (Hibasami *et al.,* 1980). Significant but marginal decreases of spermidine concentrations in rat and mouse tissues are produced by repeated intraperitoneal administration of 50–300 mg/kg body weight cyclohexylamine (Hibasami and Ito, 1981; Ito *et al.,* 1982; Porta *et al.,* 1984). Selectivity and toxicity of the compound still remain to be established. Furthermore, inhibition of spermidine synthesis by cyclohexylamine produces an increase in the concentration of putrescine (Porta *et al.,* 1984; Feuerstein *et al.,* 1985) and possibly of decarboxylated *S*-adenosylmethionine, the other substrate for spermidine synthase. These compensatory effects might account for the inability of cyclohexylamine to potentiate cytotoxic agents such as nitrosourea, in contrast to other polyamine-biosynthesis inhibitors (Feuerstein *et al.*, 1985). [In a recent paper, Batchelor *et al.* (1986) established that the spermidine synthase inhibitor tested in all of the work just cited is dicyclohexylammonium sulfate, which is cyclohexylamine, not biscyclohexylamine, as has been incorrectly reported.]

Another compound, *S*-adenosyl-1,8-diamino-3-thiooctane, has been found to be a very potent inhibitor of spermidine synthase *in vitro* and in cultured cells (Pegg *et al.*, 1982b). No *in vivo* effect of this compound has been reported so far.

At present there are no available data on inhibition of spermine synthase *in vivo*.

3. Diamine Oxidase and Polyamine Oxidase Inhibitors

While diamine oxidase catalyzes the terminal oxidation of polyamines, polyamine oxidase (PAO) is involved in the interconversion of spermidine and spermine to putrescine and spermidine, respectively (Seiler *et al.*, 1981; Pegg *et al.*, 1982d). In this respect, PAO plays an important role in the control of polyamine concentrations; inhibitors of this enzyme may displace the equilibrium levels among the three polyamines. In particular, putrescine concentrations are decreased in some rat and mouse tissues after intraperitoneal administration of N^1-methyl-N^4-2,3-butadienyl-1,4-butanediamine or N^1,N^4-bis(2,3-butadienyl)-1,4-butanediamine, two potent enzyme-activated irreversible inhibitors of PAO (Bey *et al.*, 1985; Bolkenius *et al.*, 1985). A more detailed description of the *in vivo* effects of these compounds and of DAO inhibitors is reported in the Chapter 3 of this book.

III. CONSEQUENCES OF POLYAMINE INHIBITION ON ORGAN GROWTH AND REPAIR

It is well recognized that polyamine concentrations increase concomitantly with those of proteins and nucleic acids during normal or abnormal growth of animal tissues (for reviews, see Bachrach, 1973; Russell, 1973; Jänne *et al.*, 1978). Whatever the inducers of growth, the enhancement of polyamine levels is always preceded by a striking and transient increase of ODC activity (for an exhaustive review, see Russell, 1985). Even though stimulations of ODC and of the rate of polyamine biosynthesis always appear to accompany cell growth and proliferation, it remains uncertain whether these increases are circumstantial or essential for the tissues to undergo hypertrophy and/or hyperplasia. Use of specific inhibitors of polyamine biosynthesis, and more particularly of the inhibitors of ODC, might provide answers to this problem.

A. Regenerating Rat Liver

The metabolism of polyamines and fluctuations of their concentrations have been particularly well studied in the liver in regeneration following either hepatectomy or chemical injuries (Jänne, 1967; Jänne *et al.*, 1976). It was expected that administration of an ODC inhibitor might reduce elevation of polyamines and suppress the regenerative process if these two events are causally related. The putrescine analog 1,3-diamino-2-propanol (Piik *et al.*, 1978), or the ODC competitive inhibitor DL-HAVA (Kato *et al.*, 1978), not only prevents the accumulation of putrescine and spermidine but also substantially reduces the stimulation of DNA synthesis induced by partial hepatectomy. These observations were consistent with the above proposal. However, because of the already mentioned toxicity of these drugs, and of their possible ODC-unrelated side effects (vide supra), the actual significance of these findings remains doubtful.

More recently, Pösö and Pegg (1982) reinvestigated the same problem using DFMO. This compound, injected intraperitoneally at 200 mg/kg body weight every 6 hr to partially hepatectomized rats, suppresses the large rise in putrescine concentration but affects neither the increase in spermidine concentration nor DNA synthesis in the liver. The synthesis of DNA can, however, be reduced by 70% at higher administered doses of DFMO under conditions which not only prevent putrescine elevation but reduce the hepatic putrescine level below that measured in the liver of sham-operated animals. In spite of the effect on putrescine, even high doses of DFMO have very little effect on the accumulation of spermidine. The inhibition of DNA synthesis produced by DFMO is completely reversed by the coadministration of nontoxic doses of putrescine to rats (Pösö and Pegg, 1982). These findings may suggest that putrescine is needed for DNA synthesis in this model of liver regeneration, but show clearly that only a small fraction, if any at all, of the rise in putrescine normally seen after the partial hepatectomy is essential. It is possible that the elevations of putrescine and spermidine concentrations are needed for cellular functions other than DNA synthesis per se.

B. Prostatic Growth

The regulation of polyamine synthesis in male accessory glands by circulating androgens is one of the most spectacular and best-documented examples of the close association between growth or, on the contrary, involution of a hormone-dependent tissue and polyamine metabolism (Pegg *et al.*, 1970; Williams-Ashman *et al.*, 1975a,b). The rapid involution of rat ventral prostate and seminal vesicle following castration results in a marked decrease of nucleic acid synthesis together with a striking reduction of ODC and AdoMetDC activities and a profound depletion of the three polyamines. These effects, due to the castration of the animals, can be completely and rapidly reversed by injections of testosterone to rats (Pegg *et al.*, 1978; Williams-Ashman *et al.*, 1975a; Piik *et al.*, 1977; Takyi *et al.*, 1977; Danzin *et al.*, 1979b; Käpyaho *et al.*, 1980). Administration of DFMO to castrated rats treated with testosterone results in a complete suppression of the androgen-induced increase of putrescine and spermidine in the ventral prostate and of putrescine in the seminal vesicle. Moreover, a lesser accumulation of spermine in ventral prostate and spermidine in seminal vesicle is also observed. However, limitation of polyamine synthesis is not accompanied by parallel changes in DNA accumulation in either organ studied. The testosterone-induced accumulation of RNA is slightly, but significantly, slowed by DFMO in ventral prostate but not in seminal vesicle. The most spectacular effect of DFMO is to selectively slow the androgen-induced weight gain of the ventral prostate (Danzin *et al.*, 1979b). Cytological studies (Danzin *et al.*, 1979b) and direct measurements of prostatic secretions (Käpyaho *et al.*, 1984) have shown that the effect of DFMO on the weight of ventral prostate is due to inhibition of the androgen-induced restoration of the secretion content of prostatic acini.

Administration of DFMO to intact male rats likewise inhibits the secretory activity of the ventral prostate and results in a decrease of the organ weight within 4 days. However, at variance with the effects of DFMO in the testosterone-treated castrated rat, the reduction of prostatic weight in the intact rat is accompanied by a marked reduction in the age-dependent increases of both RNA and DNA concentrations in the gland (Danzin *et al.*, 1982a). The discrepancy between the two studies, i.e., castrated versus intact animals, might be due to the use of superphysiological doses of testosterone (1 mg/animal per day) in castrated rats (Williams-Ashman *et al.*, 1975a). Thus, polyamines may exert solely a permissive role in the action of steroid androgens on the biosynthesis of nucleic acids. Such a function has been also suggested for the stimulation of casein synthesis by glucocorticoids in the mammary gland of rabbit (Houdebine *et al.*, 1978) and rat (Bolander and Topper, 1979).

More convincingly, from the studies using DFMO (Danzin *et al.*, 1979b, 1982a; Käpyaho *et al.*, 1984) and from other studies in which the relationship between ODC activity and secretory activity has been investigated (Fuller *et al.*, 1975; McKeehan *et al.*, 1982), we can conclude that polyamines are involved in the maintenance of functional secretory epithelium of the prostate.

C. Cardiac Hypertrophy

Effects of DFMO have been studied in three different models of cardiac hypertrophy in the rat: normal perinatal development of the heart and thyroxin- or isoproterenol-induced hypertrophy of the heart in adult animals.

While DFMO exposure definitively retards growth of prenatal cardiac tissue (Slotkin *et al.*, 1983), DFMO administered from day 3 up to day 17 postnatally does not produce any deleterious effect on cardiac DNA, RNA, or proteins (Slotkin *et al.*, 1984). Thyroxin-induced cardiac hypertrophy, which is normally accompanied by an increased concentration of the three polyamines, is blocked neither by the putrescine analog 1,3-diamino-2-propanol (Pegg and Hibasami, 1980) nor by DFMO (Pegg, 1981), despite the fact that the two treatments prevent the rise in polyamine concentrations usually induced by thyroxin. Regarding the third model of induced cardiac hypertrophy, Bartolome *et al.* (1980, 1982) have reported that administration of DFMO to isoproterenol-treated rats not only suppressed the accumulation of polyamines in the heart but also attenuated the cardiac weight increase. They concluded from the small but significant effect of DFMO in this model that polyamines might modulate cardiac growth in different ways, and this as a function of the growth inducers (Bartolome *et al.*, 1982). This dual mechanism remains to be determined.

D. Growth of Tissues of the Gastrointestinal System

A number of studies are concerned with the role of polyamines in the growth and adaptation of organs of the gastrointestinal system—salivary glands, gastric mucosa, intestine, and pancreas. This field is still presently under extensive investigation (for a review, see Dowling *et al.*, 1985). The highlights of these studies are summarized below.

1. Salivary Glands

Isoproterenol-stimulated growth of mouse parotid gland is accompanied by a concomitant increase of putrescine and spermidine concentrations (Inoue *et al.*, 1974). Coadministration of DL-HAVA blocks the elevation of putrescine but barely affects the accumulation of spermidine (Inoue *et al.*, 1975). DNA synthesis, but not RNA synthesis, is decreased by DL-HAVA. These findings led Inoue *et al.* (1975) to conclude that putrescine plays a role in cell proliferation in this animal model. On the contrary, administration of DFMO to isoproterenol-treated mice fails to slow DNA synthesis, although isoproterenol-induced elevations of putrescine and spermidine are totally suppressed (Claverie, 1984). The discrepancy between the two studies might be related to the previously noted lack of specificity of DL-HAVA.

2. *Gastric Mucosa*

DFMO enhances formation of gastric erosions in adult rats subjected to cold-restraint stress and delays healing subsequent to the injuries (Meeroff *et al.*, 1982). In contrast, maturation of the gastric mucosa is not prevented or delayed after chronic oral administration of DFMO to newborn rats (McNeil *et al.*, 1985).

3. *Oxyntic Gland and Intestinal Mucosa*

DFMO completely inhibits the trophic response to pentagastrin in the oxyntic gland area of the stomach and the duodenal mucosa in the adult rat (Seidel *et al.*, 1984a). Likewise DFMO suppresses the increases in putrescine and spermidine concentrations occurring in the intestinal mucosa of the newborn rat at the third week after birth, and delays the maturation of the intestinal mucosa (Luk *et al.*, 1980). Intestinal adaptation during lactation can also be abolished by administration of DFMO, resulting in a marked diminution of the adaptative increase in mucosal weight and thickness, especially in crypt depth (Yang *et al.*, 1984). In adult rats, DFMO retards intestinal mucosal recovery after chemical injury (Luk *et al.*, 1980), inhibits intestinal epithelial DNA synthesis and adaptative hyperplasia after jejunectomy (Luk and Baylin, 1984), and partially prevents the trophic response to intestinal obstruction (Seidel *et al.*, 1984b). In the dog, subacute oral or intravenous administration of DFMO produces a villous atrophy of the mucosa as visualized by light and scanning electron microscopy of the small intestine (Yarrington *et al.*, 1983).

4. *Pancreas*

The effects of DFMO have been investigated on caerulein-induced pancreatic growth (Benrezzak and Moriset, 1984). Although DFMO reduces the increases in DNA synthesis and DNA content induced by the cholecystokinin analog, it does not modify the general hypertrophic action of caerulein, i.e., the elevations in RNA, proteins, and amylase activity. DFMO neither prevents pancreatic hyperplasia induced by pancreaticobiliary diversion (Stace *et al.*, 1984; Dowling *et al.*, 1985) nor affects DNA synthesis during pancreatic regeneration in rats given ethionine (Houlihan *et al.*, 1985).

E. Kidney Hypertrophy

Consequences of the inhibition of polyamine synthesis by 1,3-diaminopropane in testosterone-induced hypertrophic kidney in castrated mouse have been reported (Cameron *et al.*, 1979). Injection of 1,3-diaminopropane prevents the accumulation of putrescine and spermidine evoked by androgen in the kidneys and even reduces renal spermine concentration. This treatment also suppresses the testosterone-induced increase in renal weight and RNA. Due to the nephrotoxicity of 1,3-diaminopropane, the relationship between prevention of renal hypertrophy and inhibition of polyamine synthesis could not be firmly established by Cameron *et al.* (1979). Using the same animal model, Goldstone *et al.* (1982) confirmed that blockade of testosterone-induced elevation of polyamine concentrations in the kidney attenuates renal hypertrophy as measured by the contents of RNA and proteins and the activities of enzymatic markers such as cytochrome *c* oxidase and lysosomal hydrolase. Furthermore, DFMO inhibits the androgen-induced increments in lysosomal enzymuria (Goldstone *et al.*, 1982) and urinary secretion of β-glucuronidase (Laitinen and Pajunen, 1983). Moreover, Kingsnorth *et al.* (1982) reported that DFMO inhibits compensatory renal hypertrophy occurring after unilateral nephrectomy in the mouse. However, this result was challenged by Etheredge *et al.* (1983).

F. Growth of Uterus

To evaluate the role of polyamines in mediating the effect of estrogens on growth of the rat uterus, Rorke and Katzenellenbogen (1984) examined the effect of DFMO on estradiol-stimulated uterine weight. The authors found that DFMO treatment concomitant with administration of estradiol resulted in a marked increase in uterine weight equivalent to that obtained with estradiol alone, although the estradiol-induced increase of spermidine concentration was totally abolished.

G. Effects of DFMO in Lung

Bleomycin induces a fibroproliferative process in the lung. Since polyamines might play a role in this aspect of differentiation, Giri *et al.* (1982) studied the effect of DFMO in this process. Not only does DFMO *not* block the bleomycin-induced accumulation of total lung insoluble collagen in hamster, but it accentuates the effect of bleomycin. This is due to a direct effect of DFMO on accumulation of insoluble collagen in lung. These results have been confirmed by Frost *et al.* (1983). Furthermore, DFMO adversely affects repair of oxygen-induced lung injury in rat (Thet *et al.,* 1984). In contrast to the above-described deleterious effects of DFMO, continuous treatment with the drug prevents the development of monocrotaline-induced pulmonary toxicity, suggesting that polyamines might be important mediators of hypertensive pulmonary vascular disease that develops in response to monocrotaline administration (Olson *et al.,* 1984).

H. Skin Proliferation

A number of experimental data have indicated a close parallel between the accumulation of polyamines and the proliferative activity of the skin (Lisiewicz and Goldsmith, 1980). Despite its marked effects on putrescine formation and polyamine turnover, DFMO administration to mice does not significantly affect the increase of DNA synthesis and epidermal cell proliferation elicited by ultraviolet irradiation (Seiler and Knödgen, 1979). This result was confirmed by Käpyaho *et al.* (1983). These authors reported, however, that MGBG partially prevented DNA synthesis in this model of skin proliferation.

IV. CONCLUSIONS

The numerous studies reviewed in this chapter demonstrate unambiguously that ornithine decarboxylase can be effectively inhibited *in vivo*. DFMO and the second generation of irreversible ODC inhibitors show better eficacy and specificity than any other means of enzyme inhibition. This is essentially due to the very nature of the inhibition exerted by these compounds. When used in sufficient doses and with appropriate frequency of administration, they can block efficiently the increases of putrescine and spermidine concentrations that occur in response to various stimuli. In a few cases, they are even able to decrease the levels of these amines below basal values. Spermine concentrations, however, remain basically unaffected in most animal tissues. Lack of effect on spermine is readily explained by the known long half-life of this polyamine in cells and by an incomplete blockade of ODC. This is particularly verified in organs in which an enormous intracellular buildup of decarboxylated *S*-adenosylmethionine permits continuous synthesis of spermine.

Suppression of the elevation of putrescine and spermidine concentrations elicited by a variety of growth inducers does not necessarily affect synthesis and accumulation of nucleic acids in animal tissues. Basal levels of polyamines allow almost normal processes of growth and proliferation in a number of examples. Therefore, in these cases, it is likely that the stimulation of polyamine synthesis may be in response to requirements for other cellular functions. In particular, it is noteworthy that DFMO, given to testosterone-treated castrated rats, impairs a return to a normal secretory activity of the ventral prostate, without affecting regrowth of the gland.

Using a very high dosage of DFMO, a reduction of putrescine and spermidine concentrations below basal levels was achieved in regenerating liver and in ventral prostate of intact rats. This profound polyamine deficiency results in a significant decrease of the nucleic acid synthesis, thus confirming the inhibition of growth observed in cells cultured in the presence of an ODC inhibitor. Effects, if any, of DFMO on growth that are unrelated to polyamines, have not been established so far. Nevertheless, before confidently attributing to polyamine deficiency the inhibition of growth of the ventral prostate, the accumulation of decarboxylated *S*-adenosylmethionine should also be considered. This metabolite can be increased 500-fold over control values in ventral prostate (Danzin *et al.*, 1982a) and may exert direct detrimental effects on prostatic cells by, for instance, inhibiting the acetylation of histones (Pegg *et al.*, 1986). Germane to this point, the 100-fold increase of decarboxylated *S*-adenosylmethionine observed in the duodenum of MAP-treated rats (Wagner *et al.*, 1985a) might also contribute to the striking effect of a potent ODC inhibitor on the proliferation of gastrointestinal tissue.

In studying the relationship between polyamines and hypertrophy/hyperplasia, the problem of polyamine interconversion has received little attention. In fact, interconversion of polyamines is induced by a certain number of stimuli known to elicit growth and repair (Pegg *et al.*, 1982d). In a situation where the retroconversion pathway is more important than *de novo* synthesis, an inhibitor of ODC would be of limited usefulness in assessing a casual relationship between the changes in polyamine concentrations and in growth rate. Rather, inhibitors of the enzymes involved in the interconversion pathway may afford a solution to this problem.

Finally, the fact that DFMO and other enzyme-activated irreversible inhibitors of ODC affect preferentially polyamines in certain tissues might constitute an advantage when considering these compounds as selective therapeutic agents and could explain their low toxicity.

REFERENCES

Abeles, R.H., and Maycock, A.L. (1976). *Acc. Chem. Res.* **9,** 313–319.

Bachrach, U. (1973). "Function of Naturally Occurring Polyamines." Academic Press, New York.

Bartholeyns, J., Mamont., P., and Casara, P. (1984). *Cancer Res.* **44,** 4972–4977.

Bartolome, J., Huguenard, J., and Slotkin, T.A. (1980). *Science* **210,** 793–794.

Bartolome, J.V., Trepanier, P.A., Chait, E.A., and Slotkin, T.A. (1982). *J. Mol. Cell. Cardiol.* **14,** 461–466.

Batchelor, K.W., Smith, R.A., and Watson, N.S. (1986). *Biochem. J.* **233,** 307–308.

Benrezzak, O., and Morisset, J. (1984). *Regul. Peptides* **9,** 143–153.

Bey, P. (1978). *In* "Enzyme-Activated Irreversible Inhibitors" (N. Seiler, M.J. Jung, and J. Koch-Weser, eds.), pp. 27–41. Elsevier, Amsterdam.

Bey, P., Gerhart F., Van Dorsselaer, V., and Danzin, C. (1983). *J. Med. Chem.* **26,** 1551–1556.

Bey, P., Bolkenius, F.N., Seiler, N., and Casara, P. (1985). *J. Med. Chem.* **28,** 1–2.

Blaschko, H. (1952). *Pharmacol. Rev.* **4,** 415–458.

Bolander, F.F., and Topper. Y.J. (1979). *Biochem. Biophys. Res. Commun.* **90,** 1131–1135.

Bolkenius, F.N., Bey, P., and Seiler, N. (1985). *Biochim. Biophys. Acta* **838,** 69–76.

Cameron, R., Henningsson, S., Persson, L., and Rosengren, E. (1979). *Acta Physiol. Scand.* **106,** 299–305.

Cannellakis, E.S., Viceps-Madore, D., Kyriakidis, D.A., and Heller, J.S. (1979). *Curr. Top. Cell Regul.* **15,** 155–202.

Casara, P., Danzin, C., Metcalf, B., and Jung, M. (1985). *J. Chem. Soc. Perkin Trans.* **1,** 2201–2207.

Claverie, N. (1984). Thèse de Doctorat d'Etat, Université Louis Pasteur de Strasbourg.

Danzin, C., Jung, M.J., Grove, J., and Bey, P. (1979a). *Life Sci.* **24,** 519–524.

Danzin, C., Jung, M.J., Claverie, N., Grove, J., Sjoerdsma, A., and Koch-Weser, J. (1979b). *Biochem. J.* **180,** 507–513.

Danzin, C., Jung, M.J., Metcalf, B.W., Grove, J., and Casara, P. (1979c). *Biochem. Pharmacol.* **28,** 627–631.

Danzin, C., Jung, M.J., Seiler, N., and Metcalf, B.W. (1979d). *Biochem. Pharmacol.* **28,** 633–639.

Danzin, C., Casara, P., Claverie, N., and Metcalf, B.W. (1981). *J. Med. Chem.* **24,** 16–20.

Danzin, C., Claverie, N., Wagner, J., Grove, J., and Koch-Weser, J. (1982a). *Biochem. J.* **202,** 175–181.

Danzin, C., Bey, P., Schirlin, D., and Claverie, N. (1982b). *Biochem. Pharmacol.* **31,** 3871–3878.

Danzin C., Casara, P., Claverie, N., and Grove, J. (1983a). *Biochem. Pharmacol.* **32,** 941–942.

Danzin, C., Casara, P., Claverie, N., Metcalf, B.W., and Jung, M.J. (1983b). *Biochem. Biophys. Res. Commun.* **116,** 237–243.

Davis, R.H., Krasner, G.N., Di Gangi, J.J., and Ristow, J.L. (1985). *Proc. Natl. Acad. Sci. U.S.A.* **82,** 4105–4109.

Dowling, R.H., Hosomi, M., Stace, N.H., Lirussi, F., Miazza, B., Levan, H., and Murphy, G.M. (1985). *Scand. J. Gastroenterol.* **20** (Suppl. 112), 84–95.

Etheredge, S.B., Lin, S.-Y., Ribstein, J., Chou, Y.-N., and Humphreys, M.H. (1983). *Clin. Res.* **31,** 428 A.

Feuerstein, B.G., Deen, D.F., and Marton, L.J. (1985). *Cancer Res.* **45,** *4950–4954.*

Frost, S.B., Rakieten, N., and Raisfeld-Danse, I.H. (1983). *Toxicol. Appl. Pharmacol.* **71,** 307–315.

Fuller, D.J.M., Donaldson, L.J., and Thomas, G.H. (1975). *Biochem. J.* **150,** 557–559.

Giri, S.N., Hyde, D.M., Schwartz, L.W., and Younker, W.R. (1982). *Am. J. Pathol.* **109,** 115–122.

Goldstone, A., Koenig, H., and Lu, C. (1982). *Biochem. Biophys. Res. Commun.* **104,** 165–172.
Grove, J., Fozard, J.R., and Mamont, P.S. (1981). *J. Chromatogr.* **223,** 409–416.
Haegele, K.D., Alken, R.G., Grove, J., Schechter, P.J., and Koch-Weser, J. (1981). *Clin. Pharmacol. Ther.* **30,** 210–217.
Harik, S.I., Hollenberg, M.D., and Snyder, S.H. (1974). *Mol. Pharmacol.* **10,** 41–47.
Heby, O. (1981). *Differentiation* **19,** 1–20.
Heby, O., and Jänne, J. (1981). *In* "Polyamines in Biology and Medicine" (D.R. Morris and L.J. Marton, eds.), pp. 243–310. Dekker, New York.
Heller, J.S., Fong, W.F., and Canellakis, E.S. (1976). *Proc. Natl. Acad. Sci. U.S.A.* **73,** 1858–1862.
Hibasami, H., and Ito, H. (1981). *Gann* **72,** 512–516.
Hibasami, H., Tanaka, M., Nagai, J., and Ikeda, T. (1980). *FEBS Lett.* **116,** 99–101.
Hölttä, E., Hannonen, P., Pispa J., and Jänne, J. (1973). *Biochem. J.* **136,** 669–676.
Hölttä, E., Korpela, H., and Hovi, T. (1981). *Biochim. Biophys. Acta* **677,** 90–102.
Hopkins, D., and Manchester, K.L. (1980). *FEBS Lett.* **109,** 299–302.
Houdebine, L.M., Devinoy, E., and Delouis, C. (1978). *Biochimie* **60,** 735–741.
Houlihan, M.J., Saluja, A., Powers, R.E., Saluja, M., Eigner, S., and Steer, M.L. (1985). *Dig. Dis. Sci.* **30,** 975.
Inoue, H., Tanioka, H., Shiba, K., Asada, A., Kato, Y., and Takeda, Y. (1974). *J. Biochem. (Tokyo)* **75,** 679–687.
Inoue, H., Kato, Y., Takigawa, M., Adachi, H., and Takeda, Y. (1975). *J. Biochem. (Tokyo)* **77,** 879–893.
Ito, H., Hibasami, H., Shimura, K., Nagai, J., and Idaka, H. (1982). *Cancer Lett.* **15,** 229–235.
Jänne, J. (1967). *Acta Physiol. Scand. (Suppl.)* **300,** 1–71.
Jänne, J., Hölttä, E., and Guha, S.K. (1976). *Prog. Liver Dis.* **5,** 100–124.
Jänne, J., Pösö, H., and Raina, A. (1978). *Biochim. Biophys. Acta* **473,** 241–293.
Jänne, J., Alhonen-Hongisto, L., Seppänen, P., and Hölttä, E. (1981). *Adv. Polyamine Res.* **5,** 85–95.
Jung, M.J., Lippert, B., Metcalf, B.W., Schechter, P.J., Böhlen, P., and Sjoerdsma, A. (1977). *J. Neurochem.* **28,** 717–723.
Kahana, C., and Nathans, D. (1985). *J. Biol. Chem.* **260,** 15390–15393.
Kallio, A. (1978). *Acta Chem. Scand. B* **32,** 759–764.
Kameji, T., Murakami, Y., and Hayashi, S. (1979). *J. Biochem. (Tokyo)* **86,** 191–197.
Käpyaho, K., Pösö, H., and Jänne, J. (1980). *Biochem. J.* **192,** 59–63.
Käpyaho, K., Lauharanta, J., and Jänne, J., (1983). *J. Invest. Dermatol.* **81,** 102–106.
Käpyaho, K., Kallio, A., and Jänne, J. (1984). *Biochem. J.* **219,** 811–817.
Karvonen, E., Kauppinen, L., Partanen, T., and Pösö, H. (1985). *Biochem. J.* **231,** 165–169.
Kato, Y., Nishihara, H., Inoue, H., Gohda, E., and Takeda, Y. (1978). *J. Biochem. (Tokyo)* **84,** 1617–1623.
Kay, J.E., and Lindsay, V.J. (1973). *Biochem. J.* **132,** 791–796.
Khomutov, R.M., Hyvönen, T., Karvonen, E., Kauppinen, L., Paalanen, T., Paulin, L., Eloranta, T., Pajula, R.-L., Andersson, L.C., and Pösö, H. (1985). *Biochem. Biophys. Res. Commun.* **130,** 596–602.
Kingsnorth, A.N., King, W.W.-K., and Malt, R.A. (1982). *Fed. Proc., Fed. Am. Soc. Exp. Biol.* **41,** 1768.
Knuutila, S., and Pohjanpelto, P. (1983). *Exp. Cell Res.* **145,** 222–226.
Koch-Weser, J., Schechter, P.J., Bey, P., Danzin, C., Fozard, J.R., Jung, M.J., Mamont,

P.S., Seiler, N., Prakash, N.J., and Sjoerdsma, A. (1981). *In* "Polyamines in Biology and Medicine" (D.R. Morris and L.J. Marton, eds.), pp. 437–453. Dekker, New York.

Kollonitsch, J., Patchett, A.A., Marburg, S., Maycock, A.L., Perkins, L.M., Doldouras, G.A., Duggan, D.E., and Aster, S.D. (1978). *Nature (London)* **274,** 906–908.

Krisch, K. (1971). *In* "The Enzymes" (P.D. Boyer, ed.), 3rd Ed., Vol. 5, pp. 43–69. Academic Press, New York.

Laitinen, S.I., and Pajunen, A.E.I. (1983). *Biochem. Biophys. Res. Commun.* **112,** 770–777.

Lisiewicz, J., and Goldsmith, L.A. (1980). *J. Invest. Dermatol.* **75,** 207–210.

Luk, G.D., and Baylin, S.B. (1984). *J. Clin. Invest.* **74,** 698–704.

Luk, G.D., Marton, L.J., and Baylin, S.B. (1980). *Science* **210,** 195–198.

McCann, P.P. (1980). *In* "Polyamines in Biomedical Research" (J.M. Gaugas, ed.), pp. 109–123. Wiley, New York.

McCann, P.P., Tardif, C., Duchesne, M.-C., and Mamont, P.S. (1977). *Biochem. Biophys. Res. Commun.* **76,** 893–899.

McKeehan, W.L., Glass, H.A., Rosser, M.P., and Adams, P.S. (1982). *Prostate* **3,** 231–246.

McNeil, N.O., Eikenburg, B.E., and Johnson, L.R. (1985). *Fed. Proc., Fed. Am. Soc. Exp. Biol.* **44,** 441.

Mamont, P.S., and Danzin, C. (1981). *Adv. Polyamine Res.* **3,** 123–135.

Mamont, P.S., Duchesne, M.-C., Joder-Ohlenbusch, A.-M., and Grove, J. (1978). *In* "Enzyme-Activated Irreversible Inhibitors" (N. Seiler, M.J. Jung, and J. Koch-Weser, eds.), pp. 43–54. Elsevier, Amsterdam.

Mamont, P.S., Bey, P., and Koch-Weser, J. (1980). *In* "Polyamines in Biomedical Research" (J.M. Gaugas, ed.), pp. 147–165. Wiley, New York.

Mamont, P.S., Joder-Ohlenbusch, A.-M., Nussli, M., and Grove, J. (1981). *Biochem. J.* **196,** 411–422.

Mamont, P.S., Danzin, C., Wagner, J., Siat, M., Joder-Ohlenbusch, A.-M., and Claverie, N. (1982a). *Eur. J. Biochem.* **123,** 499–504.

Mamont, P.S., Danzin, C., and Wagner, J. (1982b). *In* "Biochemistry of *S*-Adenosylmethionine and Related Compounds" (E. Usdin, R.T. Borchardt, and C.R. Creveling, eds.), pp. 557–565. Mcmillan, London.

Mamont, P.S., Siat, M., Joder-Ohlenbusch, A.-M., Bernhardt, A., and Casara, P. (1984). *Eur. J. Biochem.* **142,** 457–463.

Mamont, P.S., Danzin, C., Kolb, M., Gerhart, F., Bey, P., and Sjoerdsma, A. (1986). *Biochem. Pharmacol.* **35,** 159–165.

Meeroff, J.C., Fam, G., Sens, D., and Levine, J.H. (1982). *Gastroenterology* **82,** 1128.

Metcalf, B.W., Bey, P., Danzin, C., Jung, M.J., Casara, P., and Vevert, J.-P. (1978). *J. Am. Chem. Soc.* **100,** 2551–2553.

Mihich, E., (1963). *Cancer Res.* **23,** 1375–1389.

Mikles-Robertson, F., Feuerstein, B., Dave, C., and Porter, C.W. (1979). *Cancer Res.* **39,** 1919–1926.

Morris, D.R. (1981). *In* "Polyamines in Biology and Medicine" (D.R. Morris and L.J. Marton, eds.), pp. 223–242. Dekker, New York.

Murakami, Y., and Hayashi, S. (1985). *Biochem. J.* **226,** 893–896.

Olson, J.W., Hacker, A.D., Altiere, R.J., and Gillespie, M.N. (1984). *Am. J. Physiol.* **247,** H682–H685.

Pathak, S.N., Porter, C.W., and Dave, C. (1977). *Cancer Res.* **37,** 2246–2250.

Paulus, T.S., Kiyono, P., and Davis, R.H. (1982). *J. Bacteriol.* **152,** 291–297.

Pegg, A.E. (1973). *Biochem. J.* **132,** 537–540.

Pegg, A.E. (1978). *J. Biol. Chem.* **253,** 539–542.

Pegg, A.E. (1981). *J. Mol. Cell. Cardiol.* **13,** 881–887.

Pegg, A.E. (1984). *Cell Biochem. Functions* **2,** 11–15.
Pegg, A.E., and Conover, C. (1976). *Biochem. Biophys. Res. Commun.* **69,** 766–774.
Pegg, A.E., and Hibasami, H. (1980). *Am. J. Physiol.* **239,** E372–E378.
Pegg, A.E., and McCann, P.P. (1982). *Am. J. Physiol.* **243,** C212–C221.
Pegg, A.E., and McGill, S.M. (1978). *Biochem. Pharmacol.* **27,** 1625–1629.
Pegg, A.E., Lockwood, D.H., and Williams-Ashman, H.G. (1970). *Biochem. J.* **117,** 17–31.
Pegg, A.E., Corti, A., and Williams-Ashman, H.G. (1973). *Biochem. Biophys. Res. Commun.* **52,** 696–701.
Pegg, A.E., Conover, C., and Wrona, A. (1978). *Biochem. J.* **170,** 651–660.
Pegg, A.E., Pösö, H., Shuttleworth, K., and Bennet, R.A. (1982a). *Biochem. J.* **202,** 519–526.
Pegg, A.E., Tang, K.-C., and Coward, J.K. (1982b). *Biochemistry* **21,** 5082–5089.
Pegg, A.E., Pösö, H., and Bennet, R.A. (1982c). *In* "Biochemistry of *S*-adenosylmethionine and Related Compounds" (E. Usdin, R.T. Borchardt, and C.R. Creveling, eds.), pp. 547–555. Mcmillan, London.
Pegg, A.E., Seely, J.E., Pösö, H., Della Ragione, F., and Zagon, I.S. (1982d). *Fed. Proc., Fed. Am. Soc. Exp. Biol.* **41,** 3065–3072.
Pegg, A.E., Wechter, R.S., Clark, R.S., Wiest, L., and Erwin, B.G. (1986). *Biochemistry* **25,** 379–384.
Persson, L., Seely, J.E., and Pegg, A.E. (1984). *Biochemistry* **23,** 3777–3783.
Persson, L., Oredson, S.M., Anehus, S., and Heby, O. (1985). *Biochem. Biophys. Res. Commun.* **131,** 239–245.
Piik, K., Rajamäki, P., Guha, S.K., and Jänne, J. (1977). *Biochem. J.* **168,** 379–385.
Piik, K., Pösö, H., and Jänne, J. (1978). *FEBS Lett.* **89,** 307–312.
Pine, M.J., and DiPaolo, J.A. (1966). *Cancer Res.* **26,** 18–25.
Pleshkewych, A., Kramer, D.L., Kelly, E., and Porter, C.W. (1980). *Cancer Res.* **40,** 4533–4540.
Porta, R., Camardella, M., Gentile, V., and De Santis, A. (1984). *J. Neurochem.* **42,** 321–325.
Porter, C.W., Dave, C., and Mihich, E. (1981). *In* "Polyamines in Biology and Medicine" (D.R. Morris and L.J. Marton, eds.), pp. 407–436. Dekker, New York.
Pösö, H., and Pegg, A.E. (1981). *Biochem. J.* **200,** 629–637.
Pösö, H., and Pegg, A.E. (1982). *Biochim. Biophys. Acta* **696,** 179–186.
Pösö, H., Kallio, A., Scalabrino, G., and Jänne, J. (1977). *Biochim. Biophys. Acta* **497,** 288–297.
Prakash, N.J., Schechter, P.J., Grove, J., and Koch-Weser, J. (1978). *Cancer Res.* **38,** 3059–3062.
Raina, A. (1963). *Acta Physiol. Scand.* **60,** (Suppl. 218), 1–81.
Rando, R.R. (1974). *Science* **185,** 320–324.
Rando, R.R. (1984). *Pharmacol. Rev.* **36,** 111–142.
Relyea, N., and Rando, R.R. (1975). *Biochem. Biophys. Res. Commun.* **67,** 392–402.
Rorke, E.A., and Katzenellenbogen, B.S. (1984). *Biochem. Biophys. Res. Commun.* **122,** 1186–1193.
Russell, D.H. (1973). *In* "Polyamines in Normal and Neoplastic Growth" (D.H. Russell, ed.), pp. 1–13. Raven, New York.
Russell, D.H. (1985). *Drug Metab. Rev.* **16,** 1–88.
Seely, J.E., and Pegg, A.E. (1983). *Biochem. J.* **216,** 701–707.
Seidel, E.R., Tabata, K., Dembinski, A.B., and Johnson, L.R. (1984a). *Am. J. Physiol.* **249,** G16–G20.
Seidel, E.R., Haddox, M.K., and Johnson, L.R. (1984b). *Am. J. Physiol.* **246,** G649–G653.

Seiler, N., and Knödgen, B. (1979). *Biochem. Med.* **21,** 168–181.
Seiler, N., and Knödgen, B. (1980). *J. Chromatogr.* **221,** 227–235.
Seiler, N., Danzin, C., Prakash, N.J., and Koch-Weser, J. (1978a). *In* "Enzyme-Activated Irreversible Inhibitors" (N. Seiler, M.J. Jung, and J. Koch-Weser, eds.), pp. 55–71. Elsevier, Amsterdam.
Seiler, N., Jung, M.J., and Koch-Weser, J. (1978b). "Enzyme-Activated Irreversible Inhibitors." Elsevier, Amsterdam.
Seiler, N., Bolkenius, F.N., and Rennert, O.M. (1981). *Med. Biol.* **59,** 334–346.
Shirahata, A., and Pegg, A.E. (1985). *J. Biol. Chem.* **260,** 9583–9588.
Shirahata, A., Christman, K.L., and Pegg, A.E. (1985). *Biochemistry* **24,** 4417–4423.
Silverman, R.B. (1984). *Biochemistry* **23,** 5206–5213.
Sjoerdsma, A. (1981). *Clin. Pharmacol. Ther.* **30,** 3–22.
Sjoerdsma, A., and Schechter, P.J. (1984). *Clin. Pharmacol. Ther.* **35,** 287–300.
Skinner, W.A., and Johansson, J.G. (1972). *J. Med. Chem.* **15,** 427–428.
Slotkin, T.A., Seidler, F.J., Whitmore, W.L., Weigel, S.J., Slepetis, R.J., Lerea, L., Trepanier, P.A., and Bartolome, J. (1983). *Int. J. Dev. Neurosci.* **1,** 113–127.
Slotkin, T.A., Persons, D., Slepetis, R.J., Taylor, D., and Bartolome, J. (1984). *Teratology* **30,** 211–224.
Stace, N.H., Hosomi, M., Vaja, S., Murphy, G.M., and Dowling, R.H. (1984). *Gut* **25,** A1159.
Steglich, C., Grens, A., and Scheffer, I.E. (1985). *Somatic Cell. Mol. Genet.* **11,** 11–23.
Stevens, L., and Stevens, E. (1980). *In* "Polyamines in Biomedical Research" (J.M. Gaugas, ed.), pp. 167–183. Wiley, New York.
Tabor, C.W., and Tabor, H. (1984). *Annu. Rev. Biochem.* **53,** 749–790.
Tabor, C.W., Tabor, H., and Tyagi, A.K. (1983). *Adv. Polyamine Res.* **4,** 467–478.
Tabor, H., Hafner, E.W., and Tabor, C.W. (1980). *J. Bacteriol.* **144,** 952–956.
Takyi, E.E.K., Fuller, D.J.M., Donaldson, L.J., and Thomas, G.H. (1977). *Biochem. J.* **162,** 87–97.
Thet, L.A., Parra, S.C., and Shelburne, J.D. (1984). *Am. Rev. Respir. Dis.* **129,** 174–181.
Wagner, J., Danzin, C., and Mamont, P. (1982). *J. Chromatogr.* **227,** 349–368.
Wagner, J., Hirth, Y., Claverie, N., and Danzin, C. (1985a). *Am. East. Eur. Symp. Liquid Chromatogr. 5th, Budapest.*
Wagner, J., Hirth, Y., Piriou, F., Zakett, D., Claverie, N., and Danzin, C. (1985b). *Biochem. Biophys. Res. Commun.* **133,** 546–553.
Walsh, C. (1977). *Horiz. Biochem. Biophys.* **3,** 36–55.
Walsh, C. (1982). *Tetrahedron* **38,** 871–909.
Walsh, C. (1984). *Annu. Rev. Biochem.* **53,** 493–535.
Williams-Ashman, H.G., and Canellakis, Z.N. (1979). *Perspect. Biol. Med.* **22,** 421–453.
Williams-Ashman, H.G., and Schenone, A. (1972). *Biochem Biophys. Res. Commun.* **46,** 288–295.
Williams-Ashman, H.G., Tadolini, B., Wilson, J., and Corti, A. (1975a). *Vitam. Horm.* **33,** 39–60.
Williams-Ashman, H.G., Corti, A., and Sheth, A.R. (1975b). *In* "Normal and Abnormal Growth of the Prostate" (M. Goland, ed.), pp. 222–239. Thomas, Springfield, Illinois.
Williams-Ashman, H.G., Corti, A., and Tadolini, B. (1976). *Ital. J. Biochem.* **25,** 5–32.
Yang, P., Baylin, S.B., and Luk, G.D. (1984). *Am. J. Physiol.* **247,** G553–G557.
Yarrington, J.T., Sprinkle, D.J., Loudy, D.E., Diekema, K.A., McCann, P.P., and Gibson, J.P. (1983). *Exp. Mol. Pathol.* **39,** 300–316.

8

Polyamine Synthesis Inhibitors Act as both Inducers and Suppressors of Cell Differentiation

OLLE HEBY,* GORDON D. LUK,† AND JOEL SCHINDLER‡

** Department of Zoophysiology*
University of Lund
Helgonavägen 3, S-223 62 Lund, Sweden

† The Oncology Center and Department of Medicine
The Johns Hopkins University School of Medicine
Baltimore, Maryland 21205

‡ Department of Anatomy and Cell Biology
University of Cincinnati College of Medicine
Cincinnati, Ohio 45267

I. INTRODUCTION

Indirect evidence suggested many years ago that the polyamines putrescine, spermidine, and spermine play an important role in cell growth and differentiation. Experiments using polyamine auxotrophs and inhibitors of enzymes in the polyamine biosynthetic pathway eventually provided direct evidence for a role of polyamines in these events. During the last decade a large number of studies have been performed to assess the role of polyamines in cell growth and division, mainly using the inhibitor approach (for references, see Heby and Jänne, 1981; Pegg and McCann, 1982; Chapters 1–7). These studies consistently show that polyamine-deficient cells grow very slowly or not at all, and that provision of exogenous polyamines rapidly reverses the growth-inhibited state and stimulates the cells to grow at their maximal rate.

It is often difficult to distinguish proliferative processes from developmental ones, since cell division can be a prerequisite for development and differentiation. Nevertheless, studies of several experimental systems have provided convincing evidence that polyamines play a role not only in proliferative but also in differentiative processes (Heby, 1981). Such studies have exploited the availability of polyamine synthesis inhibitors, and have focused on the effects of altered polyamine levels on the expression of specific cell differentiation-related events. The studies described below fall into two broad classes, depending on the nature of the cellular response to incubation with various polyamine synthesis inhibitors. While one set of results indicates that increased polyamine levels are necessary for differentiation, a second set indicates that a reduction in polyamine levels can induce differentiation. This dichotomy only emphasizes the complex role that polyamines play in cell differentiation.

Inhibitors of ornithine decarboxylase, the initial and rate-limiting enzyme in the polyamine biosynthetic pathway, have been used in the majority of studies. α-Difluoromethylornithine (DFMO), which was developed by Metcalf *et al.* (1978), is the most potent and specific inhibitor presently available. The high specificity of DFMO is due to the fact that it can be activated exclusively by ornithine decarboxylase. It is accepted by the enzyme as a substrate and is decarboxylated to yield a highly reactive electrophilic intermediate which alkylates a nucleophilic residue within the active site, thus irreversibly inactivating the enzyme (Metcalf *et al.*, 1978). Other substrate analogs, such as α-methylornithine (MO) and α-hydrazinoornithine (HO) have been used in some studies, but these inhibitors are reversible and usually less potent (for references, see Heby and Jänne, 1981).

Prior to the availability of ornithine decarboxylase inhibitors, another type of inhibitor was discovered. Methylglyoxal bis(guanylhydrazone) (MGBG), a cancer chemotherapeutic agent with clinical activity against acute myelocytic leukemia, was found to reduce the cellular spermidine and spermine content by inhibiting the activity of *S*-adenosylmethionine decarboxylase (Williams-Ashman and Schenone, 1972). Although a potent inhibitor of this enzyme, MGBG exerts many effects that seem to be unrelated to polyamine depletion (Porter *et al.*, 1981; Heby and Jänne, 1981).

When studying early developmental processes it is particularly important to use polyamine synthesis inhibitors with a high degree of specificity. This is due to the fact that true reversal experiments, like those performed with many established cell lines, and in which the effects produced as a result of polyamine deficiency are relieved by provision of exogenous polyamines, may be inconceivable because embryos and differentiating cells gradually become restricted in their developmental potential. Prevention experiments, in which exogenous putrescine or spermidine is added at the same time as DFMO to compensate for the DFMO-mediated inhibition of polyamine synthesis, provide important information as to the specificity of the inhibitor. In such experiments there appears to be no interference with the uptake of DFMO, because the polyamines have a separate transport system. However, the possibility of interference with inhibitor uptake should be considered when spermidine is used to compensate for MGBG-mediated inhibition of spermidine synthesis, because MGBG uses the same transport carrier (Porter *et al.*, 1981).

II. GERM CELLS AND EMBRYOS

Invertebrate embryos are excellent experimental materials for studying early development. They exhibit a high degree of synchrony in their development and are usually accessible in great abundance, thus permitting biochemical analysis of early developmental processes to an extent that is not possible in vertebrate materials. Despite these advantages, invertebrate embryos have been used only in a few instances for studying the effects of polyamine synthesis inhibition on early embryonic development.

A. Polychaetes

Development may best be followed in alecithal or homolecithal eggs (without yolk or with yolk distributed evenly through the cytoplasm, respectively), as the presence of large quantities of yolk modifies the development to a greater or lesser extent. The eggs of polychaete worms, like those of most marine invertebrates, are homolecithal. At fertilization, the eggs of the polychaete *Ophryotrocha labronica* are deposited cemented together as cylinders, each containing approximately 300 eggs, in which development proceeds synchronously. During cleavage the eggs exhibited a precipitous increase in ornithine decarboxylase activity and putrescine content, resulting in peak levels at late cleavage (Heby and Emanuelsson, 1980) and at the beginning of gastrulation (Emanuelsson and Heby, 1978), respectively. Inhibition of the synthesis and accumulation of putrescine by MO or DFMO treatment resulted in developmental arrest at gastrulation (Emanuelsson and Heby, 1978; Heby and Emanuelsson, 1981). When putrescine was added 1 hr after DFMO there was no gastrular arrest. These findings suggest a role for the polyamines in gastrulation, the developmental process that initiates embryonic autonomy.

Fertilization usually starts the processes of both development and cleavage, but the eggs of *Chaetopterus variopedatus*, another polychaete worm, may develop almost normally without any cytoplasmic cleavage upon activation by KCl treatment. MO exerted no effect on such "differentiation without cleavage" (Brachet *et al.*, 1978). MO treatment during normal development did not affect the segmentation of the fertilized eggs, but at the time of hatching the trochophore larvae had no intestine and no apical tuft.

B. Sea Urchins

The eggs of echinoderms are usually homolecithal and are perhaps the favorite experimental material for studying fertilization and early development, as they are relatively easy to obtain and to culture. When fertilized sea urchin (*Hemicentrotus pulcherrimus* and *Anthocidaris crassispina*) eggs were treated with HO, development was affected during cleavage (Kusunoki and Yasumasu, 1978). Higher concentrations almost completely blocked the third cleavage. As long as the spermine concentration was maintained in HO-treated eggs, cleavage occurred, although at a considerably reduced rate. Ornithine as well as the polyamines were able to reverse the inhibition of egg cleavage.

In experiments with two other sea urchin species (*Paracentrotus* and *Arbacia*), in which MO was used instead of HO, it was found that development was arrested at a later stage, the late blastula stage (Brachet *et al.*, 1978). Polyamines reduced the inhibitory effect of MO when added at the same time as the inhibitor. Whether these discrepancies are due to species and/or inhibitor differences remains to be determined.

C. Xenopus

The full-grown amphibian oocyte possesses a vast store of structural and regulatory components, and remains in a quiescent state until stimulated by an appropriate hormone (Wasserman *et al.*, 1984). Progesterone or human chorionic gonadotropin can stimulate the oocytes to resume meiosis from their prophase arrest. The meiotic maturation culminates in the dissolution of the nuclear membrane (germinal vesicle breakdown) and the meiotic reductive divisions, preparing the egg for fertilization.

Sunkara *et al.* (1981) found that hormone treatment stimulated the ornithine decarboxylase activity in oocytes of *Xenopus laevis,* and that prevention of the increase in enzyme activity by DFMO treatment caused an inhibition of both maturation and ovulation. However, this effect was not observed if DFMO was added at the same time as the hormone (Younglai *et al.,* 1980), only if DFMO was added 4 hr in advance (Sunkara *et al.,* 1981). The fact that the cellular polyamine content had to be reduced already at the time of hormone treatment, and that putrescine was capable of reversing the effects of DFMO only if provided within 2 hr of hormone addition, suggests a role for the polyamines in the very early phase of oocyte maturation.

The development of the mesolecithal egg (with a moderate amount of yolk which displaces the nucleus toward one end) has been studied with respect to polyamine synthesis and content (Russell, 1971), but thus far the effects of inhibition of polyamine synthesis have not been investigated.

D. Chick

In the chick, fertilization takes place high in the oviduct. The egg is telolecithal (with an enormous amount of yolk which makes the cytoplasm and nucleus form a separate disc, the blastodisc). Development begins while the egg is still in the oviduct but is interrupted at laying. When development is permitted to proceed by incubation, the blastoderm contains about 80,000 cells and represents the late cleavage stage.

Upon incubation there was a marked increase in ornithine decarboxylase activity with peak activity during the primitive streak stage (gastrulation). When this increase was inhibited by DFMO treatment *in ovo,* the embryos did not develop beyond the primitive streak stage (Löwkvist *et al.,* 1980). Analyses of arrested embryos showed that mesoderm outgrowth and differentiation was drastically impaired. These findings suggested a role for the polyamines in the differentiation of epiblast cells into mesoderm cells, i.e., the fundamental developmental change occurring at gastrulation (Löwkvist *et al.,* 1983). The effects of DFMO treatment were further analyzed in primary cultures of chick embryo mesoderm (Section III,A).

E. Mouse, Rat, and Rabbit

In the rat ovary there is a preovulatory increase in ornithine decarboxylase activity (late in proestrus) which is controlled by luteinizing hormone. When this increase was suppressed by treatment with DFMO during proestrus, ovulation took place normally as did the course of pregnancies resulting from mating during DFMO treatment (Fozard *et al.*, 1980c). These data provide no support for a functional role of the preovulatory rise in putrescine synthesis in the major periovulatory events.

With the exception of the monotremes (the platypus and the echidnas), all mammals have alecithal eggs. This means that all but the very earliest nutritional requirements must be drawn from the mother's circulation. Fertilization usually takes place in the fallopian tube and attachment to the uterine wall does not occur until at the blastocyst stage. Implantation causes an experimental limitation in that it becomes more difficult to study the embryo proper. Obviously, if the whole uterus or the deciduomata (embryo plus decidual tissue) is used, biochemical analyses do not reveal whether particular events are associated with the embryo itself or with the maternal decidualizing tissue.

In the mouse uterus, ornithine decarboxylase activity was low during the first 4 to 5 days of gestation, i.e., until implantation (Fozard *et al.*, 1980a,b). It increased sharply in the deciduomata between days 6 and 7 to reach a peak on day 8, i.e., during the primitive streak stage (Fozard *et al.*, 1980b). Inclusion of DFMO in the drinking water from day 5 on abolished the increase in ornithine decarboxylase activity and caused resorption or expulsion of the embryos. Histological examination revealed implantation and decidualization to be normal, but subsequent embryonic development to be greatly reduced. Embryonic development failed to progress beyond a stage typical of days 6 to 8 of normal gestation. As early as day 9, signs of rejection and resorption were evident and on day 18 there were no signs of pregnancy. These effects of DFMO were completely prevented by the concomitant administration of putrescine, implicating inhibition of putrescine synthesis as the mechanism of the contragestational effect.

In the rat and the rabbit, DFMO exerted contragestational effects similar to those seen in the mouse (Fozard *et al.*, 1980a; Méndez *et al.*, 1983). DFMO was included in the drinking water (Fozard *et al.*, 1980a) or administered intraluminally in the uterine horns (Méndez *et al.*, 1983). As in the case of the mouse, the effects of DFMO were counteracted by provision of putrescine (Méndez *et al.*, 1983).

In the mouse, cleavage of the fertilized ovum and blastocyst formation seem to take place essentially normally during DFMO treatment. This is in agreement with the observation of Alexandre (1979) using MO to inhibit the ornithine decarboxylase activity in mouse embryos developing *in vitro*. MGBG treatment, on the other hand, blocked development during cleavage (Alexandre, 1979; Geuskens and Alexandre, 1982). The development of the embryos resumed after transfer into normal medium, but their cavitation was delayed.

Taken together, studies on embryos, invertebrate as well as vertebrate, indicate that an increase in polyamine synthesis is absolutely essential during a critical period (at about the time of gastrulation) for continued embryonal growth and differentiation.

III. ORGAN SYSTEMS

A. Primary Mesoderm: Chick Embryo

As described in Section II,D, inhibition of polyamine synthesis in early chick embryos blocked their development at gastrulation, apparently by affecting mesoderm outgrowth and differentiation. To study these effects in greater detail primary cultures of chick embryo mesoderm were used (Löwkvist *et al.*, 1987). Mesoderm from definitive streak stage chick embryos was prepared free from ectoderm and endoderm and cultured according to Keane *et al.* (1979). In culture, the mesoderm cells retained their *in ovo* outgrowth behavior and differentiation pattern.

Addition of DFMO to the culture medium delayed the attachment and outgrowth and reduced the rate of proliferation of the mesoderm cells (Löwkvist *et al.*, 1987). Furthermore, the expression of differentiated phenotypes, such as beating heart tissue, erythroid cells, chondrocytes, and adipose cells, was markedly delayed. Simultaneous addition of putrescine prevented or reduced the effects of DFMO. The data suggest that polyamine depletion decreases the rate of mesoderm cell proliferation and as a consequence delays the expression of differentiated phenotypes. This is consistent with the contention that the fates of most mesodermal cells are already determined by the end of gastrulation, and that most embryonic tissues must reach a certain mass or cell number before they can express their developmental capabilities.

B. Mammary Gland: Mouse, Rat, and Rabbit

In an extensive series of investigations the relationship between polyamines and specific cell-differentiative events in mouse mammary gland explants has been studied (Oka *et al.*, 1978, 1981). It has been shown that spermidine is necessary for milk protein synthesis. The initiation of DNA synthesis in mammary gland cells induced by insulin and glucocorticoids was blocked by MGBG-mediated inhibition of spermidine synthesis (Oka *et al.*, 1978, 1981; Sakai *et al.*, 1978). Furthermore, the synthesis of casein and α-lactalbumin following prolactin stimulation was inhibited by MGBG treatment (Oka and Perry, 1974). The MGBG-induced inhibition of DNA replication and milk protein (casein and α-lactalbumin) synthesis was abrogated by the addition of spermidine, whereas putrescine or spermine addition was less effective. Moreover, the requirement for glucocorticoids in milk protein synthesis was replaced completely for α-lactalbumin, and partially for casein, by provision of exogenous spermidine. Although the increased synthesis of polyamines, particularly putrescine, is needed for mammary cell proliferation, the requirement for spermidine in milk protein production appears to be distinct and separate from the requirement for proliferation. Only spermidine was able to substitute for the glucocorticoid needed to induce mammary gland differentiation (Oka *et al.*, 1981).

The induction of milk protein synthesis in rabbit mammary gland explants does not have an absolute requirement for glucocorticoid and spermidine (Houdebine *et al.*, 1978), but it can be enhanced by glucocorticoids (Bolander and Topper, 1979). This enhancement is dependent on spermidine, although the latter cannot replace the glucocorticoid. The rat system represents an intermediate between that of the mouse and the rabbit. It has an absolute requirement for both glucocorticoid and spermidine, but spermidine cannot substitute for the hormone (Bolander and Topper, 1979).

The addition of a combination of spermidine and a prostaglandin (or arachidonic acid, phospholipase A, or cyclic GMP) to mouse mammary gland explants stimulated casein synthesis in a prolactin-like manner (Rillema, 1976; Rillema *et al.*, 1977). The fact that MGBG abolished the stimulation of casein synthesis by prolactin, but did not abolish the effect of spermidine plus a prostaglandin, suggests a specific function for spermidine in casein synthesis.

C. Hematopoietic System: Mouse, Rat, and Human

In the normal, unperturbed adult animal, the hematopoietic system is one of the rapidly proliferating, self-renewing organ systems. The findings from studies of normal hematopoietic cell differentiation *in vitro,* and especially *in vivo,* are very complex. Polyamines may modulate the proliferation and/or differentiation of hematopoietic cells either positively or negatively (enhancement or suppression).

Hematopoietic stem cells have inherent capacity for self-replication (proliferation) as well as differentiation (commitment) to hematopoietic progenitors and finally to mature peripheral blood cell elements. The morphological identity of these stem cells and the more mature progenitors has not been characterized. However, by indirect methods clonal assays for the progeny of the stem cells have been established. The *in vivo* assay for pluripotent stem cell colony formation in the spleen of animal hosts has been termed CFU-S (colony-forming unit, spleen). The *in vitro* assays for committed erythroid progenitor cells consist of growing hematopoietic tissues in semisolid matrices for varying periods of time and examining clones of erythroid cells. Two progenitors, the earlier one termed BFU-E (burst-forming unit, erythroid) and the later one termed CFU-E (colony-forming unit, erythroid), have been described. The progenitor for granulocyte–macrophage cells has also been grown in similar *in vitro* systems and is termed the CFU-GM (Cline and Golde, 1979).

Verma and Sunkara (1982), using *in vitro* liquid cultures and semisolid agar cultures, found that the proliferation and differentiation of human granulocyte–macrophage progenitor cells exhibited remarkable polyamine dependency. With human-placenta-conditioned medium as colony-stimulating factor, there was granulocyte–macrophage colony formation and granulopoietic differentiation. These events were inhibited by the administration of DFMO. That the inhibition of proliferation and differentiation was prevented by provision of exogenous putrescine indicates that it was indeed due to polyamine deficiency. However, because colony-stimulating factor was used, it is not completely clear whether the effect of polyamine synthesis inhibition and the prevention by putrescine was an effect on the colony-stimulating factor interaction or on the actual process of differentiation.

Using other models and techniques of studying human and mouse hematopoietic progenitor cell proliferation and differentiation, rather divergent effects have been observed. Niskanen *et al.* (1983b), using the methylcellulose culture system, showed that mouse BFU-E and CFU-E were increased when the bone marrow cells were obtained from mice treated with DFMO. This stimulation could be abrogated by administration of putrescine to the mice. The stimulatory effect of DFMO *in vivo* was also reflected by an increase in CFU-DE, i.e, erythroid colonies in diffusion chambers implanted in the peritoneal cavity of mice. Sharkis *et al.* (1983) showed that the administration of DFMO to mice produced bone marrow cells that had the capacity to form significantly more colonies in the spleens of lethally irradiated mice (CFU-S). This observation was independently confirmed by Niskanen *et al.* (1983a). However, the *in vitro* effects of DFMO on CFU-E and BFU-E were rather different than those observed *in vivo*. There was a dose-dependent inhibition of BFU-E and CFU-E formation. This inhibition was not prevented by the addition of putrescine. However, addition of aminoguanidine, which blocks diamine oxidase present in fetal calf serum, increased the yield of BFU-E and CFU-E (Niskanen *et al.*, 1983a). Thus, the effect of polyamine synthesis inhibition on hematopoietic cell proliferation and differentiation does not appear to be simple or straightforward.

Sharkis *et al.* (1983) further dissected the effect of DFMO on erythroid progenitor cell proliferation and differentiation by using a coculture system in which the effect of thymocytes on bone marrow cells was observed. It was previously shown that the addition of thymocytes to erythroid progenitor cell cultures could either enhance or suppress the growth of erythroid colonies, depending on the number of thymocytes added (Sharkis *et al.,* 1980, 1982; Sieber and Sharkis, 1982). Thus, small numbers of thymocytes in coculture with the hematopoietic progenitor cells from the bone marrow produced a reduction in BFU-E and CFU-E formation, whereas large numbers of thymocytes caused an enhancement of BFU-E and CFU-E formation. It was also found that the suppressor thymocyte was more dependent on proliferation than the helper thymocyte (Sharkis *et al.,* 1982). Using this coculture system, DFMO was found to increase CFU-E formation (Sharkis *et al.,* 1983). This effect was due to an increase in helper thymocyte activity. Thus, the action of DFMO in this short-term culture system may be the acute inhibition of proliferation of the suppressor thymocyte activity, reflected by increased helper activity. This enhancing effect of DFMO was prevented by the administration of putrescine. However, when DFMO is administered on a more chronic basis, there might ensue an inhibition of proliferation of cells other than the suppressor thymocytes, including the helper thymocytes, as well as the erythroid progenitor cells directly. This would result in the hematopoietic inhibition observed *in vivo* in clinical (Abeloff *et al.,* 1984) and animal studies (Luk *et al.,* 1983).

Despite the observed stimulation of mouse CFU-E formation *in vivo,* results obtained in human studies during phase I trials of DFMO (Abeloff *et al.,* 1984) suggest that the overall *in vivo* effects of DFMO are inhibitory to the differentiation of the end-stage erythropoietic cells. In both humans and rats, the administration of DFMO for longer term periods (2–5 weeks) resulted in decreased peripheral blood cell elements, especially in the platelet count (Abeloff *et al.,* 1984; Luk *et al.,* 1983). The effects in rats were prevented by the addition of putrescine (Luk *et al.,* 1983). In humans, the effects were promptly reversed when DFMO treatment was stopped (Abeloff *et al.,* 1984).

The importance of polyamines in erythroid cell proliferation and differentiation was subsequently studied using bone marrow transplantation techniques. Congenitally aplastic mice were given marrow transplants from normal donors. When the anemic mice were exposed to DFMO, or were provided with bone marrow cells from donors that were exposed to DFMO, the results were a more rapid engraftment of the marrow (Sharkis, 1985). This again suggested that DFMO modified the proliferation of the regulatory cells of erythropoiesis.

For further elucidation of the role of polyamines in hematopoiesis, the results obtained from studies of two neoplastic cell lines may be instructive (Sections IV,D and IV,H).

D. Intestinal Mucosa: Rat

The intestinal mucosa undergoes constant renewal and cell differentiation. It may be divided into a proliferative crypt compartment and a differentiated villus compartment. During maturation of the young animal, there is rapid proliferation in the crypt compartment and rapid differentiation of the mucosal cells into differentiated villus cells. Administration of DFMO to young rats resulted in delayed maturation of the intestinal mucosa (Luk *et al.*, 1980). There was decreased cell proliferation, and the progeny produced also showed a delay in maturation, as measured by the appearance of mature, absorptive enzymes. This inhibition of intestinal mucosal differentiation was also observed when DFMO was administered to adult rats that were stressed by exposure to the cytotoxic agent 1-β-D-arabinofuranosylcytosine (Luk *et al.*, 1980).

In the normal unperturbed adult rat, prolonged administration of DFMO gradually resulted in hypoplasia of the mucosa, with decreased cell proliferation and decreased numbers of mucosal cells. It eventually resulted in mucosal atrophy and delayed cell maturation with decreased mucosal enzyme activity. When the intestinal mucosa was stressed to undergo adaptive hyperplasia, the administration of DFMO resulted in inhibition and delay of the adaptive maturation and differentiation. This was shown by a decrease in thymidine incorporation and crypt cell production, and delay in maturation of mucosal enzymes in models involving postresectional hyperplasia and hyperplasia obtained during lactation (Luk and Baylin, 1984; Yang *et al.*, 1984). Overall, it is difficult to separate the effects on cellular proliferation from those on differentiation in this organ system, as many differentiation processes require cell proliferation.

IV. PRIMARY CELL CULTURES AND ESTABLISHED CELL LINES

A. Fibroblasts: Mouse

Mouse 3T3 fibroblasts maintained *in vitro* can differentiate into adipocytes (Green and Kehinde, 1974). Two specific sublines (3T3-L1 and 3T3-L2) have been isolated which exhibit an enhanced propensity to differentiate into adipose cells (Green and Kehinde, 1974). When exposed to various compounds, including 1-methyl-3-isobutylxanthine, dexamethasone, and insulin, cells from these sublines show rapid and irreversible differentiation into adipose cells (Russell and Ho, 1976; Rubin *et al.*, 1978). To investigate the role of polyamines in this differentiative event, the influence of DFMO was analyzed (Bethell and Pegg, 1981; Erwin *et al.*, 1984). The response of induced fibroblasts to DFMO exposure was the inhibition of differentiation into adipocytes. This DFMO inhibition was temporally restricted. Concurrent addition of DFMO with the inducer was necessary to inhibit adipocyte differentiation. If DFMO was added after cells were exposed to the inducer, normal differentiation occurred. Measurement of polyamine levels following both the induction of differentiation and the DFMO-mediated inhibition of differentiation indicated that spermidine, rather than putrescine or spermine, was the important polyamine in this differentiative process. Thus, the effect of DFMO on 3T3-L1 cells is temporally restricted and reversible, and DFMO seems to affect differentiation at an essential inductive level rather than at a secondary biosynthetic level. These studies clearly suggest a role for polyamines in the differentiative process.

B. Chondrocytes: Rat and Rabbit

Ornithine decarboxylase activity and polyamines are present in cartilage (Conroy *et al.*, 1977). *In vitro* studies with both rat and rabbit cartilage cells (Suzuki *et al.*, 1976; Takigawa *et al.*, 1979) demonstrated that glycosaminoglycan synthesis can be stimulated in both cell types by parathyroid hormone. Additionally, a large and rapid increase in ornithine decarboxylase activity in rabbit costal chondrocytes resulted from parathyroid hormone stimulation (Takigawa *et al.*, 1979). The addition of exogenous polyamines to the cell cultures revealed that these molecules could induce the differentiation of chondrocytes (Takano *et al.*, 1981). Inhibition of ornithine decarboxylase activity with α-hydrazino-δ-aminovaleric acid (HAVA), a competitive inhibitor of the enzyme (Inoue *et al.*, 1975), resulted in a specific block of both morphological change and increased glycosaminoglycan synthesis (Takano *et al.*, 1983). The observations that polyamines alone can stimulate chondrocyte differentiation and that inhibition of polyamine synthesis blocks differentiation provide strong evidence that polyamines play a central role in the normal phenotypic expression of chondrocytes.

C. Myoblasts: Rat

Muscle differentiation can be studied *in vitro* by investigating events involved in the fusion of myoblasts to myotubes. One rat cell line, L6 myoblasts, can be induced to differentiate into myotubes, with a concurrent increase in the level of a muscle-specific protein, creatine kinase. Several compounds have demonstrated a stimulatory ability when added to cultures of L6 cells. Among these compounds are insulin (Florini and Ewton, 1981) and MSA (multiplication-stimulating activity), a member of the somatomedin family of hormones (Ewton and Florini, 1980, 1981).

Following induction of differentiation, the ornithine decarboxylase activity increased, but dropped to barely detectable levels in fully differentiated myotubes (Stoscheck *et al.,* 1980), suggesting a possible role for polyamines in the early inductive events. To investigate the role of polyamines in myotube formation, cultures of L6 cells were induced to differentiate in the presence of DFMO.

Initial studies with insulin-induced L6 cells showed that myoblast differentiation could be inhibited by treatment with DFMO (Erwin *et al.,* 1983). Addition of exogenous polyamines reversed this inhibition. Subsequent studies using MSA-induced L6 cells (Ewton *et al.,* 1984) demonstrated similar results. DFMO treatment blocked the induction of MSA-stimulated ornithine decarboxylase activity, lowered levels of putrescine, and blocked induction of differentiation. Exogenous polyamines again reversed this inhibition.

Myoblast differentiation involves an obligate proliferative phase. Studies with MSA-induced, DFMO-treated L6 cells indicate that the inhibition of differentiation is not simply a result of aberrant cell proliferation, but a primary response to reduced levels of polyamines. Since exogenous polyamines can block DFMO inhibition, have no effect on MSA-induced differentiation, and cannot alone stimulate differentiation, the results suggest that increased levels of polyamines are necessary, but alone are not sufficient, to induce myotube formation.

D. Friend Erythroleukemia Cells: Mouse

Friend erythroleukemia cells in culture may be induced to differentiate by a variety of compounds. Gazitt and Friend (1980) were able to separate the list of compounds into two groups, those that were shown to enhance polyamine synthesis and those that appeared to have no effect on polyamine synthesis. Compounds which appeared to induce differentiation by induction of ornithine decarboxylase included dimethyl sulfoxide and hexamethylene bisacetamide. Induction of increases in ornithine decarboxylase activity and differentiation were inhibited by polyamine synthesis inhibitors such as MO, HO, and MGBG. This inhibition was abrogated by high dosages of the polyamines putrescine, spermidine, and spermine. These studies emphasize that, even in cell systems in which polyamines appear to be necessary for cell differentiation, differentiation may nevertheless proceed or be induced in the absence of a stimulation of polyamine synthesis. Sugiura *et al.* (1984) recently showed that DFMO inhibited the differentiation of Friend erythroleukemia cells and that this inhibition was reversed by the addition of exogenous spermidine. There was a highly significant correlation between intracellular spermidine levels and differentiation of the erythroleukemia cells. These studies suggest an important role for the polyamines in the differentiation of Friend erythroleukemia cells.

E. Neuroblastoma Cells: Mouse

Mouse neuroblastoma cells can be grown *in vitro* and induced to differentiate with either cyclic AMP or 1-methyl-3-isobutylxanthine, resulting in both morphological changes leading to the formation of neurons and biochemical changes related to the synthesis of neurotransmitter (Prasad and Hsie, 1971). This differentiation involves a large decrease in ornithine decarboxylase activity, little change in the level of putrescine, but a substantial decrease in the spermidine level (Chen *et al.*, 1982). Studies with DFMO demonstrated that reduced levels of both ornithine decarboxylase activity and polyamines also induced differentiation of neuroblastoma cells (Chen *et al.*, 1983; Chen and Liu, 1983). It is particularly interesting that suboptimal concentrations of cyclic AMP can potentiate the DFMO effect, and that this potentiation is prevented by exogenous polyamines (Chen and Liu, 1983). These studies suggest that neuroblastoma differentiation involves alterations in the levels of polyamines.

F. Melanoma Cells: Mouse

Mouse melanoma cells have served as a useful model to investigate cell differentiation *in vitro*. The morphological changes which occur in the formation of melanocytes are accompanied by distinct biochemical events, increases in both tyrosinase activity and intracellular levels of melanin. Differentiation can be induced experimentally by treating cells in culture with melanotropin (melanocyte-stimulating hormone, MSH), resulting in an elevation of cyclic AMP (Wong *et al.*, 1974). Additional studies have shown that cyclic-AMP-independent mechanisms also stimulate melanocyte differentiation (Lotan and Lotan, 1980, 1981).

In order to investigate the possible role of polyamines in melanoma differentiation, Cloudman S91 mouse melanoma cells were treated with both DFMO and MGBG (Käpyaho and Jänne, 1983; Käpyaho *et al.*, 1985). Treatment with each compound alone, or the two compounds together, induced extensive differentiation as determined by both tyrosinase activity and melanin accumulation. Stimulation of these melanoma cells by MSH resulted in a decrease in the intracellular level of putrescine, comparable to the reduction observed with exposure to DFMO.

While concurrent exposure of DFMO-treated S91 cells to exogenous polyamines prevented melanocyte differentiation, treatment with putrescine alone had no effect, and treatment of MSH-stimulated S91 cells with exogenous putrescine failed to prevent tyrosinase stimulation. Additionally, DFMO potentiated the stimulatory effect of MSH. These results suggest that the mechanism of action of polyamine synthesis inhibitors is different from that of MSH in stimulating melanocyte differentiation.

Since treatment of S91 cells with either DFMO or MGBG has antiproliferative effects, it is difficult to rule out the possibility that the differentiative response is simply a reflection of altered growth characteristics.

Recent studies with another mouse melanoma cell line (B16) (Sunkara *et al.*, 1985) support the observations seen with the S91 cells. While once again one cannot rule out antiproliferative effects in the B16 cells, the aggregate studies do suggest that levels of polyamines influence the state of melanocyte differentiation.

G. Embryonal Carcinoma Cells: Mouse

Teratocarcinomas are germ cell tumors which contain a mixture of differentiated cell types and undifferentiated stem cells referred to as embryonal carcinoma (EC) cells (Graham, 1977). A large number of EC cell lines have been isolated from tumors and successfully maintained *in vitro* (Jakob *et al.*, 1973; Sherman, 1975; Martin and Evans, 1975). Cells from these lines share the same cellular morphology, yet differ in their ability to differentiate in response to various stimuli. Cell differentiation can be stimulated in some, but not all, EC cell lines (McBurney, 1976; Jakob *et al.*, 1978; Sherman and Miller, 1978; Strickland and Mahdavi, 1978; Jetten *et al.*, 1979). Once EC cells respond to any of the stimuli, the differentiated phenotype is irreversible, even after removal of the original stimulus (Jetten *et al.*, 1979).

Pluripotent EC cells are tumorigenic, but can contribute to normal development when introduced into an appropriate microenvironment. Thus, microinjection of single EC cells, derived either from cell lines or tumors, into a mouse blastocyst leads to normal differentiation in the resulting chimeric mouse (Brinster, 1974; Mintz and Illmensee, 1975). Since EC cells can participate in the normal developmental process of embryogenesis, they represent an excellent model system to study cytodifferentiation and mammalian embryogenesis.

In order to investigate the possible role of polyamines in EC cell differentiation, cells from various EC cell lines were treated with DFMO (Heby *et al.,* 1983, 1984; Schindler *et al.,* 1983, 1985). The results indicate that DFMO has a profound effect on the ability of cells from several EC lines to differentiate *in vitro*. Additionally, the effects of DFMO could be reversed if exogenous polyamines were added within the first few days of treatment. While cells from most EC lines responded very well to DFMO induction (PCC4aza1R, PCC3, OC-15, Dif(RA)$^{-}$1), one cell line (Nulli-SCC1) was refractory to DFMO-induced differentiation. In the case of this cell line, even a higher concentration of DFMO (21 versus 2.1 m*M*) had no effect on differentiation, but did influence growth kinetics, suggesting that DFMO was active in Nulli-SCC1 cells, and that induction of differentiation was not simply a result of a reduced proliferation rate. Studies with F9 cells show some differences in the extent of induced differentiation (Heby *et al.,* 1983, 1984; Jetten and Shirley, 1985; Oredsson *et al.,* 1985; Schindler *et al.,* 1985). These discrepancies are likely to be the result of differences in the actual experimental protocols employed. Importantly, however, all studies do indicate that F9 cells respond to DFMO.

While the studies above indicate a correlation between polyamine levels and the state of EC cell differentiation, both enzyme activities and polyamine levels for the differentiated cells were assessed long after differentiation had occurred. Since EC cell differentiation requires only short treatments with various inducers, it is likely that the inductive events occur early after exposure to various stimuli. Therefore, changes in polyamine levels were analyzed in a number of EC cell lines within hours of exposure to DFMO (Kelly *et al.,* 1985). In PCC4aza1R cells, a significant reduction in ornithine decarboxylase activity and putrescine and spermidine levels occurred as early as 8 hr after exposure to DFMO, long before overt morphological differentiation was detected. In Dif-(RA)$^{-}$1 cells, DFMO lowered the levels of ornithine decarboxylase and polyamines significantly within the first 8 hr of exposure, in a manner identical to that seen in PCC4aza1R cells, the cell line from which Dif-(RA)$^{-}$1 cells were isolated (Schindler *et al.,* 1981). Nulli-SCC1 cells failed to differentiate in response to DFMO, yet levels of both ornithine decarboxylase and putrescine were dramatically reduced. It is possible that the reduction seen in this cell line, while significant, is not a sufficient signal to induce differentiation.

One of the most potent inducers of EC cell differentiation is retinoic acid (RA) (Strickland and Mahdavi, 1978; Jetten *et al.,* 1979). A series of studies in a variety of experimental systems have demonstrated that retinoids can influence both ornithine decarboxylase activity and polyamine levels (Verma *et al.,* 1978, 1979; Haddox *et al.,* 1979; Russell and Haddox, 1981). Since the studies described above suggested a role for polyamines in influencing the state of EC cell differentiation, attempts were undertaken to investigate the relationship between RA-induced EC cell differentiation and the levels of both ornithine decarboxylase activity and polyamines.

Utilizing F9 cells, the influence of RA on ornithine decarboxylase and polyamine levels was determined (Jetten and Shirley, 1985; Oredsson *et al.*, 1985). The results indicated that treatment of F9 cells with RA does result in a reduction in ornithine decarboxylase activity and polyamine levels in a manner consistent with DFMO treatment. However, putrescine, which can reverse the effect of DFMO, failed to reverse RA-induced differentiation. This suggests that RA does not induce differentiation by directly interacting with ornithine decarboxylase to reduce levels of polyamines.

Results of similar studies with several other EC cell lines (Kelly *et al.*, 1985) were generally consistent with those reported for F9 cells. RA-treated PCC4aza1R cells showed a reduction in ornithine decarboxylase activity and polyamine levels that was both quantitatively and temporally comparable to that seen by DFMO treatment. Dif(RA)$^{-}$1 cells showed no RA effect, as expected, since these cells are refractory to RA-induced differentiation. Nulli-SCC1 cells treated with RA showed little change in either ornithine decarboxylase activity or putrescine level until well after (24 hr) an early inductive event should have occurred. As in the case of F9 cells, putrescine failed to reverse any inductive effect of RA.

The studies described for EC cells indicate a role for polyamines in their differentiation. The reversal studies with putrescine and RA indicate that while reduced levels of ornithine decarboxylase and polyamines are necessary for EC cell differentiation, reduction in these levels alone is not a sufficient stimulus to induce differentiation.

H. Promyelocytic Leukemia Cells: Human

The human promyelocytic leukemia cell line, HL-60, has been used extensively for the study of human hematopoietic cell differentiation (Collins *et al.*, 1977, 1980). HL-60 cells multiply continuously in simple culture medium. Morphologically and cytochemically, HL-60 cell cultures consist predominantly of progranulocytes, and less than 5–10% of the cells resemble more mature granulocytic elements. In this state, HL-60 cells have low phagocytic capacity and are tumorigenic in nude mice. The addition of a variety of substances, including dimethyl sulfoxide, *N,N*-dimethylformamide and the retinoids, causes the majority of HL-60 cells to mature into granulocytes, with morphological, cytochemical, and functional changes. These granulocytes appear to have undergone terminal differentiation, in that there is no longer continued cell proliferation. Furthermore, the cells are no longer tumorigenic in nude mice and have increased phagocytic capacity.

Exposure of HL-60 cells to phorbol esters, such as 12-*O*-tetradecanoylphorbol-13-acetate (TPA), causes them to change morphologically, cytochemically, and functionally into monocyte-like cells. These cells again appear to be terminally differentiated, and the cells are nontumorigenic in nude mice and have increased phagocytic capacity and increased adherence to plastic.

Studies investigating the response of polyamines to phorbol ester and retinoid induction of HL-60 cell differentiation show that there was a large increase in ornithine decarboxylase activity and putrescine levels accompanying differentiation (Huberman *et al.*, 1981; Luk *et al.*, 1982). Subsequently, this increase in putrescine content was shown to follow a transient increase in ornithine decarboxylase activity (Luk *et al.*, 1982; Heby *et al.*, 1983). Exposure of the HL-60 cells to DFMO, together with either retinoic acid or TPA, resulted in inhibition of ornithine decarboxylase activity and accumulation of putrescine as well as growth inhibition, but there was no effect on the induced differentiation (Luk *et al.*, 1982; Heby *et al.*, 1983). Thus, the increase in ornithine decarboxylase was said to be important for proliferation but not for differentiation of the HL-60 cell.

More recent studies have amplified upon the above observations, that the inhibition of ornithine decarboxylase activity by DFMO is not necessarily accompanied by a complete depletion of the polyamine content. When the intracellular spermidine content was reduced to a greater extent by prolonged DFMO treatment, the induction of HL-60 cell differentiation by retinoids was inhibited (Sugiura *et al.*, 1984). However, similar to the previous studies, phorbol-ester-induced HL-60 cell differentiation was not inhibited by prolonged DFMO treatment, even with extensive depletion of the intracellular spermidine content. Therefore, these studies on HL-60 differentiation show that intracellular spermidine is important in differentiation, but that cellular differentiation can again proceed by pathways apparently not involving polyamines, as in phorbol-ester-induced HL-60 differentiation.

V. SUMMARY AND FUTURE PROSPECTS

The spectrum of experimental systems outlined in the preceding sections represents a broad survey of evidence suggesting that polyamines play a role in the process of cellular differentiation. Each experimental system has its own unique biological properties. The observations that some cell types require an increase in intracellular polyamines, others require a decrease, while one (myoblasts) first requires an increase and then a decrease in intracellular polyamines during induction of differentiation suggest that one unifying mechanism to explain the role of polyamines in cell differentiation is unlikely. It is equally unlikely that all the compounds that induce differentiation in the various cell types mentioned share mechanisms with obligate alterations in the levels of intracellular polyamines. However, because of the ubiquitous nature of polyamines and the lack of any substantive evidence suggesting that alterations in polyamine levels have no influence on the differentiative capacity of a given cell, it is safe to state that polyamines have important functions in cell differentiation. Future work will have to define these functions at a molecular level.

ACKNOWLEDGMENTS

The authors' laboratories received financial support from the Swedish Natural Science Research Council and the National Institutes of Health (United States).

REFERENCES

Abeloff, M.D., Slavik, M., Luk, G.D., Griffin, C.A., Hermann, J., Blanc, O., Sjoerdsma, A., and Baylin, S.B. (1984). *J. Clin. Oncol.* **2,** 124–130.
Alexandre, H. (1979). *J. Embryol. Exp. Morphol.* **53,** 145–162.
Bethell, D.R., and Pegg, A.E. (1981). *Biochem. Biophys. Res. Commun.* **102,** 272–278.
Bolander, Jr., F.F., and Topper, Y.J. (1979). *Biochem. Biophys. Res. Commun.* **90,** 1131–1135.
Brachet, J., Mamont, P., Boloukhère, M., Baltus, E., and Hanocq-Quertier, J. (1978). *C.R. Acad. Sci. Ser. D* **287,** 1289–1292.
Brinster, R.L. (1974). *J. Exp. Med.* **140,** 1049–1056.
Chen, K.Y., and Liu, A.Y. -C. (1983). *Adv. Polyamine Res.* **4,** 743–755.
Chen, K.Y., Presepe, V., Parken, N., and Liu, A.Y.-C. (1982). *J. Cell. Physiol.* **110,** 285–290.
Chen, K.Y., Nau, D., and Liu, A.Y.-C. (1983). *Cancer Res.* **43,** 2812–2818.
Cline, M.J., and Golde, D.W. (1979). *Blood* **53,** 157–165.
Collins, S.J., Gallo, R.C., and Gallagher, R. E. (1977). *Nature (London)* **270,** 347–349.
Collins, S.J., Bodner, A., Ting, R., and Gallo, R.C. (1980). *Int. J. Cancer* **25,** 213–218.
Conroy, P.D., Simms, D.M., and Pointon, J.J. (1977). *Biochem. J.* **162,** 347–350.
Emanuelsson, H., and Heby, O. (1978). *Proc. Natl. Acad. Sci. U.S.A.* **75,** 1039–1042.
Erwin, B.G., Ewton, D.Z., Florini, J.R., and Pegg, A.E. (1983). *Biochem. Biophys. Res. Commun.* **114,** 944–949.
Erwin, B.G., Bethell, D.R., and Pegg, A.E. (1984). *Am. J. Physiol.* **246,** C293–C300.
Ewton, D.Z., and Florini, J.R. (1980). *Endocrinology* **106,** 577–583.

Ewton, D.Z., and Florini, J.R. (1981). *Dev. Biol.* **86,** 31–39.
Ewton, D.Z., Erwin, B.G., Pegg, A.E., and Florini, J.R. (1984). *J. Cell. Physiol.* **120,** 263–270.
Florini, J.R., and Ewton, D.Z. (1981). *In Vitro* **17,** 763–768.
Fozard, J.R., Part, M.-L., Prakash, N.J., Grove, J., Schechter, P.J., Sjoerdsma, A., and Koch-Weser, J. (1980a). *Science* **208,** 505–508.
Fozard, J.R., Part, M.-L., Prakash, N.J., and Grove, J. (1980b). *Eur. J. Pharmacol.* **65,** 379–391.
Fozard, J.R., Prakash, N.J., and Grove, J. (1980c). *Life Sci.* **27,** 2277–2283.
Gazitt, Y., and Friend, C. (1980). *Cancer Res.* **40,** 1727–1732.
Geuskens, M., and Alexandre, H. (1982). *Biol. Cell.* **46,** 151–160.
Graham, C.F. (1977). *In* "Concepts in Mammalian Embryogenesis" (M.I. Sherman, ed.), pp. 315–394. MIT Press, Cambridge, Massachusetts.
Green, H., and Kehinde, O. (1974). *Cell* **1,** 113–116.
Haddox, M.K., Scott, K.F.F., and Russell, D.H. (1979). *Cancer Res.* **39,** 4930–4938.
Heby, O. (1981). *Differentiation* **19,** 1–20.
Heby, O., and Emanuelsson, H. (1980). *Biochem. Biophys. Res. Commun.* **92,** 75–79.
Heby, O., and Emanuelsson, H. (1981). *Med. Biol.* **59,** 417–422.
Heby, O., and Jänne, J. (1981). *In* "Polyamines in Biology and Medicine" (D.R. Morris and L.J. Marton, eds.), pp. 243–310. Dekker, New York.
Heby, O., Oredsson, S.M., Olsson, I., and Marton, L.J. (1983). *Adv. Polyamine Res.* **4,** 727–742.
Heby, O., Oredsson, S.M., and Kanje, M. (1984). *Adv. Enzyme Regul.* **22,** 243–264.
Houdebine, L.M., Devinoy, E., and Delouis, C. (1978). *Biochimie* **60,** 735–741.
Huberman, E., Weeks, C., Herrmann, A., Callaham, M., and Slaga, T. (1981). *Proc. Natl. Acad. Sci. U.S.A.* **78,** 1062–1066.
Inoue, H., Kato, Y., Takigawa, M., Adachi, K., and Takeda, Y. (1975). *J. Biochem.* **77,** 879–893.
Jakob, H., Boon, T., Gaillard, J., Nicolas, J.-F., and Jacob, F. (1973). *Ann. Microbiol. (Inst. Pasteur)* **124B,** 269–282.
Jakob, H., Dubois, P., Eisen, H., and Jacob, F. (1978). *C.R. Acad. Sci. Ser. D* **286,** 109–111.
Jetten, A.M., and Shirley, J.E. (1985). *Exp. Cell Res.* **156,** 221–230.
Jetten, A.M., Jetten, M.E.R., and Sherman, M.I. (1979). *Exp. Cell Res.* **124,** 381–391.
Käpyaho, K., and Jänne, J. (1983). *Biochem. Biophys. Res. Commun.* **113,** 18–23.
Käpyaho, K., Sinervirta, R., and Jänne, J. (1985). *Cancer Res.* **45,** 1444–1448.
Keane, R.W., Lindblad, P.C., Pierik, L.T., and Ingram, V.M. (1979). *Cell* **17,** 801–811.
Kelly, M., McCann, P.P., and Schindler, J. (1985). *Dev. Biol.* **111,** 510–514.
Kusunoki, S., and Yasumasu, I. (1978). *Dev. Biol.* **67,** 336–345.
Lotan, R., and Lotan, D. (1980). *Cancer Res.* **40,** 3345–3350.
Lotan, R., and Lotan, D. (1981). *J. Cell. Physiol.* **106,** 179–189.
Löwkvist, B., Heby, O., and Emanuelsson, H. (1980). *J. Embryol. Exp. Morphol.* **60,** 83–92.
Löwkvist, B., Emanuelsson, H., and Heby, O. (1983). *Cell Differ.* **12,** 19–26.
Löwkvist, B., Oredsson, S.M., Holm, I., Emanuelsson, H., and Heby, O. (1987). *Cell Tissue Res.,* in press.
Luk, G.D., and Baylin, S.B. (1984). *J. Clin. Invest.* **74,** 698–704.
Luk, G.D., Marton, L.J., and Baylin, S.B. (1980). *Science* **210,** 195–198.
Luk, G.D., Civin, C.I., Weissman, R.M., and Baylin, S.B. (1982). *Science* **216,** 75–77.
Luk, G.D., Sharkis, S.J., Abeloff, M.D., McCann, P.P., Sjoerdsma, A., and Baylin, S.B. (1983). *Proc. Natl. Acad. Sci. U.S.A.* **80,** 5090–5093.

McBurney, M.W. (1976). *J. Cell. Physiol.* **89,** 441–455.
Martin, G.R., and Evans, M.J. (1975). *Proc. Natl. Acad. Sci. U.S.A.* **72,** 1441–1445.
Méndez, J.D., Diaz-Fiores, M., Durán, G., and Hicks, J.J. (1983). *Contraception* **28,** 93–98.
Metcalf, B.W., Bey, P., Danzin, C., Jung, M.J., Casara, P., and Vevert, J.P. (1978). *J. Am. Chem. Soc.* **100,** 2551–2553.
Mintz, B., and Illmensee, K. (1975). *Proc. Natl. Acad. Sci. U.S.A.* **72,** 3585–3589.
Niskanen, E., Kallio, A., McCann, P.P., and Baker, D.G. (1983a). *Blood* **61,** 740–745.
Niskanen, E.O., Kallio, A., McCann, P.P., Pou, G., Lyda, S., and Thornhill, A. (1983b). *Cancer Res.* **43,** 1536–1540.
Oka, T., and Perry, J.W. (1974). *J. Biol. Chem.* **249,** 7647–7652.
Oka, T., Sakai, T., Lundgren, D.W., and Perry, J.W. (1978). *In* "Hormones, Receptors, and Breast Cancer" (W.L. McGuire, ed.), pp. 301–323. Raven, New York.
Oka, T., Perry, J.W., Takemoto, T., Sakai, T., Terada, N., and Inoue, H. (1981). *Adv. Polyamine Res.* **3,** 309–320.
Oredsson, S.M., Billgren, M., and Heby, O. (1985). *Eur. J. Cell Biol.* **38,** 335–343.
Pegg, A.E., and McCann, P.P. (1982). *Am. J. Physiol.* **243,** C212–C221.
Porter, C.W., Dave, C., and Mihich, E. (1981). *In* "Polyamines in Biology and Medicine" (D.R. Morris and L.J. Marton, eds.), pp. 407–436. Dekker, New York.
Prasad, K.N., and Hsie, A.W. (1971). *Nature (London) New Biol.* **233,** 141–142.
Rillema, J.A. (1976). *Biochem. Biophys. Res. Commun.* **70,** 45–49.
Rillema, J.A., Linebaugh, B.E., and Mulder, J.A. (1977). *Endocrinology,* **100,** 529–536.
Rubin, C.S., Hirsch, A., Fung, C., and Rosen, O.M. (1978). *J. Biol. Chem.* **253,** 7570–7578.
Russell, D.H. (1971). *Proc. Natl. Acad. Sci. U.S.A.* **68,** 523–527.
Russell, D.H., and Haddox, M.K. (1981). *Ann. N.Y. Acad. Sci.* **359,** 281–297.
Russell, T.R., and Ho, R.-J. (1976). *Proc. Natl. Acad. Sci. U.S.A.* **73,** 4516–4520.
Sakai, T., Lundgren, D.W., and Oka, T. (1978). *J. Cell. Physiol.* **95,** 259–267.
Schindler, J., Matthaei, K.I., and Sherman, M.I. (1981). *Proc. Natl. Acad. Sci. U.S.A.* **78,** 1077–1080.
Schindler, J., Kelly, M., and McCann, P.P. (1983). *Biochem. Biophys. Res. Commun.* **114,** 410–417.
Schindler, J., Kelly, M., and McCann, P.P. (1985). *J. Cell. Physiol.* **122,** 1–6.
Sharkis, S.J. (1985). *Blood Cells* **10,** 223–231.
Sharkis, S.J., Spivak, J.L., Ahmed, A., Misiti, J., Stuart, R.K., Wiktor-Jedrzejczak, W., Sell, K.W., and Sensenbrenner, L.L. (1980). *Blood* **55,** 524–527.
Sharkis, S.J., Colvin, M.O., and Sensenbrenner, L.L. (1982). *Stem Cells* **1,** 269–273.
Sharkis, S.J., Luk, G.D., Collector, M.I., McCann, P.P., Baylin, S.B., and Sensenbrenner, L.L. (1983). *Blood* **61,** 604–607.
Sherman, M.I. (1975). *In* "Teratomas and Differentiation" (M.I. Sherman and D. Solter, eds.), pp. 189–205. Academic Press, New York.
Sherman, M.I., and Miller, R.A. (1978). *Dev. Biol.* **63,** 27–34.
Sieber, F., and Sharkis, S.J. (1982). *Blood* **60,** 845–850.
Stoscheck, C.M., Florini, J.R., and Richman, R.A. (1980). *J. Cell. Physiol.* **102,** 11–18.
Strickland, S., and Mahdavi, V. (1978). *Cell* **15,** 393–403.
Sugiura, M., Shafman, T., and Kufe, D. (1984). *Cancer Res.* **44,** 1440–1444.
Sunkara, P.S., Wright, D.A., and Nishioka, K. (1981). *Dev. Biol.* **87,** 351–355.
Sunkara, P.S., Chang, C.C., Prakash, N.J., and Lachmann, P.J. (1985). *Cancer Res.* **45,** 4067–4070.
Suzuki, F., Yoneda, T., and Shimomura, Y. (1976). *FEBS Lett.* **70,** 155–158.
Takano, T., Takigawa, M., and Suzuki, F. (1981). *Med. Biol.* **59,** 423–427.
Takano, T., Takigawa, M., and Suzuki, F. (1983). *J. Biochem.* **93,** 591–598.

Takigawa, M., Watanabe, R., Ishida, H., Asada, A., and Suzuki, F. (1979). *J. Biochem.* **85,** 311–314.

Verma, A.K., Rice, H.M., Shapas, B.G., and Boutwell, R.K. (1978). *Cancer Res.* **38,** 793–801.

Verma, A.K., Shapas, B.G., Rice, H.M., and Boutwell, R.K. (1979). *Cancer Res.* **39,** 419–425.

Verma, D.S., and Sunkara, P.S. (1982). *Cancer Res.* **42,** 3046–3049.

Wasserman, W.J., Houle, J.G., and Samuel, D. (1984). *Dev. Biol.* **105,** 315–324.

Williams-Ashman, H.G., and Schenone, A. (1972). *Biochem. Biophys. Res. Commun.* **46,** 288–295.

Wong, G., Pawelek, J., Sansone, M., and Morowitz, J. (1974). *Nature* (*London*) **248,** 351–354.

Yang, P., Baylin, S.B., and Luk, G.D. (1984). *Am. J. Physiol.* **247,** G553–G557.

Younglai, E.U., Godeau, F., Mester, J., and Baulieu, E.E. (1980). *Biochem. Biophys. Res. Commun.* **96,** 1274–1281.

9

The Contragestational Effects of Ornithine Decarboxylase Inhibition

J. R. FOZARD

Merrell Dow Research Institute
16, rue d'Ankara
F 67084 Strasbourg Cedex, France

I. INTRODUCTION

Decarboxylation of ornithine by ornithine decarboxylase (ODC; EC 4.1.1.17) is the initial and, at least in mammals, rate-limiting step in the biosynthesis of the diamine putrescine and the polyamines spermidine and spermine (Williams-Ashman *et al.*, 1972; Tabor and Tabor, 1976, 1984; Pegg, 1986). Although basal activity of ODC is generally low in most tissues, marked increases are associated with instances of rapid tissue growth (Jänne *et al.*, 1978; Heby, 1981; Pegg and McCann, 1982) and particularly with mammalian and nonmammalian embryogenesis (Raina, 1963; Russell and Snyder, 1968; Russell and McVicker, 1972; Manen *et al.*, 1977; Fozard *et al.*, 1980a,b; Heby and Emanuelsson, 1981; Luzzani *et al.*, 1982; Galliani *et al.*, 1983). The early stage of mammalian embryogenesis is therefore an obvious process wherein inhibition of ODC might be expected to result in potentially significant biological consequences. In the event, experimental data obtained from a number of species provide direct proof of an essential role for ODC in early mammalian embryogensis. To describe these data in detail is the principal objective of this review. Reflecting an author's prerogative, a number of biases will be evident. Certain of these are deliberate, for instance, embryonic development in invertebrates and nonmammalian vertebrates will not be discussed in detail except in the context of the mechanism of action of ODC inhibitors in mammals. Other biases are unavoidable; thus, a majority of the studies of the role of ODC in gestation have been carried out with DL-α-difluoromethylornithine (MDL 71782; DFMO; eflornithine), the prototype potent and selective inhibitor of ODC which acts through the mechanism of substrate-induced irreversible inhibition (Bey, 1978; Seiler *et al.*, 1978). A starring role for eflornithine is therefore inevitable. However, a number of other inhibitors of ODC have also been tested for antifertility effects in the mouse and the details of these are also included (Section III,B).

II. POLYAMINE BIOCHEMICAL CHANGES DURING GESTATION IN MAMMALS

A. General Aspects

The changes in ODC, *S*-adenosylmethionine decarboxylase (EC 4.1.1.50; AdoMetDC), and the polyamines which occur in the key tissues involved in mammalian embryonic development have been measured in rat (Russell and McVicker, 1972; Guha and Jänne, 1976; Williams and McAnulty, 1976; Maudsley and Kobayashi, 1977; Hoshiai *et al.*, 1981), mouse (Fozard *et al.*, 1980a,b), and hamster (Luzzani *et al.*, 1982). The key changes are broadly similar in each species and are illustrated by experimental data obtained in the mouse in Figs. 1 and 2.

Perhaps the most striking changes, and certainly the most relevant in functional terms (see Section III,A), occur early in gestation in whole uterus. In this tissue, ODC activity is generally low during the first 5 days after fertilization but increases rapidly between days 6 and 8 to reach a peak on day 8, which is the time following implantation when early embryonic development is taking place. Uterine putrescine concentrations follow closely the changes in ODC; the AdoMetDC and spermidine concentrations are qualitatively similar but are displaced temporally by 1 day (Figs. 1 and 2).

Placental ODC activity and putrescine concentrations increase from day 11 as gestation proceeds; in contrast, spermidine concentrations decline gradually over the same period. Fetal ODC activity is high on day 11 but declines rapidly to reach a stable low plateau by day 14 of gestation. In contrast, fetal putrescine and spermidine concentrations increase gradu-

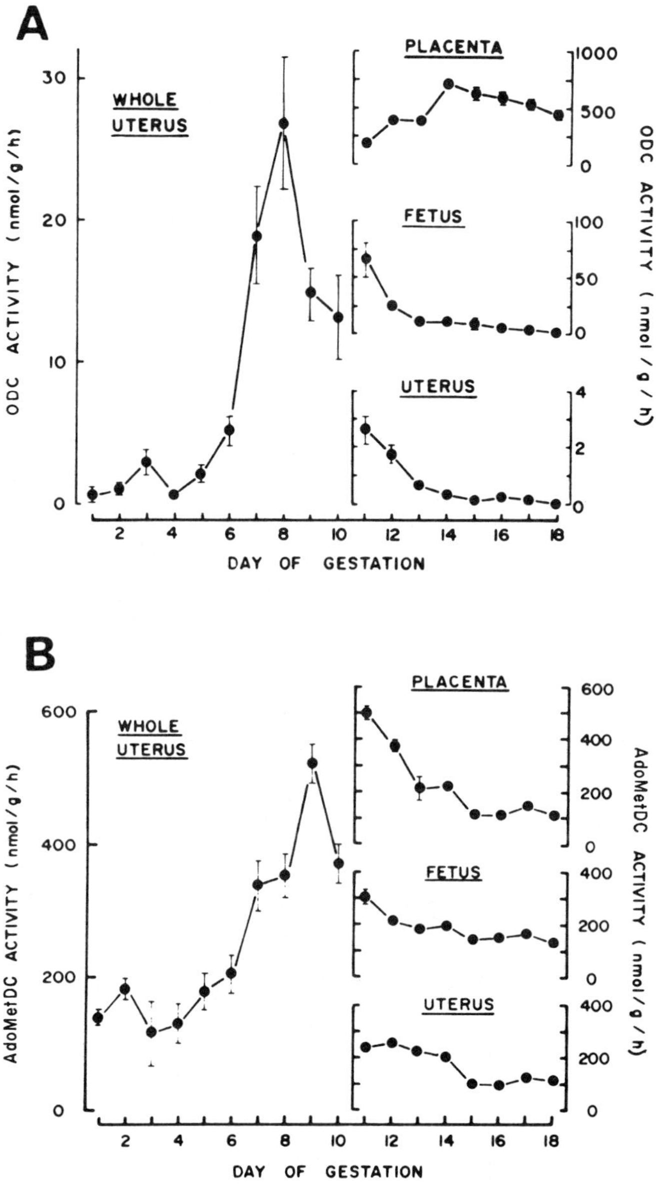

Fig. 1. Changes in ornithine decarboxylase (ODC) activity (A) and *S*-adenosyl-L-methionine decarboxylase (AdoMetDC) activity (B) in reproductive tissues of pregnant mice. Up to day 10 of gestation, whole uterus was taken for assay. From day 11, placentas, fetuses, and uteri were dissected out and assayed separately. Values represent the SEM of between 4 and 10 individual values. Reproduced from Fozard *et al.* (1980b) with permission.

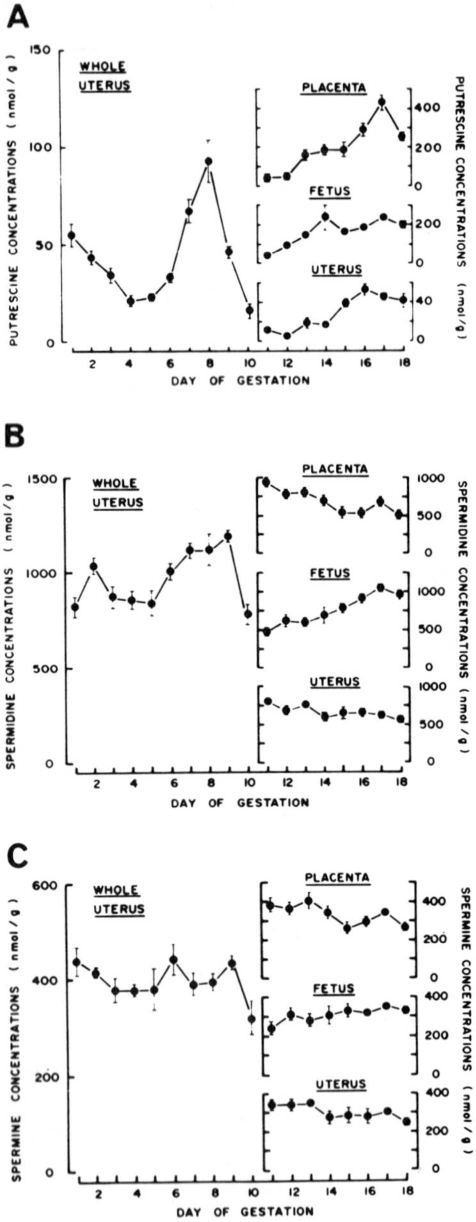

Fig. 2. Changes in putrescine (A), spermidine (B), and spermine (C) concentrations in the reproductive tissues of pregnant mice. For further details see legend to Fig. 1. Reproduced from Fozard *et al.* (1980b) with permission.

ally as parturition approaches (Fig. 2). This latter observation is the direct opposite of the situation in the rat, where fetal putrescine and spermidine concentrations decline as gestation progresses (Russell and McVicker, 1972). In the rat, the increase in tissue polyamine concentrations in the later stages of gestation is reflected in a rise in plasma spermidine (Lundgren and Oka, 1978) and a marked increase in the urinary excretion of putrescine and spermidine (Andersson *et al.*, 1978; Andersson and Henningsson, 1981). Placental tissue seems to be the source of the latter since the changes are abolished by removal of the placentas *and* fetuses but not by removal of the fetuses alone (Andersson and Heningsson, 1981).

B. Early Postimplantation Changes in Whole Uterus

By dissecting the deciduomata (decidual tissue plus embryo) from the nondecidualizing uterine tissue on day 8 and analyzing the samples separately, it was shown that the polyamine biochemical changes take place principally in the deciduomata in mouse (Fozard *et al.*, 1980b), rat (Saunderson and Heald, 1974; Hoshiai *et al.*, 1981), and hamster (Galliani *et al.*, 1983). Increases in ODC activity occur in both the decidual tissue and the developing embryo. Thus, in nonpregnant animals, decidualization of the uterus induced by intraluminal vegetable oils (Heald, 1979; Collawn *et al.*, 1981; Hoshiai *et al.*, 1981), mechanical damage (Collawn *et al.*, 1981), or systemic administration of pyrathiazine (Barkai *et al.*, 1978) is associated with a marked increase in ODC activity. Moreover, division of mouse deciduomata such that the embryo is retained intact in one of the parts revealed significantly greater ODC activity in the pure decidualizing tissue than in samples of nondecidualizing uterus (Fozard *et al.*, 1980b).

Progesterone is essential for decidualization and for the associated increase in ODC. Hence, both the functional and the biochemical changes induced in rat uterus by pyrathiazine were blocked by an antibody to progesterone (Barkai *et al.*, 1978). Furthermore, suppression of endogenous progesterone release by administration of prostaglandin $F_{2\alpha}$ or by ovariectomy prevented the rise in uterine ODC and arrested pregnancy in the hamster; these effects were completely reversed by concomitant administration of progesterone (Luzzani *et al.*, 1982).

The technique of dividing deciduomata and retaining the embryo in one part provided *indirect* evidence of a marked increase in ODC activity in the early implantation embryo. Thus, significantly more ODC activity was associated with the fraction enriched by an average of 10.2 embryos per sample than in the decidual tissue alone (Fozard *et al.*, 1980b). *Direct* evidence of an increase in embryonal ODC activity with age was provided by Huber and Brown (1982), who measured the changes in enzyme activity in rat embryos maintained in an *in vitro* culture system from days 9.5 to 11.5 of gestation.

III. EFFECTS OF INHIBITORS OF ODC ON GESTATION

A. Eflornithine

1. Biological Effects and Relationship to Inhibition of ODC

Studies in the mouse first revealed the profound effect of eflornithine on mammalian gestation (Fozard *et al.* 1980a,b). Mice were treated with eflornithine on days 1–4, 5–8, 9–12, or 13–16 of gestation by including the drug in the drinking water in an amount (2%) sufficient to ensure marked and sustained inhibition of ODC. They were killed on day 18 of gestation and the uteri and their contents were examined. The biological effects proved critically dependent on the period of treatment. For instance, animals treated with eflornithine during days 5–8 of gestation displayed no signs of pregnancy when autopsied on day 18 despite evidence of successful mating. Mice treated on days 1–4 or 9–12 contained fewer and smaller fetuses and more resorption nodules than controls. Mice treated with eflornithine between days 13 and 16 differed from controls only in having smaller fetuses (Fozard *et al.*, 1980b). Offspring from rats treated with eflornithine during the later stages of pregnancy have been reported to show organ-specific development deficits (Slotkin *et al.*, 1983), although these appear to be critically dependent on the dose (Genedani *et al.*, 1985) and schedule (Slotkin *et al.*, 1983; Bell *et al.*, 1986) of treatment.

Experiments broadly similar to those described above for the mouse have allowed definition of the critical period for intervention with an inhibitor of ODC in the rat (Fozard *et al.*, 1980a; Reddy and Rukmini, 1981; Slotkin *et al.*, 1983), hamster (Galliani *et al.*, 1983), and rabbit (Fozard *et al.*, 1980a); in each case this proved to be the period immediately following implantation, when the key stages of early embryonic development are taking place.

Histological examination of deciduomata from mice treated with eflornithine during days 5–8 of gestation reveals that decidualization takes place normally after implantation but that embryonic development fails to progress beyond a stage typical of days 6 to 7 of normal gestation. Subsequently, the arrested embryo and the surrounding decidual tissue show increasing signs of rejection, the remains being resorbed or lost from the uterus between days 16 and 18 of gestation (Fozard *et al.*, 1980a).

The contragestational effects of eflornithine are associated with marked inhibition of the increase in deciduomal ODC activity which occurs during early gestation in the mouse (Fozard *et al.*, 1980a), rat (Reddy and Rukmini, 1981), and hamster (Galliani *et al.*, 1983). A causal relationship between inhibition of ODC and contragestational activity is suggested by data from experiments designed to localize the precise time during gestation where intervention with eflornithine results in pregnancy arrest. When mice and hamsters were treated on individual days during the early part of gestation, optimal effects were achieved following drug administration on day 8 (mouse—Fig. 3) and day 7 (hamster—Galliani *et al.*, 1983). In the mouse, the period is sufficiently circumscribed that two injections of eflornithine given within 6 hr early on day 8 of gestation result in pregnancy arrest whereas drug treatment on day 9 is completely without effect (Fig. 3). In both mice and hamsters, the period most susceptible to inhibition by eflornithine corresponds exactly to the peak change in ODC activity.

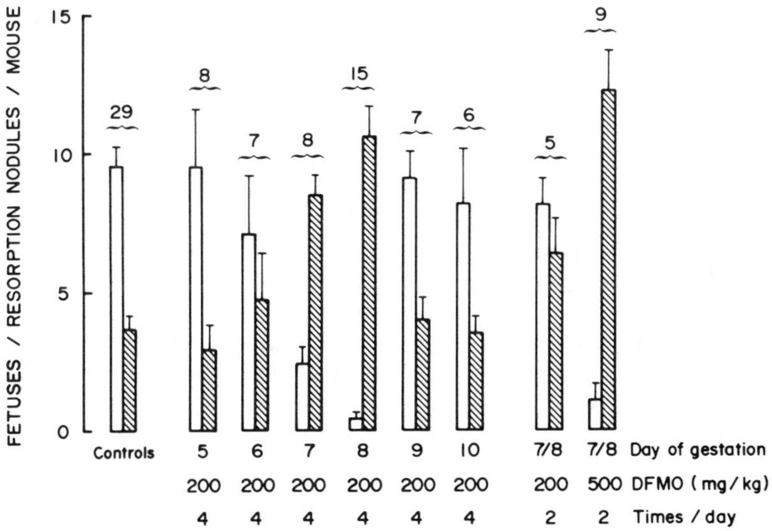

Fig. 3. Contragestational effects of eflornithine (DFMO) administered at various times during early gestation in the mouse. The histograms represent the number of fetuses (open columns) and resorption nodules (hatched columns) recovered from the uteri of mice autopsied on day 18 of gestation. Eflornithine was administered by subcutaneous injection every 6 hr starting at noon on the day indicated. The histograms on the furthest right represent the results obtained when mice received either 200 or 500 mg/kg of eflornithine as the last dose on day 7 and the first dose on day 8. Vertical bars are the standard errors of the mean values from the number of individual experiments indicated by the figures above the columns. Reproduced from Fozard *et al.* (1980b) with permission.

2. *The Mode of Action of Eflornithine as a Contragestational Agent*

As discussed above the contragestational effects of eflornithine are associated with inhibition of deciduomal ODC and optimal biological effects are closely coupled to the peak changes in ODC. The grounds for concluding that eflornithine acts exclusively as an inhibitor of ODC have been listed previously (Fozard *et al.*, 1980a). Chief among them are that eflornithine works through the principle of substrate-induced irreversible inhibition (Rando, 1974; Bey, 1978), that eflornithine does not inhibit other 1-carboxylyases (Bey, 1978; Metcalf *et al.*, 1978), and that eflornithine is both nontoxic (the acute oral LD_{50} exceeds 5000 mg/kg) and devoid of acute pharmacological activity (Fozard and Koch-Weser, 1982). Since a number of other highly selective, irreversible inhibitors of ODC are similarly contragestational (see Section III,B), there are sound reasons for concluding that the contragestational effects of eflornithine are a consequence of inhibition of ODC. A critical dependence specifically on the *de novo* synthesis of putrescine is implied by the observation that the contragestational effects of eflornithine in the mouse can be completely reversed by the concomitant administration of putrescine or the putrescine analog, α-monofluoromethyl-*trans*-dehydroputrescine (MDL 72197), which cannot be transformed to spermidine (see Table I). Similar results were reported by Mendez *et al.* (1983), who showed that the embryotoxic effects of eflornithine applied locally in rat uterus were prevented by the concomitant administration of putrescine. A key role for putrescine is also supported by the fact that the contragestational effects of eflornithine in the mouse could be partially prevented by aminoguanidine, an inhibitor of diamine oxidase (Fozard *et al.*, 1980c).

The question of the precise site in the body where inhibition of ODC results in contragestational effects may be posed. An interference with the secretion and/or actions of hormones involved in the maintenance of pregnancy could theoretically be the basis of the contragestational effects of eflornithine. However, treatment with eflornithine does not influence the reproductive cycle of nonpregnant mice (Fozard *et al.*, 1981). Moreover, a number of steps in the reproductive process which are clearly under hormonal control (Elger, 1972) are unaffected by eflornithine (Fozard *et al.*, 1980b; Genedani *et al.*, 1985). It therefore seems unlikely that interference with the hormonal control of pregnancy is an explanation of the contragestational effects of eflornithine. From the work of Mendez *et al.* (1983) it is clear that eflornithine can act locally on the uterus to produce pregnancy arrest. These authors administered eflornithine directly into one uterine horn on day 4 of gestation in amounts of 10, 50, and 100 μg. Embryonic development was interrupted selectively in the treated

TABLE I

Reversal of the Contragestational Effects of Eflornithine[a]

Treatment	Viable fetuses/ gravid female (means ± SEM)	n
Saline	10.6 ± 0.5	44
+ Putrescine (200)	9.9 ± 1.0	13
+ MDL 72197 (80)	10.5 ± 1.3	12
Eflornithine 100	5.2 ± 1.5	9
+ Putrescine (200)	10.4 ± 0.9	10
Eflornithine (200)	2.7 ±0.6	38
+ Putrescine (50)	2.2 ± 0.7	9
+ Putrescine (100)	3.5 ± 0.2	20
+ Putrescine (200)	7.0 ± 0.9	21
+ MDL 72197 (80)	9.1 ± 1.1	18

[a] Putrescine and α-monofluoromethyl-*trans*-dehydroputrescine (MDL 72197) were used. Mice were treated with the agents indicated (doses, in parentheses, in mg/kg) every 6 hr on day 8 of gestation. Eflornithine was given subcutaneously and putrescine and MDL 72197 were given intraperitoneally. Mice were autopsied on day 18 of gestation.

horn at each dose level and was complete following administration of 100 μg eflornithine. On this basis, the deciduoma would seem the likely site of action of eflornithine.

The question then arises as to whether the decidual tissue or the embryo is the site where inhibition of ODC is functionally important. The evidence is overwhelmingly in favor of an effect on the embryo. Thus, as noted above, histological evidence from the mouse shows clearly that decidualization takes place normally after implantation but that embryonic development fails to progress beyond a stage typical of days 6 to 7 of normal gestation (Fozard *et al.*, 1980a). The conclusion is supported by biochemical data obtained in both rats and mice showing changes in DNA, RNA, and protein contents of uterine "implants" consistent with normal decidualization but impaired embryonic growth (Fozard *et al.*, 1980b; Mendez *et al.*, 1983). Finally, marked inhibition of embryonic development can be produced by administration of eflornithine on day 8 of gestation (Fig. 3), at which time decidualization is essentially complete (O'Grady and Bell, 1977).

In terms of morphology, the time during mammalian gestation most

TABLE II

Comparison of Development Stages of Mouse, Rat, Rabbit, Sheep, Monkey, and Man[a]

Standard stages (Witschi)[b]	Identification of stages	Gestation age (days)					
		Mouse	Rat	Rabbit	Sheep	Monkey	Man
3	Four cells	1.5	3	0.5	1.5	1.5	2
7–8	Beginning of implantation	4–5	6	7	10	9	6.5
12	Primitive streak	7	8.5	6.5	13	19	19
16	13–20 Somite embryo	9	10.5	9	17	25	27
25	End of embryonic period	11	12.5	10	21	28	36
35–36	Birth	19	22	32	150	164	267

[a] Adapted from Table 26 in "Biology Data Handbook" *Fed. Am. Soc. Exp. Biol.*, Washington D.C., 1964.
[b] From Witschi (1956).

sensitive to inhibition by eflornithine lies between standard stages 12 and 16 (Witschi, 1956), with development failing to progress beyond the primitive streak stage (see Table II and Fozard and Part, 1983). This implies minimal disturbance of tubular egg transport, uterine spacing, or the early stages of implantation. Similarly, initial cleavage of the fertilized ovum and blastocyst formation would appear to take place normally following treatment with eflornithine. The failure to influence early cleavage accords with the observations of Alexandre (1978) and Zwierzchowski *et al.* (1986) that neither the competitive inhibitor of ODC, α-methylornithine, nor eflornithine affected the primary differentiation of mouse eggs *in vitro*. The observation is also consistent with observations made in a number of invertebrate species (Emanuelsson and Heby, 1978; Brachet *et al.*, 1978; Lundquist *et al.*, 1983) and the chick (Löwkvist *et al.*, 1980, 1983) that inhibition of ODC resulted in suppression of embryonic development at the primitive streak (or gastrulation) stage.

Many roles for the polyamines in the control of key steps in nucleic acid and protein synthesis have been suggested (Tabor and Tabor, 1976, 1984; Cohen, 1978; Jänne *et al.*, 1978; Maudsley, 1979; Williams-Ashman and Cannellakis, 1979; Pegg, 1986), most of which could be invoked to explain inhibition of the complex growth and differentiation process represented by early mammalian embryogenesis *in vivo*. Since no conclusive evidence is available in this context, detailed speculation as to the molecular basis of the mechanism of the contragestational effects of eflornithine would not be justified. However, the work of Heby and his colleagues has provided an important indication as to the key intracellular structure affected by eflornithine. Work carried out in the polychaete (*Ophryotrocha labronica*), a dipterian egg (*Caliphora erythrocephala* Meigen), and the chick suggests that inhibition of ODC and the resultant polyamine depletion results in suppression of the normal increase in nucleolar formation which occurs at gastrulation (Heby and Emanuelsson, 1978, 1981; Löwkvist *et al.*, 1980, 1983; Emanuelsson and Heby, 1983; Lundquist *et al.*, 1983). In the chick, structural changes noted in the nucleoli of eflornithine-treated embryos were closely similar to those observed in actinomycin-D-treated embryos (Löwkvist *et al.*, 1983), suggesting that polyamine deprivation results in impaired synthesis of rRNA. Löwkvist *et al.* (1983) speculate that the molecular basis for this may be the absolute requirement for polyamines of a protein kinase critically involved in rRNA gene expression in the nucleolus (Kuehn *et al.*, 1979; Atmar and Kuehn, 1981). However, results from recent experiments (Löwkvist *et al.*, 1985, 1986) indicate a more general suppression of synthesis of several RNA species in chick embryos, suggesting that eflornithine-induced gastrular arrest is due to a general inhibition of transcription.

B. Other Inhibitors of ODC

Prior to the advent of eflornithine, there were no reports in the literature of selective inhibitors of ODC having been used to inhibit mammalian gestation. Since ODC has a biological half-life of only 10–20 min and is readily induced, regulated by its reaction product and stabilized against degradation by competitive inhibitors (Maudsley, 1979), sustained *in vivo* inhibition of the enzyme proved difficult to achieve. *In vitro,* however, a number of competitive inhibitors of ODC have been used successfully to inhibit development of sea urchin eggs (α-hydrazinoornithine—Kusunoki and Yasumasu, 1978; α-methylornithine—Brachet *et al.*, 1978), *O. labronica* (Emanuelsson and Heby, 1978), and chick (Löwkvist *et al.*, 1980); in each case inhibition was reversed by concomitant addition of polyamines.

Several irreversible inhibitors of ODC, conceived and synthesized at the Merrell Dow Research Institute, Strasbourg, have been tested for their effects on gestation in the mouse using a standard procedure (Fozard and Part, 1983). With one exception, all the inhibitors of ODC suppressed gestation, although in some instances their potencies in this respect bore little relationship to their potencies to inhibit ODC *in vitro* (Table III). The single exception is α-monofluoromethyl-*trans*-dehydroputrescine (MDL 72197), which, although a powerful inhibitor of ODC *in vitro* (Table III)

TABLE III

Effects of Inhibitors of ODC on Gestation in Mice[a]

Treatment	MDL number	Dose (mg/kg sc)	Viable fetuses/ gravid female	n	Kinetic constants[b] K_i (μM)	$\tau_{1/2}$ (min)	Ref[c]
Saline	—	—	9.9 ± 0.5	63	—	—	—
Eflornithine	71782	50	7.8 ± 1.1	8	39	3.1	1
		100	5.7 ± 0.8	18			
		200	1.3 ± 0.4	54			
α-Monochloromethyl-ornithine	71660	500	6.0 ± 1.5	3	No saturation kinetics		1
α-Monofluoromethyl-ornithine	71919	50	9.1 ± 1.5	8	75	1.6	2
		100	12.7 ± 0.9	7			
		200	9.2 ± 1.5	10			
		400	5.4 ± 1.5	11			
(2*RS*,5*RS*)6-Heptyne-2,5-diamine	72110	50	7.5 ± 1.3	8	13	1.8	3
		100	6.6 ± 1.1	12			
		200	9.5 ± 1.5	11			
(2*R*,5*R*)6-Heptyne-2,5-diamine	72175	10	7.2 ± 2.0	6	3	1.7	3
		20	7.3 ± 1.1	12			
		200	0.1 ± 0.1	9			
(2*R*,5*S*)6-Heptyne-2,5-diamine	72174	10	7.1 ± 1.9	7	1200	2.2	4
		20	8.8 ± 1.2	12			
		200	9.8 ± 1.3	8			
α-Monofluoromethyl-*trans*-dehydroputrescine	72197	5	10.4 ± 1.2	7	42	0.2	5
		20	11.2 ± 0.6	15			
		80	9.6 ± 1.9	8			
α-Monofluoromethyl-*trans*-dehydroornithine	72246	5	9.3 ± 1.8	7	2.7	2.6	5
		20	10.6 ± 0.6	8			
		80	3.3 ± 1.0	8			
α-Monofluoromethyl-*trans*-dehydroornithine methylester	72403	50	12.0 ± 0.5	10	No direct inhibition		6
		200	1.0 ± 0.6	10			

[a] Mice were treated subcutaneously (sc) with the agents indicated every 6 hr on day 8 of gestation. Mice were autopsied on day 18 of gestation (for full details of experimental technique, see Fozard and Part, 1983).

[b] Obtained using partially purified rat liver ODC. For full details see individual references.

[c] References: 1, Metcalf *et al.* (1978); 2, Bey (1978); 3, Danzin *et al.* (1983); 4, Casara *et al.* (1985); 5, Bey *et al.* (1983); 6, Mamont *et al.* (1986).

and deciduomal ODC *in vivo* (J.R. Fozard, unpublished), is not able to arrest gestation. The data in Table I suggest MDL 72197 has "putrescine-like" activity which may compensate for the deficit in endogenous putrescine arising from inhibition of ODC. Worthy of emphasis is the comparison between (2 *R*,5 *R*)6-heptyne-2,5-diamine (MDL 72175) and its isomer, (2 *R*,5 *S*)6-heptyne-2,5-diamine (MDL 72174), which is only weakly active on ODC; only the active isomer (MDL 72175) inhibited gestation. Also noteworthy is the fact that the methyl ester of α-monofluoromethyl-*trans*-dehydroornithine (MDL 72403) inhibits gestation despite having no direct inhibitory effects on ODC per se. Presumably hydrolysis of the ester takes place *in vivo* and it is the free amino acid (MDL 72246) which causes the functional response (see Mamont *et al.*, 1986).

IV. CLINICAL SIGNIFICANCE OF THE CONTRAGESTATIONAL EFFECTS OF ODC INHIBITION

The interruption of gestation by inhibition of a critical enzyme step in early embryogenesis provides the basis for consideration of potent and selective inhibitors of ODC as an alternative and entirely novel approach to the interruption of early human pregnancy. From animal studies, the underlying concept behind the use of such compounds to terminate pregnancy is that treatment should take place during a key stage in embryogenesis when the embryo has an absolute requirement for putrescine formed *de novo* from ornithine. This would lie between standard stages 8 and 16 as defined by Witschi (1956) (see Table II), which for the human extends from the sixth or seventh day after fertilization to day 27. However, the more precise localization of the susceptible time of gestation of mice and hamsters to day 7 or 8 (see Section II,A,1) suggests that Witschi Standard Stages 12–16 would adequately cover this period. This corresponds to days 19–27 following fertilization in human gestation. Similar considerations could be used to define treatment schedules appropriate to these species in which the consequences of veterinary misalliance need to be avoided.

A means of termination of early pregnancy which does not surgically invade the cavity of the uterus, does not require hospitalization, and can be administered with the minimum of medical dexterity or supervision would be a valuable addition to the current family-planning therapeutic modalities (Editorial, 1983). The therapeutic potential of inhibitors of ODC as postcoital contraceptives, inducers of menstruation, or first-trimester abortifacients is clear; it remains for the future to determine whether the potential for clinical usefulness will translate into therapeutic reality.

V. CONCLUSION

Experiments with eflornithine, a highly selective, enzyme-activated inhibitor of ODC, have established that a marked increase in ODC leading to a rapid rise in putrescine concentrations is essential for the progression of early mammalian embryonic growth. Experiments in invertebrates and chick, mouse, rat, hamster, and rabbit have established that the halt to development occurs at gastrulation. Although the molecular mechanism of action of eflornithine as a contragestational agent remains to be established, a decrease in the accelerated formation of nucleoli which takes place at gastrulation seems to be a critical factor. These observations raise the possibility that potent and long-acting inhibitors of ODC could find clinical application as a novel approach to fertility control.

REFERENCES

Alexandre, H. (1978). *C.R. Acad. Sci. Paris* **286,** 1215–1217.

Andersson, A.-C., and Henningsson, S. (1981). *Acta Endocrinol. (Copenhagen.)* **98,** 456–463.

Andersson, A.-C., Henningsson, S., and Rosengren, E. (1978). *J. Physiol. (London)* **285,** 311–324.

Atmar, V.J., and Kuehn, G.D. (1981). *Proc. Natl. Acad. Sci. U.S.A.* **78,** 5518–5522.

Barkai, U., Shaham, Y., and Kraicer, P.F. (1978). *Life Sci.* **23,** 679–687.

Bell, J.M., Whitmore, W.L., and Slotkin, T.A. (1986). *Neuroscience* **17,** 399–407.

Bey, P. (1978). *In* "Enzyme-Activated Irreversible Inhibitors" (N. Seiler, M.J. Jung, and J. Koch-Weser, eds.), pp. 27–41. Elsevier, Amsterdam.

Bey, P., Gerhart, F., Van Dorsselaer, V., and Danzin, C. (1983). *J. Med. Chem.* **26,** 1551–1556.

Brachet, J., Mamont, P., Bolonkheŕe, M., Baltus, E., and Hanocq-Quertier, J. (1978). *C.R. Acad. Sci. Paris* **287,** 1289–1292.

Casara, P., Danzin, C., Metcalf, B., and Jung, M. (1985). *J. Chem. Soc. Perkin. Trans.* **1,** 2201–2207.

Cohen, S.S. (1978). *Nature (London)* **271,** 209–210.

Collawn, S.S., Rankin, J.C., Ledford, B.E., and Baggett, B. (1981). *Biol. Reprod.* **24,** 528–535.

Danzin, C., Casara, P., Claverie, N., Metcalf, B.W., and Jung, M.J. (1983). *Biochem. Biophys. Res. Commun.* **116,** 237–243.

Editorial (1983). *Lancet* **1,** 855–856.

Elger, W. (1972). *Ergeb. Physiol.* **67,** 69–168.

Emanuelsson, H., and Heby, O. (1978). *Proc. Natl. Acad. Sci. U.S.A.* **75,** 1039–1042.

Emanuelsson, H., and Heby, O. (1983). *Cell Differ.* **12,** 277–285.

Fozard, J.R., and Koch-Weser, J. (1982). *Trends Pharmacol. Sci.* **2,** 107–110.

Fozard, J.R., and Part, M.-L. (1983). *In* "Methods in Enzymology" (H. Tabor and C.W. Tabor, eds.), Vol. 94, pp. 213–216. Academic Press, New York.

Fozard, J.R., Part, M.-L., Prakash, N.J., Grove, J., Schechter, P.J., Sjoerdsma, A., and Koch-Weser, J. (1980a). *Science* **208,** 505–508.

Fozard, J.R., Part, M.-L., Prakash, N.J., and Grove, J. (1980b). *Eur. J. Pharmacol.* **65,** 379–391.

Fozard, J.R., Grove, J., Part, M.-L., and Prakash, N.J. (1980c). *Br. J. Pharmacol.* **69,** 335P–336P.

Fozard, J.R., Prakash, N.J., and Grove, J. (1981). *Life Sci.* **27,** 2277–2283.

Galliani, G., Columbo, G., and Luzzani, F. (1983). *Contraception* **28,** 159–170.

Genedani, S., Bernardi, M., and Bertolini, A. (1985). *Neurobehav. Toxicol. Teratol.* **7,** 57–65.

Guha, S.K., and Jänne, J. (1976). *Biochim. Biophys. Acta* **437,** 244–252.

Heald, J.P. (1979). *Biol. Reprod.* **20,** 1195–1199.

Heby, O. (1981). *Differentiation* **19,** 1–20.

Heby, O., and Emanuelsson, H. (1978). *Cell Tissue Res.* **194,** 103–114.

Heby, O., and Emanuelsson, H. (1981). *Med. Biol.* **59,** 417–422.

Hoshiai, H., Lin, Y.C., Loring. J.M., Perelle, B.A., and Villee, C.A. (1981). *Placenta* **2,** 105–116.

Huber, B.E., and Brown, N.A. (1982). *In Vitro* **18,** 599–605.

Jänne, J., Pösö, H., and Raina, A. (1978). *Biochim. Biophys. Acta* **473,** 241–293.

Kuehn, G.D., Affolter, H.-U., Atmar, V.J., Seebeck, T., Gubier, U., and Braun, R. (1979). *Proc. Natl. Acad. Sci. U.S.A.* **76,** 2541–2545.

Kusunoki, S., and Yasumasu, I. (1978). *Dev. Biol.* **67,** 336–345.

Löwkvist, B., Heby, O., and Emanuelsson, H. (1980). *J. Embryol. Exp. Morphol.* **60,** 83–92.

Löwkvist, B., Emanuelsson, H., and Heby, O. (1983). *Cell Differ.* **12,** 19–26.

Löwkvist, B., Emanuelsson, H., and Heby, O. (1985). *J. Exp. Zool.* **234,** 375–382.

Löwkvist, B., Emanuelsson, H., Egyházi, E., Sjöberg, J., Langström, E., and Heby, O. (1986). *Dev. Biol.* **116,** 291–301.

Lundgren, D.W., and Oka, T. (1978). *Am. J. Physiol.* **234,** E451–E456.

Lundquist, A., Löwkvist, B., Linden, M., and Heby, O. (1983). *Dev. Biol.* **95,** 253–259.

Luzzani, F., Colombo, G., and Galliani, G. (1982). *Life Sci.* **31,** 1553–1558.

Mamont, P.S., Danzin, C., Kolb, M., Gerhart, F., Bey, P., and Sjoerdsma, A. (1986). *Biochem. Pharmacol.* **35,** 159–165.

Manen, C.-A., Hadfield, M.G., and Russell, D.H. (1977). *Dev. Biol.* **57,** 454–459.

Maudsley, D.V. (1979). *Biochem. Pharmacol.* **28,** 153–161.

Maudsley, D.V., and Kobayaski, Y. (1977). *Biochem. Pharmacol.* **26,** 121–124.

Mendez, J.D., Diaz-Flores, M., Duran, G., and Hicks, J.J. (1983). *Contraception* **28,** 93–98.

Metcalf, B.W., Bey, P., Danzin, C., Jung, M.J., Casara, P., and Vevert, J.P. (1978). *J. Am. Chem. Soc.* **100,** 2551–2553.

O'Grady, J.E., and Bell, S.C. (1977). *In* "Development in Mammals" (M.H. Johnson, ed.), Vol. 1, pp. 165–243. North-Holland Publ., Amsterdam.

Pegg, A. (1986). *Biochem. J.* **234,** 249–262.

Pegg, A., and McCann, P.P. (1982). *Am. J. Physiol.* **243,** C212–C221.

Raina, A. (1963). *Acta Physiol. Scand. (Suppl.)* **218,** 1–81.

Rando, R. (1974). *Science* **185,** 320–324.

Reddy, P.R.K., and Rukmini, V. (1981). *Contraception* **24,** 215–220.

Russell, D.H., and McVicker, T.A. (1972). *Biochim. Biophys. Acta* **259,** 247–258.

Russell, D.H., and Snyder, S.H. (1968). *Proc. Natl. Acad. Sci. U.S.A.* **60,** 1420–1427.
Saunderson, R., and Heald, J.P. (1974). *J. Reprod. Fertil.* **39,** 141–143.
Seiler, N., Danzin, C., Prakash, N.J., and Koch-Weser, J. (1978). *In* "Enzyme-Activated Irreversible Inhibitors" (N. Seiler, M.J. Jung, and J. Koch-Weser, eds.), pp. 55–71. Elsevier, Amsterdam.
Slotkin, T.A., Seidler, F.J., Whitmore, W.L., Weigel, S.J., Slepetis, R.J., Lerea, L., Trepanier, P.A., and Bartoleme, J. (1983). *Int. J. Dev. Neurosci.* **1,** 113–127.
Tabor, C.W., and Tabor, H. (1976). *Annu. Rev. Biochem.* **45,** 285–386.
Tabor, C.W., and Tabor, H. (1984). *Annu. Rev. Biochem.* **53,** 749–790.
Williams, P.G., and McAnulty, P.A. (1976). *J. Embryol. Exp. Morphol.* **35,** 545–559.
Williams-Ashman, H.G., and Canellakis, Z.N. (1979). *Perspect. Biol. Med.* **22,** 421–453.
Williams-Ashman, H.G., Jänne, J., Coppac, G.L., Geroch, M.E., and Schenone, A. (1972). *Adv. Enzyme Regul.* **10,** 225–245.
Witschi, E. (1956). "The Development of Vertebrates." Sauders, Philadelphia.
Zwierzchowski, L., Czlonkowska, M., and Guszkiewicz, A. (1986). *J. Reprod. Fertil.* **76,** 115–121.

10

Modulation of Antineoplastic Drug Action by Inhibitors of Polyamine Biosynthesis

CARL W. PORTER* AND JUHANI JÄNNE†

** Grace Cancer Drug Center*
Roswell Park Memorial Institute
666 Elm Street
Buffalo, New York 14263

† Department of Biochemistry
University of Helsinki
SF-00170 Helsinki 17, Finland

I. INTRODUCTION

Relative to antimicrobial chemotherapy, the use of single agents in antineoplastic chemotherapy has been only marginally successful. With the exception of cures of Burkitt's lymphoma using cyclophosphamide and of choriocarcinoma using methotrexate, complete remissions of lasting duration are infrequently and inconsistently obtained among the 100 or so other malignancies. Our general inability to develop highly selective drugs has weighed heavily in this dilemma. Cytotoxic agents abound but their usefulness extends only so far as their specificity for tumor cells or tissues. Whereas near-absolute selectivity has been attained with the antimicrobials, similar achievements have not been forthcoming with antineoplastics. Tumor cells derive genetically from their host tissue and tend, therefore, to express nearly identical metabolic and antigenic phenotypes. Accordingly, differences targeted for chemotherapeutic exploitation usually amount to distinctions in degree and invariably lead to untoward host effects.

During the past 20 years, the limited success of single-agent chemotherapy has been largely offset by advances made with drug combinations. A significant percentage of complete remissions was initially observed in hematogenous malignancies such as acute lymphocytic leukemia using the VAMP regimen [*v*incristine, *a*rabinosylcytosine (ara-C), *m*ethotrexate, and *p*rednisone] conceived by Freireich *et al.* (1964) and in Hodgkins disease using the MOPP regimen (nitrogen *m*ustard, *O*ncovin (vincristine), *p*rednisone, and *p*rocarbazine) devised by DeVita *et al.* (1972). Impressive response rates were also obtained in solid tumors such as advanced breast cancer with the five-drug regimen of Cooper (1969). Today, combination chemotherapy, or polychemotherapy as it is sometimes referred to, is known to be superior to single-drug chemotherapy in incidence of use, in degree of response, and in prolongation of survival. While it is encouraging to consider that about a dozen cancers, in their disseminated form, may now be treated with curative intent using multidrug regimens (Table I; Berger, 1986; Wittes and Goldin, 1986), the most com-

TABLE I

Successful Drug Combinations Based Mainly on Different Mechanisms of Drug Action and/or Different Host Toxicities[a]

Acronym	Regimen	Disease	Reference
VAMP	Vincristine, amethopterin mercaptopurine, prednisone	Acute lymphocytic leukemia	Freireich *et al.* (1964)
COAP	Cyclophosphamide, arabinosylcytosine, oncovin, prednisone	Acute myelocytic leukemia	Bodey *et al.* (1970)
MOPP	Mechlorethamine, oncovin, prednisone, procarbazine	Hodgkin's disease	DeVita *et al.* (1970)
ABVD	Adriamycin, bleomycin, vinblastine, dacarbazine	Hodgkin's disease	Santoro *et al.* (1982)
CHOP	Cyclophosphamide, hydroxydaunomycin, oncovin, prednisone	Histiocytic lymphoma	Armitage *et al.* (1980)
CAF	Cyclophosphamide, adriamycin, fluorouracil	Breast cancer	Smalley *et al.* (1977)
FAM	Fluorouracil, adriamycin, mitomycin	Gastric cancer	Macdonald *et al.* (1980)
PVB	Platinum, vinblastine, bleomycin	Testicular cancer	Einhorn and Donahue (1977)

[a] From a recent review by Berger (1986).

mon and life-threatening malignancies (i.e., lung and colon) are largely unaffected by any form of chemotherapy.

While early drug combinations were selected on empirical considerations relevant to single-drug effectiveness and host toxicity, present day drug combinations are more likely to be made on the basis of mechanistic rationale. Accordingly, those experimental agents for which the mode of action is well understood are best suited for such studies. In this respect, α-difluoromethylornithine (DFMO) is an ideal candidate. Consistent with its initial design as an irreversible inhibitor of ornithine decarboxylase (Metcalf *et al.*, 1978), DFMO has thus far withstood the test of time as a highly specific inhibitor of polyamine biosynthesis. Several other characteristics of DFMO indicating its suitability for combination chemotherapy are presented in Table II and will be considered separately throughout the text. Because DFMO is the only specific polyamine inhibitor undergoing clinical evaluation at the present time, the main theme of this review will be presented from the perspective of DFMO as a modulator of antineoplastic drug activity. Merrell Dow Research Institute has recently synthesized new and more potent polyamine inhibitors modeled after DFMO (Sjoerdsma and Schecter, 1984) which are likely to follow the same clinical path as DFMO. Should they prove to be more clinically effective than DFMO as single agents, they undoubtedly, will also be more useful in combination with other drugs. At a minimum, such inhibitors may find clinical application in retardation of tumor regrowth following exposure to cytotoxic agents or to radiation.

Although the present chapter is restricted in its breadth to DFMO in combination with antineoplastic agents, the reader should be aware that DFMO has also been used to achieve impressive therapeutic effects in combination with antiparasitic agents. These are reviewed in Chapters 15

Table II

Characteristics of DFMO Favoring Its Utilization in Combination Chemotherapy

1.	Highly specific and apparently singular mode of action
2.	Modulates a major metabolic pathway in proliferating cells
3.	Growth inhibitory as a single agent
4.	Wide tumor spectrum of effectiveness *in vitro*
5.	Induces increased cellular uptake of certain cationic drugs
6.	Induces specific cell cycle changes in certain tumors
7.	Minimally host toxic
8.	Displays an atypical profile of dose-limiting toxicities
9.	Innate resistance is unusual
10.	Immunoenhancement effects

and 16 of this volume. Readers interested in a more in-depth consideration of the current status of combination chemotherapy in general are referred to a well-focused and timely review recently published by Wittes and Goldin (1986).

II. GOALS OF COMBINATION CHEMOTHERAPY

The progress made in the clinical management of cancer with combination chemotherapy is directly related to the ability of those drug combinations to achieve one or more of the following goals (Table III): (1) to optimize tumor cell kill, (2) to minimize host toxicity (thereby increasing effective drug selectivity), (3) to provide broader range coverage of innately resistant cells in the typically heterogeneous tumor cell population, and (4) to reduce the incidence of acquired resistance within a treated tumor cell population.

The first of these goals is heavily dependent upon the guiding principles for cell kill and chemotherapy established by Skipper *et al.* (1964) from studies with murine L1210 leukemia. One particularly relevant principle of this hypothesis states that a given drug dosage will destroy a constant fraction, *not number,* of cells regardless of the tumor burden present at the time of therapy. Following first-order kinetics, a tumor of 10^9 cells might be reduced to 10^4 cells or a tumor of 10^7 cells to 100 cells by drug therapy at the same given dose. The relationship between tumor cell kill and dose appears to be log/linear (DeVita, 1982), so that if the log kill dose is doubled, the cell kill increases to two logs or 99% of the cells. On the basis of these considerations and because the growth fraction of a tumor (i.e., those cells probably responsive to chemotherapy) is inversely related to tumor size (Frindel *et al.,* 1970), it is obviously desirable to initiate chemotherapy with the smallest tumor cell number present and to administer the highest tolerated drug dose. As an alternative to the latter, more than one drug might be used to significantly increase the fraction of cells killed. The combination of two drugs which are independently capable of producing a log or 90% cell kill is additive when a 2 log or 99% kill is

TABLE III
Objectives of Drug Combinations in Anticancer Chemotherapy

1. Optimizing tumor cell destruction
2. Minimizing host toxicity (thereby increasing drug selectivity)
3. Maximizing coverage of heterogenous cell types
4. Limiting the development of acquired resistance

obtained, synergistic when more than a 2 log kill takes place, and antagonistic when less than a single log or 90% kill occurs.

Maximizing tumor cell kill serves little purpose if, at the same time, the dose-limiting toxicity of either drug is also intensified. Thus, even additive drug interactions can be particularly advantageous when the agents express toxicity toward different normal host tissues. This allows destruction of the tumor cells to be selectively enhanced with no significant increase in damage to any single host tissue–toxicity is diffused over several host tissues. This is a major consideration in the COAP regimen used in treating acute myelocytic leukemia (Bodey *et al.*, 1970), where *c*yclophosphamide and *a*rabinosylcytosine are differently myelosuppressive, *O*ncovin is neurologically toxic, and *p*rednisone induces an endocrine imbalance. Relative to other anticancer agents, DFMO-induced host toxicities, which include mile thrombocytopenia, anorexia, and auditory acuity loss (Abeloff *et al.*, 1984; Maddox *et al.*, 1985), are minimal and atypical among known anticancer agents. On this basis, therefore, DFMO qualifies as a very suitable agent for combination studies intent on minimizing host toxicities.

When the precise mode of action of an anticancer agent is known, selectivity may be increased with agents which prevent or reverse the host tissue toxicity of the cytotoxic agent. The use of citrovorum factor (Leukovorin) to rescue host tissues but not the tumor tissues from the toxicity of high-dose methotrexate is one such example (Goldin *et al.*, 1974). Because host toxicity is minimal with DFMO it is unlikely that such strategies will evolve for its use, although, if necessary, putrescine or spermidine could possibly be used to prevent or reverse toxicity so long as tumor tissues were not similarly benefited. Such a "polyamine rescue strategy" has been considered (Kallio *et al.*, 1983) with another more toxic inhibitor of polyamine biosynthesis, methylglyoxal bis(guanylhydrazone) (MGBG, also known as methyl-GAG or mitoguazone).

The last two goals of combination chemotherapy deal with the problems of drug resistance and selectivity, which together comprise the two major obstacles to successful anticancer chemotherapy. Drug resistance can be classified as being innate or acquired. Innate resistance exists naturally among tumor cell populations and is not dependent on prior drug exposure. Acquired resistance takes place once chemotherapy is initiated and is secondary to genetic or epigenetic phenomena. With either form of resistance, drug-induced reductions in populations of sensitive tumor cells exert a selection pressure which invites repopulation by drug-resistant cell types. Given these considerations, it is not unexpected that optimal responses have been obtained with mixtures of drugs having different mechanisms of action.

To a significant extent, innate resistance in solid tumors is attributable to the presence of a slowly cycling or noncycling (G_0) cells. Most of the available anticancer agents kill cells by altering DNA synthesis and/or function and, as a consequence, affect mainly the cycling cells of a tumor population. Quiescent cells, by contrast, are usually refractory to these drugs (as well as to DFMO). Following elimination of proliferating cells, these resting cells are recruited into cycle. In addition, it has been shown repeatedly that the cycling cells themselves vary widely in their sensitivity to cytotoxic agents. Barranco *et al.* (1972) showed that four permanent cell lines derived from the same human melanoma nodule displayed very different sensitivities to ara-C. Similarly, Talmadge *et al.* (1984) reported that murine B_{16} melanoma cells cloned in animals could be separated into cell populations with varying sensitivities to different chemotherapeutic agents. Tumor cell heterogeneity, whether defined by drug sensitivity or a wide variety of other parameters, generally exceeds that exhibited by normal tissues and appears to be due to a genetic instability associated with the neoplastic state. The influences of such heterogeneity with respect to drug sensitivity may be favorably modulated with drug combinations. Cell populations inherently resistant to one agent are likely to be sensitive to another, particularly if its mode of action is substantially different.

Acquired resistance may also be minimized by combination chemotherapy. The phenomenon is wholly dependent on mutagenic events which, through a variety of mechanisms, impart a drug-insensitive phenotype to emergent cells. The frequency with which these events occur varies according to the drug, with values in the range of 10^{-6} to 10^{-9} being average for most antineoplastics (Ling, 1985). Since an acute leukemia in relapse may comprise a tumor burden of up to 10^{12} cells ($\sim$1 kg), the probability of it developing a significant fraction of multiply resistant cells is significant. When two or more drugs are used, the probability of mutations resulting in double or triple drug resistance becomes much less likely. It has been noted, however, that multiple resistance seems to occur at a greater incidence than would be expected from the product of the mutation frequency for each drug. To some extent, this may be because cells resistant to one drug grow to a significant number and then, through mutation, became resistant to the second drug. Alternatively, certain forms of resistance, such as those involving the plasma membrane P-glycoprotein (P for permeability), may confer pleiotropic (multi-) drug resistance to several agents having seemingly unrelated modes of action (Ling, 1975).

Murine cells with acquired resistance to DFMO have been isolated previously and shown to contain an increased ornithine decarboxylase activity secondary to overproduction by gene amplification (Alhonen-

Hongisto *et al.*, 1985a,d; Pohjanpelto *et al.*, 1985; McConlogue and Coffino, 1983), to an apparent change in enzyme half-life (Choi and Scheffler, 1983), or to changes related to the enzyme catalytic site (Pohjanpelto *et al.*, 1985). Although it is generally agreed that human cell lines, including tumor-derived lines, are genetically more stable than murine cells, it is apparent that human tumor cells could develop resistance to DFMO by increasing the gene dosage of ornithine decarboxylase and overproducing the enzyme (Leinonen *et al.*, 1986). This detracts considerably from the potential of DFMO as a single agent but, because of the considerations raised above, has less importance in its suitability for drug-combination studies.

III. APPROACHES TO COMBINATION CHEMOTHERAPY

While early clinical trials with drug combinations were based on relatively straightforward principles involving drug–tumor specificity and host toxicity profiles, modern day strategies have become extremely complex, encompassing biochemical, pharmacological, toxicological, cytokinetic, and immunological considerations. These and various other approaches have been thoroughly reviewed in several monographs (Sartorelli and Creasy, 1982; Mihich and Grindey, 1977; Keiser and Capizzi, 1977). In the present discussion, consideration will be given to those approaches which seem applicable, in theory at least, to polyamine inhibitors, particularly DFMO. More detailed accounts of specific types of drug interactions will be provided in the following section on targeted combinations.

By the biomedical approach, drug combinations are selected to focus a multiple drug assault on a specific metabolic pathway, although cell functions, organelles, or structures can also be targeted. The basic strategies have developed out of attempts to affect DNA biosynthesis and integrity but were applicable to polyamine biosynthesis as well. The first strategy is the *sequential blockade* (Potter, 1951) and involves the inhibition of two sites along the same biosynthetic pathway. The synergism that occurs in the growth of L1210 leukemia with the combination of ara-C, an inhibitor of DNA polymerase, and hydroxyurea, an inhibitor of ribonucleotide reductase, results from a sequential blockade of DNA synthesis (Schabel, 1975). The combination of an inhibitor of ornithine decarboxylase (i.e., DFMO) with one of spermidine synthase provides an example of the sequential blockade with polyamine inhibitors (Pegg *et al.*, 1982b). The second strategy, termed *concurrent blockade* (Elion *et al.*, 1954), relates to inhibition of two sites along alternative pathways involved in the synthesis of the same product. While there are no clear examples of this strategy in inhibition of DNA biosynthesis, the use of the ornithine decarboxylase inhibitor, α-methylornithine, in combination with MGBG, an inhibitor of *S*-adenosylmethionine (AdoMet) decarboxylase (Williams-Ashman and Schenone, 1972), represents an early application in polyamine biosynthesis (Morris *et al.*, 1977). Recently, Porter *et al.* (1984b) have applied this strategy in combining DFMO with the methionine-analog inhibitor of AdoMet synthetase, L-2-amino-4-methoxy-*cis*-but-3-enoic acid (L-*cis*-AMB). If one considers the catabolic conversion of spermine to spermidine via the "intraconversion pathway" (Pegg *et al.*, 1986), a concurrent blockade could be established with DFMO and certain butanediamine inhibitors of polyamine oxidase (Bolkenius *et al.*, 1985), presuming, of course, that spermidine, not spermine, is essential for cell proliferation. In the *complementary blockade* strategy, one drug is used to affect macromolecular structure and a second to inhibit its biosynthesis. Typically, an antimetabolite such as 5-fluorouracil (5-FU) is paired with a DNA-damaging agent such as 1,3-bis(2-chloroethyl)-1-nitrosourea (BCNU). Since the role of newly synthesized polyamines in the context of cell proliferation is uncertain, it is difficult to devise an analogous strategy using DFMO. The use of DFMO to enhance BCNU-induced cytotoxicity and DNA damage (Hung *et al.*, 1981; Cavanaugh *et al.*, 1984; Seidenfeld and Komar, 1985) is similar but the drugs are applied in reversed sequence. Presumably, the polyamine biosynthetic block imposed by DFMO leads to cellular depletion of spermidine and renders DNA, which is otherwise thought to be stabilized by polyamines, more vulnerable to BCNU.

A major consideration in the pharmacological approach to combination chemotherapy involves the delivery of the cytotoxic agent across various membranes to the appropriate target site. Certain drugs can facilitate this process by increasing the cellular uptake of another—an event which invariably leads to enhanced cytotoxicity. Vincristine increases the uptake and toxicity of methotrexate by interfering with the efflux mechanisms of the latter (Creasy, 1975). Likewise, the calcium channel blockers, such as verapamil, are thought to overcome membrane-mediated multidrug resistance to a variety of antineoplastic agents through uncertain mechanisms governing intracellular drug concentrations (Tsuruo *et al.*, 1983). In the case of polyamine inhibitors, it is well known that DFMO increases the uptake of polyamines and agents that utilize polyamine transport mechanisms (Alhonen-Hongisto *et al.*, 1980; Seppänen *et al.*, 1981a,b). Foremost among these is MGBG and its analogs, but cytotoxic polyamine analogs (Porter *et al.*, 1985) would also be similarly affected. Evidence is accumulating that DFMO, like vincristine with methotrexate, acts by decreasing the efflux of MGBG.

Altering drug metabolism is also among the pharmacological approaches. The antibiotic 2-deoxycoformycin—an inhibitor of adenosine deaminase (Woo *et al.,* 1974)—has no antitumor activity itself but has proved useful in limiting the deamination of adenosine analogs, such as arabinosyladenosine (ara-A) and 3-deoxyadenosine (cordycepin), thereby increasing their therapeutic effectiveness. Since various serum oxidases are particularly active in metabolizing polyamines, it is useful to combine polyamine analogs intended as antiproliferative agents with aminoguanidine, an inhibitor of serum diamine oxidase (Shore and Cohn, 1960). This represents a routine strategy in tissue culture studies utilizing polyamines or polyamine analogs (Porter *et al.,* 1985), but its general usefulness *in vivo* has yet to be demonstrated.

Certain biochemical phenomena, particularly those associated with the fabrication and replication of DNA, occur at specific phases of the cell cycle. According to the cytokinetic approach, drug combinations are devised with the purpose of using one drug to influence the cycling characteristics of a tumor cell population in order to optimize the cytotoxic effects of a second drug, thereby achieving a measure of drug cooperativity. Ideally, it should be possible, with the use of cell cycle inhibitors, to partially synchronize the proliferating cells with a tumor such as at the G_1/S interface, and then, with the proper scheduling, to assault these cells as they progress to the S phase with an S-phase-specific agent. It has been shown with L1210 leukemia that ara-C followed by the alkylating agent BCNU is an effective regimen in treating L1210 leukemia (Tyrer *et al.,* 1967). By design, ara-C is used to kill cells in S phase and produce a G_1/S blockade which renders the cells maximally sensitive to BCNU. Such also was the rationale in combining DFMO with ara-C by Sunkara *et al.* (1980, 1981). The strategy becomes particularly appealing if it can be demonstrated that the syncronizing agent affects normal cell cycling differently from that of tumor cells, as has been thought to be the case with MGBG (Rupniak and Paul, 1978) and DFMO (Sunkara *et al.,* 1981). It was reported that "normal cell" lines treated with polyamine inhibitors are arrested in G_1 phase while transformed cell lines are halted in S phase. A major limitation of the aforementioned strategy is that, in most solid tumors, only a fraction of the cells are cycling and the remainder are either incapable of cycling or exist as resting cells which may be stimulated to proliferate when the tumor cell population is altered. Schabel (1969) has suggested the use of alkylating agent to destroy the cycling cells of a tumor followed by an antimetabolite to halt G_0 cells as they mobilize for proliferation. A second problem is that, while it is relatively easy to achieve cell synchronization *in vitro,* it is exceedingly more difficult in animals.

The toxicological approach has already been briefly considered previously under goals of combination chemotherapy. Two useful strategies have emerged: (1) combining drugs which have nonoverlapping tissue toxicity profiles, thereby diffusing host toxicity while increasing tumor cell kill, and (2) preventing or reversing the toxicity manifested by an active anticancer agent in normal tissues with a second compound while not reducing toxic effects to the tumor. Examples of both strategies were presented earlier. With respect to polyamines, DFMO is ideally suited to the former strategy since, relative to most anticancer agents, it is minimally toxic and has a rather unusual toxicity profile which includes mild thrombocytopenia, anorexia, and ototoxicity, all of which are reversible (Abeloff *et al.*, 1984; Maddox *et al.*, 1985; Meyskins *et al.*, 1986). Because DFMO is relatively atoxic, it seems unnecessary to consider means to prevent or reverse host tissue toxicities. It is well known from tissue culture studies that putrescine or spermidine can prevent or reverse DFMO-induced cytostasis. Whether this can be achieved selectively in host tissues has not yet been studied, although it has been attempted, with minimal success, in the case of MGBG toxicity (Mihich, 1963a,b; Kallio *et al.*, 1983).

The final and most recent approach to be applied to combination chemotherapy is immunological. Increasingly it has become recognized that various components of the immune system play a major role in tumor cell destruction (Ehrke and Mihich, 1985a,b). Because a majority of anticancer agents are immunosuppressive, it is worthwhile considering the probable effect of drug combinations on the specific elements of the immune system and, as in the case of host tissue toxicity, how to minimize untoward immunosuppressive effects. Interestingly, we have found that DFMO has very mild immunosuppressive properties and in fact, appears to stimulate certain subsets of splenic lymphocytes known to participate in antitumor activity (Ehrke *et al.*, 1986). Similarly, MGBG markedly inhibits the development of antibody-forming cells in mice but not the function of cytotoxic effector cells and phagocytes (Bennett *et al.*, 1977).

Given the successes of combination chemotherapy over the past two decades, it is not surprising that, during their development as potential antineoplastics, new agents routinely undergo early evaluation for their ability to potentiate the activity of clinically established agents. *In vitro* screens based on isobologram analyses have been devised for such testing and sophisticated means for data interpretation based on mathematical models have evolved (Chou and Talalay, 1984). The difficulty, however, comes in translating *in vitro* findings obtained with such models to *in vivo* model systems, where pharmacokinetic parameters such as drug clearance, partitioning, metabolism, etc., loom as critical factors in achieving appropriate drug interactions. Even once favorable drug cooperativity is identified *in vivo,* there is little assurance, without accompanying biochemical or pharmacological studies, that it is taking place via the mechanism determined by *in vitro* studies.

IV. DFMO EFFECTS ON POLYAMINE BIOSYNTHESIS

Owing to its high specificity of action, its effectiveness in depleting putrescine and spermidine, and the generality of its effect in various *in vitro* and *in vivo* systems, DFMO has achieved benchmark status among inhibitors of polyamine biosynthesis. While it has been extremely valuable in delineating the involvement of polyamines in various cellular processes, its usefulness as a single agent in cancer chemotherapy has been largely disappointing. Unlike most experimental anticancer agents, however, the limitations of DFMO have not been in the realm of intolerable host toxicities but rather in its general inability to effect a cytotoxic antitumor response. Typically, DFMO induces a depletion of putrescine and spermidine pools, an increase in those of spermine, and a subsequent cessation of cell growth (Mamont *et al.,* 1978). It has been speculated that this incomplete depletion of total polyamines might be responsible for its cytostatic, as opposed to cytotoxic, effects. This remains to be demonstrated and efforts are underway in numerous laboratories to develop inhibitors (or inhibitor combinations) capable of total polyamine depletion.

As will be detailed in Section IV, a meaningful criterion in selecting a drug for use in combination strategies is a thorough knowledge of its mode of action. It has been said that the longer a specific inhibitor is studied, the less specific it becomes. Such has not been the case with DFMO thus far. Since its introduction into the literature in 1978 as an irreversible inhibitor of ornithine decarboxylase (Mamont *et al.,* 1978), there have been no published indications of other sites of action. Nonetheless, by virtue of regulatory mechanisms associated with the polyamine pathway and through interrelationships between polyamine biosynthesis and intermediate metabolism, a number of biochemical responses secondary to depletion of putrescine and spermidine have been reported. Because several of these may be relevant to the other agents used in combination with DFMO, it is important that they be briefly reviewed. Foremost among these is the significant increase in AdoMet decarboxylase consequential to spermidine depletion. This, in part, may be responsible for the rise (as opposed to decline) of spermine pools associated with DFMO treatment. Concomitant with the increase in the enzyme is a substantial (i.e., 300- to 400-fold) increase in decarboxylated AdoMet (Pegg *et al.,* 1982a; Mamont *et al.,* 1982). Whether this alters transmethylation reactions by depleting available AdoMet remains to be determined, but it has recently been shown that decarboxylated AdoMet may serve as a substrate for N-acetylation, leading to the appearance of this unusual derivative in DFMO-treated cells (Wagner *et al.*, 1985; Pegg *et al.*, 1986). Contrary to expectations, pooling of decarboxylated AdoMet does not decrease cellular ADP or ATP pools. In fact, Heby *et al.* (1984) have shown that polyamine inhibition causes a significant increase in the cellular content of purine and pyrimidine ribonucleotides in general, possibly reflecting the inhibition of cell division and DNA biosynthesis.

While the direct biochemical effects of DFMO per se are fairly well known, the cellular consequences of DFMO-induced polyamine depletion are not (except at the phenomenological level, such as inhibition of cell growth, etc.) This is because the precise function of polyamines has yet to be determined. There exists a large body of literature to indicate that, on the basis of electrostatic considerations and experimental evidence, they are involved in interactions with nucleic acids. The combination studies described in Section V,D with DFMO and the DNA alkylating agent BCNU provide the best available indication that such interactions may occur in the context of an intact cell (Tofilon *et al.*, 1983; Cavanaugh *et al.*, 1984). More detailed biochemical studies using DFMO and other polyamine inhibitors and/or analogs will undoubtedly lend insight into the cellular role of polyamines. As such information becomes available, it will become possible to design more effective drug combinations.

A final consideration which bears on the potential of DFMO in drug combination studies regards the concentrations and doses required for its utilization. Because it penetrates cells ineffectively via a nonactive process (Erwin and Pegg, 1982), it is typically used at millimolar concentrations in *in vitro* systems. In mice, it is usually administered as a 2% solution in drinking water which, assuming the consumption of 3 ml per day, is equivalent to about 3 g/kg/day or 210 g for a 70-kg man. Even if it is assumed that only 50% of the DFMO is absorbed into the bloodstream, the amount (~100 g/day) is still far and above what can be realistically achieved in patients via intravenous injection or infusion. Clearly, on this basis alone, animal findings cannot be readily translated into clinical protocols. Thus, it is worthwhile to view drug combinations involving DFMO as means to lower its effective clinical dose into a realistic range. Alternatively, Merrell Dow Research Institute (Sjoerdsma and Schecter, 1984) and others (reviewed by Porter and Sufrin, 1986) have been working on the development of new and more effective polyamine inhibitors and analogs. Already, some of these have been described (Bey *et al.*, 1983; Danzin *et al.*, 1982) and certain ones have now been found to be effective in inhibiting polyamine biosynthesis in animals at doses as low as 25 mg/

kg/day (Mamont *et al.*, 1986). Most likely, these compounds, particularly when used in combination with other agents, will be of more practical usefulness than DFMO in future clinical studies.

V. TARGETED DRUG INTERACTIONS USING POLYAMINE INHIBITORS

A. Polyamine Biosynthesis (Table IV)

The use of more than one polyamine inhibitor as a means to achieve near-total blockade of the biosynthetic pathway was first applied by Morris *et al.* (1977). Having determined previously (Fillingame *et al.*, 1975) that MGBG-induced decreases in spermidine and spermine pools of activated lymphocytes were complicated by marked increases in putrescine pools, the authors sought to minimize putrescine biosynthesis. Thus, cells were treated concomitantly with the reversible inhibitor of ornithine decarboxylase, α-methylornithine, and MGBG. The combination accentuated polyamine depletion and inhibition of DNA synthesis but had no effect on total protein or RNA synthesis (Morris *et al.*, 1977; Seyfreid and Morris, 1979). Ultimately, their findings contributed to the growing realization at that time that polyamines were critically involved in and essential for cell proliferation.

Interest in the obvious potential of the DFMO and MGBG combination as an antiproliferative strategy developed rapidly in the early 1980s at the preclinical and clinical levels in United States and Europe largely as a result of findings made by Jänne and colleagues (Alhonen-Hongisto *et al.*, 1980; Seppänen *et al.*, 1981a,b). The numerous reports that appeared in this area can be divided into two categories that differ according to their mechanistic objectives. In the first, championed by the Finnish group, the drugs are administered sequentially (i.e., DFMO followed by MGBG) with the intent of increasing the uptake and cytotoxicity of MGBG as a consequence of DFMO-induced polyamine depletion (Alhonen-Hongisto *et al.*, 1980; Seppänen *et al.*, 1981a,b). Though since determined to be of limited clinical usefulness, this particular strategy has attracted the greatest worldwide interest of any DFMO–drug combination thus far studied. In the second category, championed largely by investigators at Memorial Sloan Kettering, the DFMO and MGBG are administered concomitantly with the intent of achieving a more effective and complete inhibition of polyamine biosynthesis and, hence, cell proliferation. Eventually, leads uncovered by this approach were pursued further by utilizing the increased drug transport concept. The latter approach is reviewed below while the former is considered in Section V,B.

TABLE IV

The Effects of DFMO in Combination with other Inhibitors of Polyamine Biosynthesis[a]

Drug	Cell system (*in vitro* unless indicated)	Growth inhibition[b]	Reference
AdoDato	SV-3T3 cells	Increased	Pegg *et al.* (1982b)
L-*cis*-AMB (methionine analog)	L1210 leukemia	None	Porter *et al.* (1984b)
Cycloleucine (methionine analog)	L1210 leukemia	None	Porter *et al.* (1984b)
MGBG	Renal adenocarcinoma (*in vivo*)	None	Duzendorfer *et al.* (1983a)
MGBG	Murine renal adenocarcinoma (*in vivo*)	Increased	Herr *et al.* (1984a)
MGBG	Dunning R-3327-G prostatic adenocarcinoma (*in vivo*)	Increased	Duzendorfer *et al.* (1983b); Herr and Kleinert (1984); Herr *et al.* (1984b)
MGBG	L1210 leukemia (*in vivo*)	Increased	Burchenal *et al.* (1981); Warrell and Burchenal (1983)
MGBG and stilbamadine	L1210 leukemia (*in vivo*)	Increased (60% survivors)	Burchenal *et al.* (1981); Warrell and Burchenal (1983)
MGBG and mitomycin C	Human stomach carcinoma in nude mice (*in vivo*)	Increased	Fujimoto *et al.* (1985)
MGBG	Phase I trial (22 patients)	None, toxicity increased	Warrell *et al.* (1983)
N^4-Spermidine analogs	L1210 leukemia	Decreased	Porter *et al.* (1985)
N^1,N^8-Spermidine analogs	L1210 leukemia	None	Porter *et al.* (1985)

[a] See also Table IV.

[b] Increased = additive and/or synergistic drug interactions; none = subadditive or no effect; decreased = antagonistic.

It should be noted that while it was the intention of a large proportion of the studies reviewed in this section to effect growth inhibition through a combined inhibition of polyamine biosynthesis by DFMO and MGBG, it is highly unlikely that this was actually the mechanism responsible. In no instance is there sufficient indication from polyamine pool-size determinations that such perturbations might be responsible. In fact, for reasons to be discussed in Section V,B, the combination of DFMO with MGBG makes such a possibility less likely than with DFMO alone. Rather, enhanced growth inhibition with DFMO and MGBG, whether administered sequentially or concomitantly, is almost certainly due to increased cellular uptake of MGBG and hence to an intensification of its many and varied cellular manifestations, including profound effects on mitochondrial function (Mikles-Robertson *et al.,* 1979; Porter *et al.,* 1979). These considerations notwithstanding, certain of these studies, because of their original intent, are reviewed here as being targeted at polyamine biosynthesis (Table III).

Among the first of such studies were those of Burchenal *et al.* (1981) in which impressive *in vivo* findings were reported. While DFMO alone produced no significant therapeutic effect in L1210-bearing mice, the combination of MGBG plus DFMO, administered concomitantly, achieved a greater than 100% increase in the median life-span over untreated animals. Interestingly, when the antiparasitic agent stilbamidine was included in the combination, median survival was extended even more, with 60% of the animals surviving long term. The potentiating effect of stilbamidine, which is otherwise used in the treatment of South American blastomycosis, on MGBG was described previously (Burchenal *et al.,* 1963) but the basis for it is not understood.

The combination of DFMO and MGBG has undergone phase I clinical trials under a schedule designed to evaluate the effects of increasing doses of MGBG with fixed doses of DFMO given on a continuous daily basis (Warrell *et al.,* 1983). Among 22 patients entering the protocol, only 1 patient responded. While DFMO was well tolerated, the drug combination caused severe hematological and gastrointestinal toxicity that was greater than observed previously with MGBG alone (Warrell *et al.,* 1981). In view of the impressive activity of the combination of DFMO, MGBG, and stilbamidine against L1210 leukemia, and since stilbamidine is already approved for use as an antiparasitic agent, clinical trials may be warranted with the three-drug combination.

Significant potentiation of MGBG activity by DFMO has also been observed in transplantable rat prostate carcinoma lines (Duzendorfer *et al.,* 1983b; Herr *et al.,* 1984b; Herr and Kleinert, 1984). Herr and colleagues at Memorial Sloan Kettering found that, used singly, neither

DFMO nor MGBG was particularly effective against the Dunning R-3327-G line *in vivo* but the combination produced a significant percentage (50–60%) of tumor-free animals. Relative to other chemotherapeutic findings with this model tumor (Block *et al.,* 1977; Muntzing *et al.,* 1979), the DFMO–MGBG combination appears to be superior, although its usefulness is still limited by a very narrow chemotherapeutic index. The mechanism of the apparent synergistic effect exerted by DFMO and MGBG on prostatic cancer is uncertain since Heston *et al.* (1984b) observed that pretreatment of prostate-tumor-bearing animals with DFMO selectively enhanced the uptake of the natural polyamines but, unexpectedly, not MGBG. Possibly, prostate tumors are unique among cell types thus far studied, in being vulnerable in their growth to a dual inhibition of the polyamine pathway.

Another urological tumor model which appears to be responsive to the DFMO–MGBG combination is renal adenocarcinoma in mice (Duzendorfer *et al.,* 1983a; Herr *et al.,* 1984a). Initial studies were prompted by an earlier observation that human renal adenocarcinomas contain unusually high levels of spermidine (Duzendorfer and Russell, 1978), and by a report that DFMO alone had a significant effect on this tumor (Kingsnorth *et al.,* 1983). The combination of inhibitors reduced local tumor growth, prevented metastases in some animals, reduced metastases in others, and increased host survival over that seen with either drug alone. While encouraging, the findings were not quite so impressive as those with the rat prostate carcinoma line. Because the prostate has unique characteristics relative to polyamine biosynthesis and uptake, it may be the preferred malignancy for evaluating the potential of combination chemotherapy with polyamine inhibitors. Certainly the data with the fast-growing Dunning R-3327-G rat carcinoma suggest that clinical trials might be worthwhile, particularly with hormone-resistant prostatic cancer. This lead is being pursued clinically by Sher and colleagues at Memorial Sloan Kettering in a protocol using DFMO pretreatment to increase the cellular uptake of MGBG. Initial attempts in this area encountered severe toxicity with no therapeutic responses (Herr *et al.,* 1986).

DFMO and MGBG have also been used in combination with other cancer chemotherapeutic agents. Fujimoto *et al.* (1985) found that although the combination of DFMO and MGBG exerted a clear-cut growth inhibitory effect on human stomach cancer xenotransplanted into nude mice, there was a rapid regrowth of the tumor after cessation of the therapy. When mitomycin C was added to the regimen, this regrowth was effectively suppressed (Fujimoto *et al.,* 1985).

The studies reviewed above had as their main objective, in combining DFMO with MGBG, a dual inhibition of polyamine biosynthesis. While MGBG is regarded as a reversible inhibitor of AdoMet decarboxylase, it is known to have a variety of actions which seem to be unrelated to polyamines and probably account for its potent cytotoxicity as a single agent (Pleshkewych *et al.*, 1980) and for the efficacy achieved by the drug combination. Other studies using DFMO in combination with more specific polyamine inhibitors have been reported. Pegg *et al.* (1982b) treated SV-3T3 cells with DFMO and *S*-adenosyl-1,8-diamino-3-thiooctane (AdoDato), a transition-state inhibitor of spermidine synthase, and achieved a reduction of spermine greater than that seen with DFMO alone. Growth inhibition by the two drugs, however, was not remarkably different from that obtained by DFMO alone.

Porter and colleagues (Porter *et al.*, 1982, 1985; Porter and Bergeron, 1983; Casero *et al.*, 1984) carried out extensive studies with the combination of DFMO and various spermidine analogs with the expectation that growth inhibition by DFMO might be enhanced by the presence of nonfunctional analogs of spermidine. Conceivably, the latter might displace spermidine from critical binding sites in the cell and thereby contribute to its more rapid depletion. It was found that alkyl derivatives of spermidine with substituents at the central amine were capable of preventing DFMO cytostasis, presumably by substituting for authentic spermidine in cell functions related to growth (Porter *et al.*, 1985). Thus, the interaction of these particular analogs with DFMO was antagonistic. By contrast, alkyl spermidine derivatives with substituents at the terminal (N^1 and N^8) amines failed to prevent DFMO cytostasis but, contrary to expectations, did not enhance DFMO cytostasis. Certain of these derivatives, such as N^1,N^8-bis(ethyl)spermidine, were themselves potent inhibitors of polyamine biosynthesis and seemed to act by regulating the enzyme activity of ornithine decarboxylase (Porter *et al.*, 1985). Nonetheless they too failed to increase DFMO cytostasis presumably because, under *in vitro* conditions, the blockade established by either agent alone is sufficiently complete to effect efficient growth inhibition. The combination may have greater potential in *in vivo* model systems, where it is exceedingly more difficult to establish and maintain a blockade of the polyamine pathway. Thus, having an inhibitor and regulator focused on the same enzyme (ornithine decarboxylase) might yield more meaningful drug cooperativity in terms of both inhibition of polyamine synthesis and tumor growth in animals (Porter and Sufrin, 1986).

A final drug combination that has been attempted (Porter *et al.*, 1984b) involved the methionine analog inhibitors of AdoMet biosynthesis, cycloleucine and L-2-amino-4-methoxy-*cis*-but-3-enoic acid. Both analogs inhibit AdoMet synthetase preparations and, in L1210 cells, lead to substantial decreases in AdoMet pools (Porter *et al.*, 1984b). Despite the critical involvement of AdoMet in polyamine biosynthesis, spermidine and spermine pools were not reduced apparently because of a compensatory increase in AdoMet decarboxylase. In combination with DFMO, the analogs exhibited only a slight increase in growth inhibition. However, in combination with MGBG, growth inhibition was significantly enhanced, possibly because AdoMet metabolism is more completely blocked through MGBG inhibition of AdoMet decarboxylase. *In vivo* studies with MGBG and cycloleucine using a renal adenocarcinoma have been reported (Duzendorfer *et al.*, 1983a) to be ineffective. In our own experience with using cycloleucine in animals (J.R. Sufrin, A.J. Bernacki, and C.W. Porter, unpublished data), we have found that scheduling is of critical importance. Thus, the combination of MGBG and methionine analogs may still have therapeutic potential when administered under optimal conditions.

B. Cellular Uptake (Table V)

The discovery by Alhonen-Hongisto *et al.* (1980) that DFMO-induced polyamine depletion of Ehrlich ascites cells elicits a striking increase in the uptake of exogenous polyamines has been of basic importance to antiproliferative strategies utilizing inhibitors of polyamine biosynthesis. Although the enhanced uptake of polyamines by cells treated with polyamine inhibitors is obviously disadvantageous in single-agent therapy, it offered significant potential in drug combinations. As polyamine-depleted cells attempt to restore the intracellular pools from exogenous sources, the natural compensatory reaction can be "misused" by substituting cytotoxic agents for the polyamines. Foremost among these is MGBG, which has been known for a long time to utilize the polyamine transport system (Dave and Caballes, 1973). Thus, in association with their initial observation that DFMO-treated cells demonstrate enhanced uptake characteristics toward polyamines, Alhonen-Hongisto *et al.* (1980) also found that MGBG transport into cells was markedly increased. More importantly, it was determined that "priming" of the tumor with DFMO followed by exposure to MGBG resulted in a much enhanced antiproliferative effect. These simple but fundamental observations initiated a chain of reactions among numerous laboratories which ultimately lead to worldwide clinical trials employing the sequential combination of DFMO and MGBG.

Before reviewing the combination studies employing this strategy (Table IV), the mechanism underlying the enhanced cellular accumulation of polyamines and MGBG following DFMO treatment will be considered. Ultimately, a detailed biochemical and molecular characterization of the

TABLE V

DFMO-Induced Effects on Growth Inhibition via Increased Cellular Uptake

Drug	Cell system (*in vitro* unless indicated)	Growth inhibition[a]	Reference
Aziridinylputrescine	PC-3 human prostate carcinoma	Increased	Heston *et al.* (1985)
MGBG	Ehrlich ascites cells	Increased	Alhonen-Hongisto *et al.* (1980); Seppänen *et al.* (1981a)
MGBG	HeLa	Increased	Sunkara *et al.* (1983a)
MGBG	9L rat glioblastoma	Increased	Alhonen-Hongisto *et al.* (1985b)
MGBG	V79 hamster lung cells	Increased	Alhonen-Hongisto *et al.* (1985b)
MGBG	ST-2 human stomach carcinoma	Increased	Fujimoto *et al.* (1985)
MGBG	Ehrlich ascites (*in vivo*)	Increased	Seppänen *et al.* (1981b)
MGBG	L1210 leukemia (*in vivo*)	Increased	Seppänen *et al.* (1983); Kramer *et al.* (1985)
MGBG	ALL and AML patients (5)	Increased	Siimes *et al.* (1981)
MGBG	CLL and lymphoma patients (6)	None, toxicity increased	Porter *et al.* (1983)
MGBG	Phase I trials (28 patients)	None, toxicity increased	Splinter and Romijin (1986)
MGBG	Brain tumor patients	Increased	Wilson and Gutin (1984)

[a] Increased = additive and/or synergistic drug interactions; none = subadditive or no effect; decreased = antagonistic.

polyamine transport system will be required since, at present, very little is known about it. As late as 1981, Seppänen reported that the system was rather nonspecific since, in addition to putrescine, spermidine, and spermine, other nonphysiological diamines and polyamines (with a backbone longer than three carbons) were readily taken up by tumor cells and were also subject to enhanced accumulation with polyamine depletion. Porter *et al.* (1984b) determined that putrescine, spermidine, and spermine all utilize the same carrier system in L1210 cells and that the transport specificity is greatest for molecules having the same aliphatic chain length as spermidine and spermine. Interestingly, the greatly enhanced cellular uptake of polyamines was not found to be due to changes in the activation energy or affinity for the carrier but rather to an apparent increase in maximum velocity of transport (Seppänen, 1981).

Subsequent studies with analogs of MGBG are consistent with the likelihood that the polyamine-depletion-induced enhanced accumulation of bis(guanylhydrazones) is related to an increased intracellular binding and retention of the compounds rather than to an activation of the putative polyamine carrier. Among the derivatives of MGBG, the cellular accumulation of the parent compound, glyoxal bis(guanylhydrazone), is stimulated by DFMO in a manner similar to MGBG (Seppänen *et al.*, 1984a), while that of ethylglyoxal bis(guanylhydrazone) (EGBG) is not (Seppänen *et al.*, 1984b). However, when tested as competitive inhibitors for spermidine uptake, all three compounds showed roughly comparable affinities for the putative polyamine carrier, even though their retention by the cells was quite different (Alhonen-Hongisto *et al.*, 1984c). Thus, while certain alkylated derivatives of MGBG are better inhibitors of AdoMet decarboxylase, their poor retention by tumor cells makes them less suitable as antiproliferative agents for combined use with DFMO (Jänne and Morris, 1984; Igarashi *et al.*, 1984; Igarashi and Morris, 1984; Jänne *et al.*, 1986).

It appears, therefore, that the enhanced accumulation of bis(guanylhydrazones) in polyamine-deprived tumor cells might be based on a decreased efflux rather than on an enhanced influx of these compounds (Alhonen-Hongisto *et al.*, 1984c). This view is strongly supported by studies with human cell variants resistant to the antiproliferative effects of MGBG. Porter and colleagues (Wiseman *et al.*, 1983; Kramer *et al.*, 1983) described several variants of a Simian-virus-40-transformed human cell line that were up to 20-fold more resistant to the antiproliferative effects of by MGBG than the parent cell line. The resistance was apparently based upon poor accumulation of the drug in the variant cells, yet the short-term uptake characteristics of the parental and resistant cells were practically identical. Subsequent efflux studies clearly indicated that the drug-resistant cells released MGBG into the medium much more rapidly than did the sensitive parental cells (Wiseman *et al.*, 1983).

Like DFMO, MGBG also appears to stimulate the cellular accumulation of the polyamines (Alhonen-Hongisto *et al.*, 1982), thus producing a possible "autostimulation" of its own accumulation. Recently Wallace and Keir (1986) have found that administration of various polyamine inhibitors, including MGBG, prevents efflux of polyamines (and presumably agents which utilize this transport system) in baby hamster kidney cells.

The combined effects of DFMO and MGBG on cell growth are further complicated by certain antagonistic drug interactions. There are observations indicating that inhibitors of AdoMet decarboxylase, such as MGBG or its close derivative, 1,1′-[(methylethanediylidene)dinitrilo]bis(3-aminoguanidine) (MBAG), not only abolish DFMO-induced growth inhibition but also result in a restoration of the normal intracellular polyamine pattern in experimental animal tumor models (Prakash *et al.*, 1980; Bartholeyns and Koch-Weser, 1981; Alhonen-Hongisto *et al.*, 1982). This finding is most likely attributable to the strong inhibition exerted by these compounds toward diamine oxidase (Hölttä *et al.*, 1973; Pegg and McGill 1978; Jänne and Morris, 1984). In fact, when aminoguanidine, a known inhibitor of diamine oxidase having no effect on the biosynthetic enzymes of polyamines, is administered together with DFMO, it brings about a restoration of intracellular polyamines similar to that seen with MGBG (Kallio and Jänne, 1983). This restoration phenomenon is never seen in cultured cells, but occurs in tumor-bearing animals where polyamines can be derived from the circulation. It has been directly shown that in mice treated with combined DFMO and MGBG, radioactive putrescine administered into the intestine with the aid of a stomach tube was taken up by tumor cells (via circulation) at greatly enhanced rate in comparison with untreated or DFMO-treated animals (Kallio *et al.*, 1984). The above experimental data strengthen the view that the profound cytotoxicity exerted by MGBG in various animal tumor screens (for references, see Mihich, 1975; Porter *et al.*, 1981) is, in all likelihood, not related to depletion of spermidine and spermine but to its various other cellular effects. Understandably, this would also be expected to hold true in the combined use of MGBG with DFMO.

The initial experiments employing the sequential combination of DFMO and MGBG in tumor cell cultures clearly indicated that DFMO served as a "primer" to enhance the accumulation of MGBG. The antiproliferative effect was due solely to the cytotoxic effects of MGBG since there exists no indication that growth inhibition is due to polyamine depletion brought about by the combined inhibition of ornithine (DFMO) and AdoMet (MGBG) decarboxylases (Seppänen *et al.*, 1980). The uptake of MGBG by polyamine-deprived cells is so effective that the cells virtually explode after clearing the medium of the drug and reaching a more than

5000-fold concentration gradient across the cell membrane (Seppänen *et al.*, 1980b; Mandel and Flintoff, 1978). This accumulation of MGBG appears to be directly related to the extent of putrescine and spermidine depletion established by DFMO, since the greater the depletion the more effectively MGBG is accumulated (Seppänen *et al.*, 1981a; Jänne *et al.*, 1981a).

The cell culture studies with the sequential combination were rapidly extended to animal experiments. The "priming" of non-tumor-bearing mice with DFMO followed by MGBG injections rather selectively increased the accumulation of the latter drug in rapidly proliferating tissues, such as bone marrow and small intestine, but did not influence drug accumulation in most other tissues (Kallio *et al.*, 1983). In tumor-bearing animals the situation appears to be different. In mice bearing Ehrlich ascites carcinoma cells (Seppänen *et al.*, 1981b) or L1210 leukemia cells (Seppänen *et al.*, 1983) pretreatment with DFMO selectively enhanced the accumulation of MGBG in the tumor cells relative to normal tissues, including rapidly proliferating tissues (Seppänen *et al.*, 1981b, 1983; Jänne *et al.*, 1983). This apparent increase in MGBG selectivity by DFMO may be related to the route of injection of MGBG. In the above studies it was introduced intraperitoneally into the same tissue compartment as the tumor cells. When MGBG is introduced by intravenous injection, the DFMO enhancement of MGBG accumulation is similar in both tumor and normal tissues (Kramer *et al.*, 1985). Whether DFMO enhances the selectivity of MGBG may also be determined by the tumor type since Herr and Kleinert (1984) found that in rats with transplantable prostate tumors, DFMO seemed to reduce host toxicity.

Studies by Siimes *et al.* (1981) with leukemia patients showed that the "priming" concept [i.e., treatment with DFMO followed by relatively low doses (500–700 mg/m^2) of MGBG] was clinically applicable. Four children with advanced lymphoblastic leukemia and one with myeloblastic leukemia were treated. The concentration of MGBG in circulating blast cells was increased three- to fourfold by DFMO priming while the blood level of these cells fell dramatically. Side effects were absent or mild and all patients achieved a remission. The study aroused a great deal of clinical interest. Phase II trials of the sequential combination were initiated at Roswell Park Memorial Institute (Porter *et al.*, 1983). In six lymphoma patients (all adults) the combination produced three partial responses and intensified the host toxicity relative to MGBG alone. The response rate was similar to that seen by Warrell *et al.* (1981) with MGBG alone (50%). A phase I study of the combination has recently been reported (Splinter and Romijin, 1986). In 28 evaluable patients with a wide variety of malignancies, two partial remissions were seen. The toxicity of MGBG was enhanced and side effects were observed which were not apparent with either drug above. Thus, although the cell culture data and results obtained with tumor-bearing animal models suggest usefulness for the DFMO–MGBG sequential combination, available clinical data do not appear to fulfill these expectations. While it is possible that children may be more tolerant of the sequential DFMO and MGBG combinations (Siimes *et al.*, 1981), it seems apparent from the above studies and those reported in Section V,A that, in adults, DFMO priming does not increase the selectivity of MGBG for tumor tissue. This is supported by recent animal studies (Kramer *et al.*, 1985). Again, this conclusion may be dependent on the age of the patient and on the specific malignancy. Finally, an anecdotal report has appeared describing a dramatic effect of the DFMO–MGBG combination in brain tumor patients but no details of the study are available (Wilson and Gutin, 1984).

In addition to various neoplastic conditions, the combination of DFMO and MGBG has also been used, both systemically and topically, to treat epidermal hyperproliferation induced in mouse skin by UV irradiation. It was found that systemic or topical treatment with DFMO alone not only lowered epidermal putrescine and spermidine levels but also greatly enhanced the epidermal accumulation of intraperitoneally administered MGBG (Käpyaho *et al.*, 1982). A sequential topical application of these two drugs profoundly inhibited epidermal DNA and protein synthesis in UV-irradiated mouse skin (Käpyaho *et al.*, 1983; Jänne *et al.*, 1983). In this connection, a rescue concept was developed so that, in case of high epidermal MGBG concentrations, topically applied spermidine was used to effectively lower drug levels (Käpyaho *et al.*, 1983; Jänne *et al.*, 1983). Recently, McCullough *et al.* (1985) quantitated the dose dependence of the DFMO–MGBG combination and optimized the vehicle for topical delivery. In doing so, they were able to establish conditions whereby profound effects on polyamines and DNA biosynthesis could be achieved with minimal cutaneous irritation. While topical DFMO has been used as a single agent in the treatment of psoriasis in small-scale clinical trials (Grosshans *et al.*, 1980; Henry *et al.*, 1981), there are no published data on the combined use of DFMO and MGBG in human psoriasis.

The same sequential administration regimen has also been utilized employing compounds other than MGBG in combination with DFMO. Heston *et al.* (1985a) synthesized aziridinylputrescine, which itself exerts antiproliferative effects on a human prostatic carcinoma cell line PC-3. The growth inhibitory effect was greatly enhanced by a prior treatment of the carcinoma cells with DFMO, as indicated by clonogenic assay. The observed synergism was apparently based on the fact that this putrescine derivative, like the bis(guanylhydrazones), utilizes the polyamine carrier for its entry into cells.

C. Cell Cycle (Table VI)

Inhibition of polyamine biosynthesis affects the cell cycle traverse in different manners depending on the cell type or line used (for review see Heby and Jänne, 1981) and it is difficult to make generalizations with respect to the behavior of normal and transformed cells. However, if anything, most nontransformed cells, brought to polyamine depletion, appear to halve their growth rate before entering the S phase of cell cycle (Heby and Jänne, 1981). Sunkara *et al.* (1980, 1981) observed that, in certain *in vitro* systems, DFMO arrests tumor cells in S phase and normal cycling cells in G_1 phase. They subsequently attempted to exploit this finding by combining DFMO with an S-phase-specific drug, ara-C. In the case of cultured HeLa cells, this combination resulted in an apparent synergistic inhibition of cell growth (Sunkara *et al.*, 1981). In the only *in vivo* evaluation of a DFMO and ara-C combination, Prakash and Sunkara (1983) reported that, while being almost totally ineffective itself, DFMO in combination with ara-C increased the survival time of mice bearing L1210 leukemia by 8.5 days as compared to 5.0 days for ara-C alone. Uncertainty about the generality of this effect was created by the finding that pretreatment of 9L rat glioblastoma cells with DFMO actually decreases the cytotoxicity of ara-C (Oredsson *et al.*, 1983b). In the human colon adenocarcinoma cell line, the effects of DFMO and 5-FU, another S-phase-specific agent, were found to be additive (Kingsnorth *et al.*, 1983).

It is possible that differing cytokinetic responses to polyamine depletion described above might account for the apparent discrepancies. Cell

TABLE VI

DFMO Effects on Cell-Cycle-Specific Agents

Drug	Cell system (*in vitro* unless indicated)	Growth inhibition[a]	Reference
Arabinosylcytosine	HeLa	Increased	Sunkara *et al.* (1980)
Arabinosylcytosine	WI-38 normal fibroblasts	Decreased	Sunkara *et al.* (1981)
Arabinosylcytosine	SV-WI-38 transformed fibroblasts	Increased	Sunkara *et al.* (1981)
Arabinosylcytosine	L1210 leukemia (*in vivo*)	Increased	Prakash and Sunkara (1983)
Arabinosylcytosine	9L rat glioblastoma	Decreased	Oredsson *et al.* (1983b)
5-Fluorouracil	COLO 205 human colon adenocarcinoma	Increased	Kingsnorth *et al.* (1983)

[a] Increased = additive and/or synergistic drug interactions; none = subadditive or no effect; decreased = antagonistic.

types which respond to DFMO by a decreased cell cycle traverse, as opposed to arresting at a drug-sensitive phase of the cycle, are prone to decreased cell kill by phase-specific agents since fewer cells enter the vulnerable portion of the cycle. By contrast, those which arrest at a particular phase are likely to exhibit increased sensitivity to agents specific for that phase. It remains to be seen whether possible cytokinetic interactions between polyamine antimetabolites and conventional cancer chemotherapeutic agents will find a useful place in combination chemotherapy.

D. DNA Structure and Integrity (Table VII)

It is now well established that pretreatment of cells with DFMO renders them more vulnerable to the cytoxic effects of several, but not all, DNA-reactive agents (Table VI). In many systems, this can be induced with concentrations of DFMO which, given alone, have no effect on cell growth, thus qualifying the phenomenon as one of true potentiation. The rationale for the earliest of such studies originated with Seidenfeld *et al.* (1980), who reasoned that, since polyamines are known to interact electrostatically with nucleic acids and, in fact, are capable of protecting them from a variety of physical and chemical influences, polyamine depletion of cells with DFMO might increase cytotoxic effects related to drug-induced DNA damage. In essence, the strategy is a mirror image of the complementary inhibition rationale (Elion *et al.*, 1954) in which one drug is used to induce macromolecular damage while the other inhibits its synthesis and repair, such as with BCNU and ara-C, respectively.

The initial test of the hypothesis set forth by Seidenfeld *et al.* (1980) involved the use of α-methylornithine, a reversible inhibitor of ornithine decarboxylase, DFMO, or MGBG as potential "chemosensitizing" agents in augmenting the cytotoxic effects of X-irradiation. Although this attempt failed to alter the response curve of X-irradiation, subsequent studies using various alkylating agents upheld expectations (Hung *et al.*, 1981). In particular, the cytotoxic activity of the nitrosourea BCNU against cultured rat brain gliosarcoma cells was increased by dose enhancement ratios of 1.2 to 1.3. Subsequently it was demonstrated, with the use of 9L spheroids as a model for long-term drug effects (Deen *et al.*, 1980), that, when two courses of DFMO–BCNU are administered, growth inhibition is even more dramatically increased (Sano *et al.*, 1984). In addition to its chemotherapeutic potential, DFMO enhancement of BCNU activity provides the only instance of polyamines modulating a DNA-related phenomenon in the context of the intact cell, thus providing insight into their possible cellular distribution and function.

TABLE VII

DFMO Effects on Agents Affecting DNA Structure and/or Integrity

Drug	Cell system (*in vitro* unless indicated)	Growth inhibition[a]	Reference
m-AMSA	L1210 murine leukemia	None	Zwelling *et al.* (1985)
Adriamycin	EMT_6 (*in vivo*)	Increased	Bartholeyns and Koch-Weser (1981)
Adriamycin	L1210 leukemia (*in vivo*)	Increased	Bartholeyns and Koch-Weser (1981)
Adriamycin	HTC (*in vivo*)	Increased	Bartholeyns and Koch-Weser (1981)
Aziridinylbenzoquinone	9L rat glioblastoma	Decreased	Alhonen-Hongisto *et al.* (1984a)
BCNU	9L rat glioblastoma	Increased	Hung *et al.* (1981; Tofilon *et al.* (1982)
BCNU	9L/BCNU resistant	None	Oredsson *et al.* (1983c)
BCNU	L1210 murine leukemia	Increased	Cavanaugh *et al.* (1984)
BCNU	HuTu 80 human stomach carcinoma	Increased	Seidenfeld and Komar (1985)
BCNU	HT-29 human colon carci- noma	Increased	Seidenfeld and Komar (1985)
BCNU	ME-180 human cervical carcinoma	Increased	Seidenfeld and Komar (1985)
BCNU	MCF7 human breast carci- noma	Increased	Seidenfeld and Komar (1985)
BCNU	A-427 human lung carcinoma	Increased	Seidenfeld and Komar (1985)
BCNU	9L rat glioblastoma spheroids	Increased	Sano *et al.* (1984)
BCNU	Murine glioma 26, 92 (*in vivo*)	Increased	Marton *et al.* (1981)

1,3-Bis(*trans*-4-hydroxycyclohexyl-1-nitrosourea	9L rat glioblastoma	Decreased	Oredsson *et al.* (1983a)
Bleomycin	9L rat glioblastoma	None	Oredsson and Marton (1984)
1-(2-Chloroethyl)-3-*trans*-4-methylcyclohexyl-1-nitrosourea	9L rat glioblastoma	Increased	Oredsson *et al.* (1984)
Chlorozotocin	9L rat glioblastoma	Increased	Oredsson *et al.* (1983a)
cis-Platinum	9L rat glioblastoma	Decreased	Oredsson *et al.* (1982); Tofilon *et al.* (1982)
cis-Platinum	PANC-1 human pancreatic adenocarcinoma	Increased	Chang *et al.* (1984)
cis-Platinum	COLO 357 human pancreatic adenocarcinoma	Increased	Chang *et al.* (1984)
cis-Platinum	WD PaCa hamster pancreatic adenocarcinoma	Increased	Chang *et al.* (1984)
cis-Platinum	PD PaCa hamster pancreatic adenocarcinoma	Increased	Chang *et al.* (1984)
Cyclophosphamide	EMT_6 (*in vivo*)	Increased	Prakash *et al.* (1980)
Cyclophosphamide	R-3327 MAT-La rat prostate tumor (*in vivo*)	None	Heston *et al.* (1982)
Ellipticine	9L rat glioblastoma	None	Oredsson and Marton (1984)
N-Ethyl-*N*-nitrosourea	9L rat glioblastoma	None	Oredsson *et al.* (1983a)
N-Methyl-*N*-nitrosourea	9L rat glioblastoma	None	Oredsson *et al.* (1983a)
Mitomycin C	ST-2 human stomach tumor	Increased	Fujimoto *et al.* (1985)
X-Irradiation	9L rat glioblastoma	None	Seidenfeld *et al.* (1980)

[a] Increased = additive and/or synergistic drug interactions; none = subadditive or no effect; decreased = antagonistic.

Because 9L cells are somewhat atypical with regard to certain aspects of polyamine uptake and biosynthesis, it is important to note that this finding has been confirmed in several other cell lines. The enhancement effect has since been demonstrated in cultured L1210 leukemia cells (Cavanaugh *et al.*, 1984) and in a series of five human carcinoma lines, including those of lung, duodenum, breast, colon, and cervix, in order of sensitivity to BCNU (Seidenfeld and Komar, 1985). Interestingly, the most resistant of the human carcinoma lines demonstrated the greatest responsiveness to cytotoxicity enhancement by DFMO. The ability of DFMO to overcome resistance, however, has been tested directly in BCNU-insensitive 9L rat brain cells and was found to be ineffective (Oredsson *et al.*, 1983c), even though polyamine depletion was comparable to that seen in sensitive 9L cells. This could be dependent on the mechanism of resistance which, in the case of these particular cells, is not certain.

The combination of DFMO and BCNU was found to be advantageous in animals bearing either murine glioma 26 or rat 9L gliosarcoma intracerebral tumors (Marton *et al.*, 1981). Pretreatment with DFMO prior to BCNU administration had a modest potentiating influence on BCNU activity on the basis of animal survival without an apparent increase in toxicity. As a single agent, DFMO had little or no activity against either tumor. Interest in the clinical applicability of the DFMO–BCNU combination was further heightened by the finding that polyamine depletion potentiated BCNU cytotoxic effects in all phases of the cell cycle (Bjerkvig *et al.*, 1983). Thus, nondividing cells, which are typically inherently resistant to most chemotherapeutic agents, would be susceptible to the effect (see Section II,C). The various experimental studies with DFMO and BCNU have culminated in the clinical trials in brain tumors which are presently being conducted at the University of California at San Francisco.

The mechanistic aspects of DFMO potentiation of BCNU activity have generated considerable interest. Initially, it was determined that, with 9L cells, the phenomenon was unrelated to cell cycle perturbations since, under the conditions of the assay, they were not observed to take place (Hung *et al.*, 1981). In these studies and in all subsequent studies, the effect was fully reversed or prevented by exogenous putrescine or spermidine. Moreover, Alhonen-Hongisto *et al.* (1984b) found that the magnitude of DFMO potentiation correlated temporally with the extent of total polyamine depletion. In prevention studies using varying concentrations of spermidine to titrate the intracellular content of spermidine, Cavanaugh *et al.* (1984) was able to make the interesting observation that 30% of the normal spermidine pool was sufficient to prevent DFMO potentiation. Since this was the same amount required to prevent DFMO cytostasis (Casero *et al.*, 1984), it appeared that the spermidine pool required for cell proliferation might be the same pool associated with DNA, but further studies are needed to substantiate this possibility. Thus far, only the effects of putrescine and spermidine depletion on BCNU activity are recognized. The consequences of spermine or spermidine and spermine depletion pose intriguing possibilities since spermine, more than putrescine or spermidine, has been shown to be associated with the cell nucleus (Mach *et al.*, 1984) and since, being a tetramine, spermine might have a greater influence on DNA structure than the triamine, spermidine. Appropriate studies now await the discovery of a specific inhibitor of spermine synthase or AdoMet decarboxylase.

The cellular cytotoxicity of BCNU has been shown to be directly related to DNA strand damage resulting from alkylation and cross-link formation (Kohn, 1977). Indication that DFMO-enhancement of BCNU cytotoxicity was the result of increased DNA damage was first provided by studies using sister chromatid exchange (SCE) methodology. Tofilon *et al.* (1982) found that DFMO increased SCE incidence by BCNU by about twofold, but by itself did not increase the incidence over that of control. While BCNU-induced SCE in 9L cells has been correlated with cell survival (Tofilon *et al.*, 1983), the molecular basis for SCE formation is not yet understood, although there exists some evidence to indicate that it might involve DNA damage mediated by drug interference with topoisomerase II (Nelson *et al.*, 1984). More direct evidence for DNA damage was provided later by Tofilon *et al.* (1983) using alkaline elution techniques. DFMO-induced polyamine depletion of rat brain tumor cells increased BCNU cross-link formation in DNA. Using a different methodology based on enhancement of bis(benzamide) fluorescence, Cavanaugh *et al.* (1984) correlated DFMO enhancement of cytotoxicity in L1210 cells with increases in single-strand DNA breaks. It appears that DNA damage is not due to inhibitory effects on nucleic acid repair secondary to polyamine depletion, since Oredsson *et al.* (1984) found that the activities of three enzymes capable of removing alkylation damage from DNA were not affected by DFMO pretreatment.

Nitrosoureas rapidly decompose under physiological conditions to form highly reactive isocyamates and carbonium ion intermediates (Cheng *et al.*, 1972; Montgomery *et al.*, 1967). The carbonium ions alkylate available nucleophiles in nucleic acids and proteins (Cheng *et al.*, 1972; Montgomery *et al.*, 1967; Wheeler, 1976). In the case of BCNU, cytotoxicity is thought to result from an initial monoalkylation of DNA and the subsequent formation of DNA-interstrand and DNA-protein cross-links (Erikson *et al.*, 1980; Kohn, 1977). In an attempt to delineate the relative sensitivity of alkylation, cross-linking, and carbamoylation to DFMO enhancement by polyamine depletion, Oredsson *et al.* (1983a) compared four nitrosoureas having different reactive decomposition products (see Table VII). Of the four, only the activity of chlorozotocin was potentiated, indicating that polyamine depletion increased the cytotoxicity of agents (such as BCNU) which alkylate and form cross-links, but not those which alkylate or carbamoylate only. Not unexpectedly, in view of these findings, the cytotoxicity of 1-(2-chloroethyl)-3-*trans*-4-methylcyclohexyl-1-nitrosourea, which has a mechanism of action similar to BCNU (Wheeler, 1976), was also found to be potentiated by DFMO pretreatment (Oredsson *et al.*, 1984). Overall, the precise mechanism by which polyamine depletion leads to enhanced DNA cross-linking is not yet known. It has been suggested (Oredsson *et al.*, 1982) that removal of spermidine from DNA might alter the helical structure slightly so that the configuration of the nucleophilic groups would favor cross-link formation. Attempts to confirm this theory will also shed considerable light on the localization and function of polyamines in cells.

cis-Diaminedichloroplatinum(II) (*cis*-platinum) reacts with DNA to form single-strand adducts and ultimately intrastrand cross-links (Alvarez *et al.*, 1978; Roberts and Pascoe, 1972). Unlike BCNU, the cytotoxicity of *cis*-platinum was not potentiated in rat 9L cells—in fact, it was antagonized (Oredsson *et al.*, 1982). Consistent with this finding, it was subsequently found that both sister chromatid exchange incidence (Tofilon *et al.*, 1982) and DNA cross-link formation (Tofilon *et al.*, 1983) by *cis*-platinum were also adversely affected by DFMO-induced polyamine depletion in 9L cells. Although the generality of DFMO potentiation of BCNU has been established in several cell lines, the antagonism observed with DFMO and *cis*-platinum has not. In fact, in four pancreatic adenocarcinoma cell lines, DFMO increased the cytotoxicity of *cis*-platinum (Chang *et al.*, 1984), indicating that such phenomena might be tumor-type specific.

Recently DFMO has been shown to enhance the activity of at least one agent which has as its primary target DNA topoisomerase II (Zwelling *et al.*, 1985). The latter is responsible for the cleavage and rejoining cycle of single DNA strands which ultimately allows for the passing of one strand through another (Liu *et al.*, 1983). Although the biological function of the DNA topoisomerases is unknown, the enzyme has recently been implicated as being involved in DNA breaks and protein cross-link formation resulting from treatment with a variety of intercalating agents (Nelson *et al.*, 1984; Liu *et al.*, 1983). In particular, the potent anticancer agent 4′-(9-acridinylamino)methanesulfon-*m*-anisidide (*m*-AMSA) has been suspected as having DNA topoisomerase II as its primary target (Micetich *et *al.*, 1982). Inasmuch as spermidine has been shown to alter topoisomerase-II-mediated reactions (Brown and Cozzarelli, 1981; Kreuzer and Cozzarelli, 1980), Zwelling *et al.* (1985) examined the effects of DFMO-induced polyamine depletion on DNA effects of *m*-AMSA and found they were substantially enhanced. Pretreatment of L1210 cells with DFMO for 24 hr increased *m*-AMSA single-strand DNA breaks by about 40%. The authors speculate that polyamine depletion might either increase *m*-AMSA intercalation into DNA or increase the DNA binding affinity or site number for the topoisomerase enzyme. Unfortunately, DFMO was not found to increase the cytotoxicity of *m*-AMSA. Since the DNA effects were more dramatic at low concentrations of *m*-AMSA, the possibility exists that its antitumor activity might be increased *in vivo* where saturation of such effects is less likely to occur.

Other DNA-reactive agents which have also been found to be unaffected in their cytotoxicity by polyamine depletion are aziridinylbenzoquinone (AZQ), despite expectations that it might alkylate and cross-link DNA (Alhonen-Hongisto *et al.*, 1984a); the strand scission agent, bleomycin (Oredsson and Marton, 1984); and the intercalating agent, ellipticine (Oredsson and Marton, 1984). Given the discrepancy between rat brain 9L cells and pancreatic adenocarcinoma cells, with respect to DFMO effects on *cis*-platinum activity, it might be worthwhile to study the above agents in other cell lines before their insensitivity to polyamine depletion is considered to be a generality.

During its early preclinical evaluation, DFMO was studied *in vivo* in combination with certain widely used antitumorals. These studies were not based on rationale beyond single-drug effectiveness nor were they preceded by *in vitro* findings to indicate the possible success of the combinations. Prakash *et al.* (1980) found that DFMO given concomitantly with cyclophosphamide to mice bearing EMT_6 mammary tumor resulted in a slight increase in activity (based on tumor weight) over that seen with cyclophosphamide alone. Interestingly, this particular tumor appears to be very sensitive to polyamine depletion since 3% DFMO alone was comparable in antitumor activity to cyclophosphamide alone. In the case of R-3327 MAT-Lu rat prostate tumor, DFMO in combination with cyclophosphamide was no more effective than cyclophosphamide alone, based on tumor volume determinations (Heston *et al.*, 1982b). As a single agent, DFMO was far less effective than cyclophosphamide against this particular tumor.

Investigators at Merrell Dow (Bartholeyns and Koch-Weser, 1981) found that DFMO in combination with adriamycin substantially increased the overall survival time of animals bearing L1210 leukemia, EMT_6 mammary tumor, or HTC tumor. Similar findings were also obtained with the combination of DFMO and the mitotic poison vindesine in these model systems (See Section V,G). Despite the success of these combinations, additional studies indicating development toward clinical application have not appeared. Since the efficacy of these particular drug combinations has not yet been demonstrated *in vitro* so that various host factors might be excluded and since they involve drugs with widely different modes of action, it is difficult to speculate on how polyamine depletion might account for the apparent drug cooperativity.

E. DNA Biosynthesis (Table VIII)

There are no reported instances of DFMO-induced polyamine depletion being useful in enhancing the effect of an antagonist of DNA biosynthesis per se. Rather, since many DNA antimetabolites exhibit varying degrees of cell cycle specificity, their combination with DFMO has usually been applied as a cytokinetic strategy in which polyamine depletion is used to render a tumor cell population more susceptible to a particular agent by increasing the proportion of cells in the sensitive phase of their cycle. The rationale for such studies stems from the observation (Sunkara *et al.*, 1980, 1981) that DFMO treatment arrests certain tumor cells in S phase while halting normal cells in G_1 phase. Attempts to exploit this finding with drug combinations involving the S-phase-specific DNA antimetabolites ara C and 5-FU are reviewed in detail in Section V,C.

TABLE VIII

DFMO Effects on Agents Affecting DNA Biosynthesis

Drug	Cell system (*in vitro* unless indicated)	Growth inhibition[a]	Reference
Ara-A	Dunning R-3327-G rat prostatic adenocarcinoma (*in vivo*)	None	Duzendorfer *et al.* (1983b)
Ara-C	HeLa	Increased	Sunkara *et al.* (1980)
Ara-C	SV-WI-38 transformed fibroblasts	Increased	Sunkara *et al.* (1981)
Ara-C	WI-38 normal fibroblasts	Decreased	Sunkara *et al.* (1981)
Ara-C	9L rat glioblastoma	Decreased	Oredsson *et al.* (1983b)
Ara-C	L1210 leukemia (*in vivo*)	Increased	Prakash and Sunkara (1983)
5-FU	COLO 205 human colon adenocarcinoma	Increased	Kingsnorth *et al.* (1983)
Methotrexate	9L rat glioblastoma	Decreased	Alhonen-Hongisto *et al.* (1984d)

[a] Increased = additive and/or synergistic drug interactions; none = subadditive or no effect; decreased = antagonistic.

Overall, since DFMO itself is not known to induce DNA damage, there exists little rationale for combining it with inhibitors of DNA biosynthesis once cell cycle considerations are excluded. In fact, because DFMO treatment tends to slow cell growth and, hence, DNA biosynthesis, it might be predicted that polyamine depletion would be antagonistic to DNA antimetabolites. The few studies which have attempted such combinations do, in fact, bear out this reasoning. Marton's group combined DFMO with methotrexate, an inhibitor of dihydrofolate reductase, and hence, of thymidylic acid (DNA biosynthesis) and inosinic acid (RNA biosynthesis). They found that the activity of methotrexate was decreased by polyamine depletion in rat 9L glioblastoma cells (Alhonen-Hongisto *et al.*, 1984d). Thus, the clinical potential of drug combinations utilizing polyamine inhibitors with DNA antimetabolites does not seem to be promising unless cytokinetic strategies are taken into account.

F. Biological Response Modifiers (Table IX)

One of the most interesting combinations involving biological response modifiers is that of DFMO and interferon. Sunkara *et al.* (1983b) reported that DFMO administered together with mouse type I interferon was highly toxic to cultured B_{16} mouse melanoma cells and resulted in a practically total supression of tumor growth in melanoma-bearing mouse. The possible therapeutic synergism between these agents is somewhat difficult to judge, however, because melanoma cells in general (Käpyaho *et al.*, 1985) appear to be exceptionally sensitive to DFMO (Sunkara *et al.*, 1983b; Meyskins *et al.*, 1985). Nonetheless, it is noteworthy that no tumors were found in 20% of animals receiving the combined drug regimen while in groups receiving DFMO or interferon alone, all of the animals had tumors (Sunkara *et al.*, 1983b). These studies prompted clinical trials evaluating the effectiveness of DFMO and interferon in melanoma patients. Results from these studies, which are currently ongoing, have not yet been made available.

The findings with melanoma were soon confirmed in a variety of other systems. Heston *et al.* (1984a) observed a synergism between DFMO and human recombinant interferon against human renal cell adenocarcinoma as judged by clonogenic assay *in vitro* and an additive therapeutic effect in this same tumor xenografted into mice. Kovach and Svingen (1985) found a synergistic antiproliferative effect between DFMO and human interferon against a variety of human tumor cell lines having widely different sensitivities to interferon. By isobologram analysis, Rosenblum and Gutterman (1984) assessed true synergy with this drug combination in cultured human lymphoblastoid (Daudi) cells.

TABLE IX

DFMO Effects on Biological Response Modifiers

Drug	Cell system (*in vitro* unless indicated)	Growth inhibition[a]	Reference
Mouse fibroblast interferon	B_{16} melanoma	Increased	Sunkara *et al.* (1983b)
Mouse fibroblast interferon	B_{16} melanoma (*in vivo*)	Increased	Sunkara *et al.* (1983b)
Leukocyte interferon	Lymphoblastoid (Daudi cells)	Increased	Rosenblum and Gutterman (1984)
Recombinant human interferons	HM7 (human melanoma)	Increased	Wolf *et al.* (1985)
Recombinant human interferons	MDA-MB-231 human breast carcinoma	Increased	Wolf *et al.* (1985)
Recombinant human interferon	JDF-1 renal cell adenocarcinoma	Increased	Heston *et al.* (1984a)
Recombinant human interferon	JDF-1 (*in vivo*)	Increased	Heston *et al.* (1985)
Recombinant leukocyte type A human interferon	Seven different human cell lines of various tissue origin	Increased	Kovach and Svingen (1985)
Poly(I) : poly(C) (interferon inducer)	B_{16} melanoma (*in vivo*)	Increased	Sunkara *et al.* (1984)
Poly(I) : poly(C) (interferon inducer)	Lewis lung carcinoma (*in vivo*)	Increased	Sunkara *et al.* (1984)
Tilorone (interferon inducer)	B_{16} melanoma (*in vivo*)	Increased	Sunkara *et al.* (1984)
Tilorone (interferon inducer)	Lewis lung carcinoma (*in vivo*)	Increased	Sunkara *et al.* (1984)
Tuftsin (interferon inducer)	B_{16} melanoma (*in vivo*)	Increased	Sunkara *et al.* (1983c)
Corynebacterium parvum	B_{16} melanoma	Increased	Bowlin *et al.* (1985)

[a] Increased, additive and/or synergistic drug interactions; none, subadditive or no effect; decreased, antagonistic.

Sunkara *et al.* (1983c, 1984) subsequently extended their studies by combining DFMO with the known inducers of natural interferon, tilorone and polyriboinosinic: polyribocytidilic acid, and obtained an apparent enhancement effect. Again, the possible potentiation of the antitumor activity of DFMO by interferon inducers is somewhat difficult to judge due to the near-complete tumor growth suppression exerted by DFMO alone which had a profound inhibition on the formation of metastatic foci (Sunkara *et al.*, 1983c, 1984).

The mechanism of drug cooperativity between DFMO and interferon is still completely open. Although the interferons have been reported to suppress decarboxylase activity (Sreevalsan *et al.*, 1979; Sekar *et al.*, 1983), Sunkara *et al.* (1983b) found no greater polyamine depletion with the combination than with DFMO alone. The view that the interaction would be based solely on the host defense system is unlikely since the synergism between DFMO and interferon is also observed in *in vitro* systems (Sunkara *et al.*, 1983b; Heston *et al.*, 1984a; Rosenblum and Gutterman, 1984). It is likely that both agents would exert their antiproliferative effects independently. This view is supported by the finding that, when doxorubicin is included in the DFMO and interferon regimen, a greater antiproliferative effect than would be expected from the additive effects is observed (Kovach and Svingen, 1985).

In addition to interferon, Sunkara's group has examined the effect of combining DFMO with other immunomodulating agents. The intention in doing so was to potentiate the antitumor activity of DFMO by immunologically mediated mechanisms. Both tuftsin (Sunkara *et al.*, 1983c), a known immunoadjuvant (Nishioka, 1979), and *Corynebacterium parvum* (Bowlin *et al.*, 1985), a lymphoreticular-stimulating agent, substantially augmented the antitumor activity against B_{16} melanoma, particularly at low doses of DFMO. This latter finding has especially significant implications since the doses of DFMO required to achieve antitumor effects in mice are far above those attainable in humans (see Section IV), and such drug combinations would lower the effective dose to a realistic range. In addition, Bowlin *et al.* (1985) also reported that DFMO gives no indication of being immunosuppressive. This is in agreement with a very recent study by Ehrke *et al.* (1986) which finds that, depending on schedule and dose, DFMO can actually stimulate certain immune responses related to antitumor defenses.

G. Miscellaneous Combinations (Table X)

DFMO has been studied in combination with several agents having widely diverse modes of action as antiproliferative agents (Table IX).

TABLE X

DFMO Effects on Miscellaneous Agents

Drug	Cell system (*in vitro* unless indicated)	Growth inhibition[a]	Reference
3,4-Dihydroxybenzylamine	B_{16} melanoma	Increased	Prakash *et al.* (1985)
3,4-Dihydroxybenzylamine	B_{16} melanoma (*in vivo*)	Increased	Prakash *et al.* (1985)
Tamoxifen	*N*-Methyl-*N*-nitrosourea-induced rat mammary tumor	None	Manni and Wright (1983, 1984)
Vincristine	9L rat glioblastoma	Decreased	Alhonen-Hongisto *et al.* (1984d)
Vindesine	EMT6 (*in vivo*)	Increased	Bartholeyns and Koch-Weser (1981)
Vindesine	L1210 leukemia (in vivo)	Increased	Bartholeyns and Koch-Weser (1981)
Vindesine	HTC (*in vivo*)	Increased	Bartholeyns and Koch-Weser (1981)
Vitamin A analogs	Rat glioma (clone 6)	Increased	Chapman (1980)
Vitamin A analogs	C-1300 murine neuroblastoma	Increased	Chapman (1980)

[a] Increased, additive and/or synergistic drug interactions; none, subadditive or no effect; decreased, antagonistic.

Perhaps best known among these are the vinca alkaloid, vincristine, and the synthetic analog, vindesine. Both are regarded as mitotic poisons which appear to act by interfering with the polymerization of microtubules, particularly those involved in the formation of the mitotic spindle, causing cells to accumulate in metaphase. Bartholeyns and Koch-Weser (1981) found that the combination of DFMO and vindesine (given concomitantly) doubled survival time of mice bearing L1210 leukemia, completely arrested hepatoma tumor cell growth in rats, and inhibited tumor growth of solid EMT6 tumors in mice. All of these activities were substantially greater than that seen with either drug alone. While the basis for this drug cooperativity has yet to be established, it is interesting to note that mutants unable to synthesize polyamines lose their actin filaments and microtubules when deprived of exogenous polyamines (Pohjanpelto *et al.*, 1981). Although the DFMO–vindesine combination was found by Bartholeyns and Koch-Weser (1981) to be effective in three model tumor systems, the phenomenon is apparently not a generality since, in rat 9L glioblastoma, DFMO decreased the activity of vincristine (Alhonen-Hongisto *et al.*, 1984d). The authors attributed this lack of effect to a DFMO-induced decrease in cell cycle traverse, thereby reducing the number of vincristine-sensitive cells at mitosis.

Vitamin A and its analogs (retinoids) are known to induce a variety of biological and biochemical effects in cells which seem to vary according to the cell type. Among these are their ability to prevent the chemical induction and growth of tumors in model systems, to induce differentiation in a variety of cell lines (Breitman *et al.*, 1980), and to exert an antiproliferative effect in certain cell lines. At the biochemical level retinoids have been shown to markedly reduce the cellular level of ornithine decarboxylase (Haddox and Russell, 1979; Haddox *et al.*, 1979). On this basis, Chapman (1980) applied DFMO in combination with retinoids in the treatment of cultured neuroblastoma and glioma cells. Growth inhibition by the retinoids alone correlated with decreases in ornithine decarboxylase activity. Both effects were enhanced by DFMO. It is interesting to note that growth inhibition of murine embryonal F9 cells by DFMO can be prevented by putrescine but growth inhibition by the retinoids cannot be prevented (Jetten and Shirley, 1985), indicating that the retinoid-induced growth inhibition is probably not mediated by polyamine depletion despite decreases in ornithine decarboxylase activity.

One tumor type that has recently been shown to be uniquely sensitive to DFMO is B_{16} melanoma under either *in vitro* or *in vivo* conditions (Sunkara *et al.*, 1985). Growth inhibition has associated with it an increase in tyrosinase activity and accumulation of melanin, suggesting induction of differentiation secondary to DFMO-mediated polyamine depletion. As a follow-up to this observation these same authors (Prakash *et al.*, 1985) have tested DFMO in combination with 3,4-dihydroxybenzylamine, a tyrosinase-dependent melanolytic agent with a therapeutic effect superior to that of other catechols against B_{16} melanoma (Wick, 1972). An enhanced cell kill was observed under both *in vitro* and *in vivo* conditions which the authors believed to be due to the increased tyrosinase activity induced by DFMO. Although the *in vivo* therapeutic effect was only marginally greater than that predicted for additivity, the low host toxicities of both agents at the doses employed indicate the potential usefulness of such a drug combination.

There are recent reports to indicate that polyamines may be involved as mediators of estrogen action of hormone-dependent experimental breast cancers. Indirect evidence to support this view was provided by Manni and Wright (1984), who found that DFMO blocked the growth-promoting effect of estradiol on a nitrosourea-induced rat mammary tumor grown in soft agar. Although exogenous polyamine did not have any effect on the growth of the tumor, they effectively abolished the inhibitory effect of DFMO on the tumor. Consistent with this finding is a report by Lima and Shiu (1985) showing that DFMO blocked the estradiol-induced proliferation of an estrogen-responsive subline of human breast cancer. This same study also contained the important finding that DFMO abolished the estradiol-induced growth in hormone-responsive cell lines but not in unresponsive cells.

A further indirect piece of evidence suggesting that the polyamines might be needed for the mediation of estrogen action in mammary cancer cells was provided previously by Manni and Wright (1983). They compared the antiproliferative effect of the antiestrogen, tamoxifen, with that exerted by DFMO in carcinogen-induced mammary tumor cells grown in soft agar. Both compounds exerted similar growth inhibition on the tumor cell and the effects were not additive when combined. The latter observation was taken as a circumstantial indication that the growth-inhibitory effects of DFMO and tamoxifen may be mediated by the same system, i.e., through inhibition of polyamine biosynthesis (Manni and Wright, 1984). More direct evidence to support the involvement of polyamines in the action of estrogens and antiestrogens is available in the finding that the inhibitory effect of tamoxifen on the colony formation of breast cancer was reversed in a dose-dependent fashion by putrescine, spermidine, or spermine (Manni and Wright, 1984). The polyamines did not affect colony formation when added alone.

Although the above-cited experimental work fits very well with the proposed role of the polyamines as mediators of estrogen and antiestrogen action, important control experiments must be performed before the role of the polyamines in this process is unambiguously resolved. As correctly pointed out by Manni and Wright (1984), the reversal of tamoxifen-induced antitumor effect by exogenous putrescine, spermidine, or spermine could be based on a decreased cellular uptake of the antiestrogen. Furthermore, these studies were not supported by determinations of the intracellular polyamine pools under the experimental conditions. In any event, the potentially central role of the polyamines in the growth of hormone-dependent tumors, including others in addition to mammary cancer, remains as an exciting possibility and one which certainly is worthy of further study.

VI. SUMMARY

The basic principles and approaches which govern the design and implementation of drug combinations in anticancer chemotherapy have been reviewed. Owing to a number of unusual characteristics (summarized in Table I), DFMO offers promise as a modulator of antineoplastic drug action. Since its introduction in 1978 as a specific inhibitor of ornithine decarboxylase, DFMO has been widely used with various antineoplastics in combinations directed at polyamine biosynthesis, cellular uptake of drugs, cell cycle progression, DNA structure and integrity, DNA biosynthesis, biological responses, and others. These are summarized in Table XI. Several of these combinations have progressed from basic observations made in experimental systems to evaluation by clinical trials. At present, the combinations which, on the basis of preclinical data, show promise of clinical potential include DFMO with interferon in melanoma patients and DFMO with MGBG in patients with urological carcinomas. Apparently, the former of these is being actively pursued clinically since phase I trial data for the combination have recently been reported (Talpaz *et al.*, 1986). Conditions of acceptable toxicity were established and partial responses were noted in 2 of 12 metastatic melanoma patients.

Numerous groups, and particularly those at Merrell Dow Research Institute are presently involved in identifying new and more effective polyamine inhibitors and analogs while, at the same time, others are using available inhibitors, such as DFMO, to delineate the role of polyamines in cell proliferation. Undoubtedly, progress made in both of these areas will contribute immensely to the exploitation of polyamine biosynthesis as a target site in cancer chemotherapy.

TABLE XI

Summary of DFMO Interactions with Various Antineoplastic Agents

Class	Rationale	Compound[a,b]
DNA-reactive agents	Destabilization of DNA, exposure of reactive sites	*m*-AMSA, adriamycin, AZQ, BCNU, bleomycin, MeCCNU, chlorozotocin, cyclophosphamide, ellipticine, ENU, MNU, mitomycin C
DNA antimetabolites	Combined antiproliferative effects	Ara-A, ara-C, 5-FU, methotrexate
Cell cycle agents	Cell cycle perturbations favoring phase-specific drugs	Ara-C, 5-FU
Polyamine inhibitors	Dual inhibition of pathway and/or increased uptake	AdoDato, MGBG, methionine analogs
Polyamine analogs	Increased uptake and analog competition for spermidine sites	N^4-Spermidine and N^1,N^8-spermidine analogs, aziridinylputrescine
Biological response modifiers	Slowing of tumor cell growth	Interferon, interferon inducers, *Corynebacterium parvum*
Miscellaneous	Variable	3,4-Dihydroxybenzylamine, tamoxifen, vincristine, vitamin A analogs

[a] Underscoring indicates DFMO/drug combinations which have undergone clinical testing or which have shown promising *in vivo* findings.

[b] Abbreviations: AdoDato, *S*-adenosyl-1,8-diamino-3-thiooctane, *m*-AMSA, 4′-(9-acridinylamino)-methanesulfon-*m*-aniside; ara-A (or -C), arabinosyladenosine (or cytosine); AZQ, aziridinylbenzoquinone; BCNU, 1,3-bis(2-chloroethyl)-1-nitrosourea; ENU, *N*-ethyl-*N*-nitrosourea; 5-FU, 5-fluorouracil; MeCCNU, 1-(2-chloroethyl)-3-*trans*-4-methylcyclohexyl-1-nitrosourea; MGBG, methylglyoxyl bis(guanylhydrazone); MNU, *N*-methyl-*N*-nitrosourea.

REFERENCES

Abeloff, M.D., Slavik, M., and Luk, G.D. (1984). *J. Clin. Oncol.* **2,** 124–130.
Alhonen-Hongisto, L., Seppänen, P., and Jänne, J. (1980). *Biochem J.* **192,** 941–945.
Alhonen-Hongisto, L., Seppänen, P., and Jänne, J. (1982). *FEBS Lett.* **145,** 182–186.
Alhonen-Hongisto, L., Deen, D.F., and Marton, L.J. (1984a). *Cancer Res.* **44,** 39–42.
Alhonen-Hongisto, L., Deen, D.F., and Marton, L.J. (1984b). *Cancer Res.* **44,** 1819–1822.
Alhonen-Hongisto, L., Fagerstrom, R., Laine, R., Elo, H., and Jänne, J. (1984c). *Biochem. J.* **221,** 273–276.
Alhonen-Hongisto, L., Hung, D.T., Deen, D.F., and Marton, L.J. (1984d). *Cancer Res.* **44,** 4440–4442.
Alhonen-Hongisto, L., Kallio, A., Sinervirta, R., Seppänen, P., Kontula, K.K., Jänne, O.A., and Jänne, J. (1985a). *Biochem. Biophys. Res. Commun.* **126,** 734–740.
Alhonen-Hongisto, L., Levin, V.A., and Marton, L.J. (1985b). *Cancer Res.* **45,** 509–514.
Alhonen-Hongisto, L., Sinervirta, R., Jänne, O.A., and Jänne, J. (1985d). *Biochem. J.* **232,** 605–607.
Alvarez, M.V., Cobreros, G., Heras, A., and Lopes Zumel, C. (1978) *Br. J. Cancer* **37,** 68–72.
Armitage, J.G., Fyfe, M.A.E., and Lewis, J. (1984). *J. Clin. Oncol.* **2,** 898–902.
Barranco, S.C., Ho, D.H.W., and Drewinko, B. (1972). *Cancer Res.* **32,** 2733–2736.
Bartholeyns, J., and Koch-Weser, J. (1981). *Cancer Res.* **41,** 5158–5161.
Bennett, F., Ehrke, J., Dave, C., and Mihich, E. (1977). *Biochem. Pharmacol.* **26,** 723–728.
Berger, N.A. (1986). *J. Clin. Invest.* 1131–1135.
Bey, P., Gerhart, F., Van Dorsselaer, V., and Danzin, C. (1983). *J. Med. Chem.* **26,** 1551–1556.
Bjerkvig, R., Oredsson, S.M., Marton, L.J., Linden, M., and Deen, D.F. (1983). *Cancer Res.* **43,** 1497–1500.
Block, N.L., Camuzzi, F., Denefrio, J., Troner, M., Clafin, A., Stover, G., and Politano, V.A. (1977). *Oncology* **34,** 110–113.
Bodey, G.P., Rodriguez, F., Whitecar, J., Hart, J., and Freireich, E.J. (1970). *In* "Leukemia-Lymphoma," pp. 333–345. Year Book Medical Publ., Chicago.
Bolkenius, F.N., Bey, P., and Seiler, N. (1985). *Biochim. Biophys. Acta* **838,** 69–76.
Bowlin, T.L., Rosenberger, A.L., and Sunkara, P.S. (1985). *Cancer Immunol. Immunother.* **20,** 214–218.
Breitman, T.R., Selonick, S.E., and Collins, S.J. (1980). *Proc. Natl. Acad. Sci. U.S.A.* **77,** 2936–2940.
Brown, P.O., and Cozzarelli, N.R. (1981). *Proc. Natl. Acad. Sci. U.S.A.* **78,** 843–847.
Burchenal, J.H., Purple, J.R., Bucholz, E., and Straub, P.W. (1963). *Cancer Chemother. Rep.* **29,** 85–89.
Burchenal, J.H., Lokys, L., Smith, R., Cartmell, R., and Warrell, R.P., Jr. (1981). *Proc. AACR/ASCO* **22,** 191.
Casero, R.A., Bergeron, R.J., and Porter, C.W. (1984). *J. Cell. Physiol.* **121,** 476–482.
Cavanaugh, P.F., Jr., Pavelic, Z.P., and Porter, C.W. (1984). *Cancer Res.* **44,** 3856–3861.
Chang, B.K., Black, O., Jr., and Gutman, R. (1984). *Cancer Res.* **44,** 5100–5104.
Chapman, S.K. (1980). *Life Sci.* **26,** 1359–1366.
Cheng, C.J., Fujimura, S., Grunberger, D., and Weinstein, I.B. (1972). *Cancer Res.* **32,** 22–27.
Choi, J.H., and Scheffler, I.E. (1983). *J. Biol. Chem.* **258,** 12601–12608.
Chou, T.-C., and Talalay, P. (1984). *Vol. Enzyme Regul.* **22,** 27–55.
Cooper, R.G. (1969). *Proc. Am. Assoc. Cancer Res.* **10,** 15.

Creasy, W.A. (1975). *Handb. Exp. Pharmacol.* **38,** 670–694.

Danzin, C., Bey, P., Schirlin, P., and Claverie, N. (1982). *Bioch. Pharmacol.* **31,** 3871–3878.

Dave, C., and Caballes, L. (1973). *Fed. Proc., Fed. Am. Soc. Exp. Biol.* **32,** *736.*

Deen, D.F., Hoshino, T., Williams, M.E., Murooka, I., Knebel, K.D., and Barker, M. (1980). *J. Natl. Cancer Inst.* **64,** 1373–1382.

DeVita, V.T. (1982). *In* "Cancer: Principles and Practice of Oncology" (V.T. DeVita, Jr., S. Hellman, and S.A. Rosenberg, eds.), Chap. 8, pp. 132–155. Lippincott, Philadelphia.

DeVita, V.T., Canellos, G.P., and Moxley, J.H. (1972). *Cancer* **30,** 1495–1504.

Duzendorfer, U., and Russell, D.H. (1978). *Cancer Res.* **38,** 2321–2324.

Duzendorfer, U., Kleinert, E., and Whitmore, W.F., Jr. (1983a). *Urol. Int.* **38,** 162–165.

Duzendorfer, U., Relyea, N.M., Kleinert, E., Balis, M.E., and Whitmore, W.F., Jr. (1983b). *Oncology* **40,** 57–62.

Ehrke, M.J., and Mihich, E. (1985a). *Trends Pharmacol. Sci.* **6,** 412–417.

Ehrke, M.J., and Mihich, E. (1985b). *In* "The Reticuloendothelial System: A Comprehensive Treatise" (J.W. Hadden and A. Szentivany, eds.), Vol. 8., pp. 309–347. Plenum, New York.

Ehrke, M.J., Porter, C.W., Eppolito, C., and Mihich, E. (1986). *Cancer Res.* **46,** 2798–2803.

Einhorn, L.H., and Donohue, J. (1977). *Ann. Intern. Med.* **87,** 293–298.

Elion, G.B., Singer, S., and Hitchings, G.H. (1954). *J. Biol. Chem.* **208,** 477.

Erickson, L.C., Bradley, M.O., Ducore, J.M., Ewig, R.A.G., and Kohn, K.W. (1980). *Proc. Natl. Acad. Sci. U.S.A.* **77,** 467–471.

Erwin, B.G., and Pegg, A.E. (1982). *Biochem. Pharmacol.* **31,** 2820–2823.

Fillingame, R.H., Jorstad, C.M., and Morris, D.R. (1975). *Proc. Natl. Acad. Sci. U.S.A.* **72,** 4042–4045.

Freireich, E.J., Karon, M., and Frei, E., III (1964). *Proc. Am. Assoc. Cancer Res.* **5,** 20.

Frindel, E., Vassort, F., and Tubiana, M. (1970). *J. Radiat. Biol.* **17,** 329–337.

Fujimoto, S., Igarashi, K., Shrestha, R.D., Miyazaki, M., and Okui, K. (1985). *Int. J. Cancer* **35,** 831–825.

Fuller, D.J.M., and Gerner, E.W. (1982). *Cancer Res.* **42,** 5046–5049.

Goldin, A., Venditti, J.M., and Mantel, N. (1974). *Handb. Exp. Pharmacol.* **38,** 411–448.

Grosshans, E., Henry, M., Tell, G., Schechter, P., Bohlen, P., Grove, J., and Koch-Weser, J. (1980). *Ann. Dermatol. Venereol. (Paris)* **107,** *377–387.*

Haddox, M.K., and Russell, D.H. (1979). *Cancer Res.* **39,** 2476–2480.

Haddox, M.K., Scott, K.F.F., and Russell, D.H. (1979). *Cancer Res.* **39,** 4930–4938.

Heby, O., and Jänne, J. (1981). *In* "Polyamines in Biology and Medicine" (D.R. Morris and L.J. Marton, eds.), pp. 243–310. Dekker, New York.

Heby, O., Oredsson, S.M., and Kanje, M. (1984). *Adv. Enzyme Regul.* **22,** 243–264.

Henry, M., Juillard, J., Tell, G., Schechter, P.J.S., Grove, J., Koch-Weser, J., and Grosshans, E. (1981). *Br. J.Dermatol.* **105,** (Suppl. 20), 33–34.

Herr, H.W., and Kleinert, E.L. (1984). *Prostate* **5,** 439–444.

Herr, H.W., Kleinert, G.L., Conti, P.S., Burchenal, J.H., and Whitmore, W.F., Jr. (1984a). *Cancer Res.* **44,** 4382–4385.

Herr, H.W., Kleinert, L.L., Relyea, N.M., and Whitmore, W.F., Jr. (1984b). *Cancer* **53,** 1294–1298.

Herr, H.W., Warrell, R.P., and Burchenal, J.H. (1986). *Urology* **28,** 508–511.

Heston, W.D.W., Kadmon, D., and Fair, W.R. (1982). *Cancer Lett.* **16,** 71–79.

Heston, W.D., Fleischmann, J., Tackett, R.E., and Ratliff, T.L. (1984a). *Cancer Res.* **44,** 3220–3225.

Heston, W.D., Kadmon, D., Covey, D.F., and Fair, W.R. (1984b). *Cancer Res.* **44,** 1034–1040.

Heston, W.D.W., Uy, L., Fair, W.R., and Covey, D.F. (1985). *Biochem. Pharmacol.* **34,** 2409–2410.
Hölttä, E., Hannonen, P., Pispa, J., and Jänne, J. (1973). *Biochem. J.* **136,** 669–676.
Hung, D.T., Deen, D.F., Seidenfeld, J., and Marton, L.J. (1981). *Cancer Res.* **41,** 2783–2785.
Igarashi, K., and Morris, D.R. (1984). *Cancer Res.* **44,** 5332–5337.
Igarashi, K., Porter, C.W., and Morris, D.R. (1984). *Cancer Res.* **44,** 5326–5331.
Jänne, J., and Morris, D.R. (1984). *Biochem. J.* **218,** 947–951.
Jänne, J., Alhonen-Hongisto, L., Seppänen, P., and Hölttä, E. (1981a). *Adv. Polyamine Res.* **3,** 85–95.
Jänne, J., Alhonen-Hongisto, L., Seppänen, P., and Siimes, M. (1981b). *Med. Biol.* **59,** 448–457.
Jänne, J., Alhonen-Hongisto, L., Käpyaho, K., and Sëppänen, P. (1983a). *Adv. Polyamine Res.* **4,** 17–32.
Jänne, J., Hölttä, E., Kallio, A., and Käpyaho, K. (1983b). *Spec. Top. Endocrinol. Metabol.* **5,** 227–293.
Jänne, J., Alhonen-Hongisto, L., Nikula, P., and Elo, H. (1986). *Adv. Enzyme Regul.* **24,** 125–139.
Jetten, A.M., and Shirley, J.E. (1985). *Exp. Cell Res.* **156,** 221–230.
Kallio, A., and Jänne, J. (1983). *Biochem. J.* **212,** 895–898.
Kallio, A., Seppänen, P., Alhonen-Hongisto, L., and Jänne, J. (1983). *Cancer Res.* **43,** 324–327.
Kallio, A., Nikula, P., and Jänne, J. (1984). *Biochem. J.* **218,** 641–644.
Käpyaho, K., Liinamaa, K., and Jänne, J. (1982). *J. Invest. Dermatol.* **78,** 391–394.
Käpyaho, K., Lauharanta, J., and Jänne, J. (1983). *J. Invest. Dermatol.* **81,** 102–106.
Käpyaho, K., Sinervirta, R., and Jänne, J. (1985). *Cancer Res.* **45,** 1444–1448.
Keiser, W.L., and Capizzi, R.L. (1977). *In* "Cancer: A Comprehensive Treatise; (F.F. Becker, ed.), Vol. 5, pp. 163–190. Plenum, New York.
Kingsnorth, A.N., Russell, W.E., McCann, P.P., Diekema, K.A., and Malt, R.A. (1983). *Cancer Res.* **43,** 4035–4308.
Kohn, K.W. (1977). *Cancer Res.* **37,** 1450–1454.
Kovach, J.S., and Svingen, P.A. (1985). *Cancer Treat. Rep.* **69,** 97–103.
Kramer, D.L., Zychlinski, L., Wiseman, A., and Porter, C.W. (1983). *Cancer Res.* **43,** 5943–5950.
Kramer, D.L., Paul, B., and Porter, C.W. (1985). *Cancer Res.* **45,** 2512–2515.
Kreuzer, K.N., and Cozzarelli, N.R. (1980). *Cell* **20,** 245–254.
Leinonen, P., Alhonen-Hongisto, L., Laine, R., Jänne, O.A., and Jänne, J. (1987). *Biochem. J.* **242,** 199–203.
Lima, G., and Shiu, R.P. (1985). *Cancer Res.* **45,** 2466–2470.
Ling, V. (1975). *Can. J. Genet. Cytol.* **17,** 503–510.
Ling, V. (1985). *In* "Drug and Hormone Resistance in Neoplasia" (N. Brochosvsky and J.H. Goldie, eds.), Vol. 1, pp. 1–19. CRC Press, Boca Raton, Florida.
Liu, L.F., Rowe, T.C., Yang, L., Tewey, K.M., and Chen, G.L. (1983). *J. Biol. Chem.* **258,** 15365–15370.
McConlogue, L., and Coffino, P. (1983). *J. Biol. Chem.* **258,** 8384–8388.
McCullough, J.L., Peckham, P., Klein, J., Weinstein, G.D., and Jenkins, J.J. (1985). *J. Invest. Dermatol.* **85,** 518–521.
MacDonald, J.S., Schein, P.S., Wooley, P.V., Smythe, T., Ueno, W., Hoth, D., Smith, F., Boiron, M., Gisselbrecht, C., Bruent, R., and Lagarde, C. (1980). *Ann. Intern. Med.* **93,** 533–536.
Mach, M., Ebert, P., Popp, R., and Ogilvie, A. (1982). *Biochem. Biophys. Res. commun.* **104,** 1327–1334.

Maddox, A.M., Keating, M.F., McCredie, K.E., Estey, E., and Freireich, E.J. (1985). *Invest. New Drugs* **3,** 287–292.
Mamont, P.S., Duchesne, M.-C., Grove, J., and Tardif, C. (1978). *Exp. Cell Res.* **115,** 387–393.
Mamont, P.S., Danzin, C., Kolb, M., Gerhart, F., Bey, P., and Sjoerdsma, A. (1986). *Biochem. Pharmacol.* **35,** 159–165.
Mamont, P.S., Danzin, C., Wagner, J., Slat, M., Joder-Ohlenbusch, A.-M., and Claverie, N. (1982). *Eur. J. Biochem* **123,** 499–504.
Mandel, J.L., and Flintoff, W.F. (1978). *J. Cell. Physiol.* **97,** 335–344.
Manni, A., and Wright, C. (1983). *Cancer Res.* **43,** 1084–1086.
Manni, A., and Wright, C. (1984). *J. Natl. Cancer Inst.* **73,** 511–514.
Marton, L.J., Levin, V.A., Hervatin, S.J., Koch-Weser, J., McCann, P.P., and Sjoerdsma, A. (1981). *Cancer Res.* **41,** 4426–4431.
Metcalf, B.W., Bey, P., Danzin, C., Jung, M.J., Casara, P., and Vevert, J.P. (1978). *J. Am. Chem. Soc.* **100,** 2551–2553.
Meyskins, F.L., Kingsley, E.M., Glattke, T., Loescher, L., and Booth, A. (1986). *Invest. New Drugs* **4,** 257–262.
Micetich, K.C., Zwelling, L.A., Gormley, P., and Young, R.C. (1982). *Cancer Treat. Rep.* **66,** 1813–1817.
Mihich, E. (1975). *Handb. Exp. Pharmacol.* **37,** 766–788.
Mihich, E., and Grindey, G.B. (1977). *Cancer* **40,** 534–543.
Mikles-Robertson, F., Feuerstein, B.F., Dave, C., and Porter, C.W. (1979). *Cancer Res.* **39,** 1919–1926.
Montgomery, J.A., James, R., McCaleb, G.S., and Johnston, T.P. (1967). *J. Med. Chem.* **10,** 668–674.
Morris, D.R., Jorstad, C.M., and Seyfried, C.E. (1977). *Cancer Res.* **37,** 3168–3172.
Muntzing, J., Kirdani, R.Y., Murphy, G.P., and Sandberg, A.S. (1979). *Invest. Urol.* **17,** 37–41.
Nelson, E.M., Tewey, K.M., and Liu, L.F. (1984). *Proc. Natl. Acad. Sci. U.S.A.* **81,** *1361–1365.*
Nishioka, K. (1979). *Br. J. Cancer* **39,** 342–345.
Oredsson, S.M., and Marton, L.J. (1984). *Prog. Exp. Tumor Res.* **28,** 102–117.
Oredsson, S.M., Deen, D.F., and Marton, L.J. (1982). *Cancer Res.* **42,** 1296–1299.
Oredsson, S.M., Deen, D.F., and Marton, L.J. (1983a). *Cancer Res.* **43,** 4606–4609.
Oredsson, S.M., Gray, J.W., Deen, D.F., and Marton, L.J. (1983b). *Cancer Res.* **43,** 2541–2544.
Oredsson, S.M., Tofilon, P.J., Feurstein, B.G., Deen, D.F., Rosenblum, M.L., and Marton, L.J. (1983c). *Cancer Res.* **43,** 3576–3578.
Oredsson, S.M., Pegg, A.E., Alhonen-Hongisto, L., Deen, D.F., and Marton, L.J. (1984). *Eur. J. Cancer Clin. Oncol.* **20,** 535–542.
Pegg, A.E., and McGill, S.M. (1978). *Biochem. Pharmacol.* **27,** 1625–1629.
Pegg, A.E., Päsö, H., Shuttleworth, K., and Bennett, R.R. (1982a). *Biochem J.* **202,** 519–526.
Pegg, A.E., Tang, K.C., and Coward, J.K. (1982b). *Biochemistry* **21,** 5082–5089.
Pegg, A.E., Wechter, R.S., Clark, R.S., Wiest, L., and Erwin, B. (1986). *Biochemistry* **25,** 379–384.
Pleshkewych, A., Kramer, D.L., and Porter, C.W. (1980). *Cancer Res.* **40,** 4533–4540.
Pohjanpelto, P., Virtanen, I., and Hölttä, E. (1981). *Nature (London)* **293,** 475–477.
Pohjanpelto, P., Hölttä, C., Jänne, O.A., Knutilu, S., and Alitalo, K. (1985). *J. Biol. Chem.* **260,** 8532–8537.

Porter, C.W., and Bergeron, R.J. (1983). *Science* **219,** 1083–1085.
Porter, C.W., and Sufrin, J.R. (1986). *Anticancer Res.* **6,** 525–542.
Porter, C.W., Mikles-Robertson, F., Kramer, D., Rustum, Y.M., and Dave, C. (1979). *Cancer Res.* **39,** 2414–2421.
Porter, C.W., Dave, C., and Mihich, E. (1981). *In* "Polyamines in Biology and Medicine" (D.R. Morris and L.J. Marton, eds.), pp. 407–436. Dekker, New York.
Porter, C.W., Bergeron, R.J., and Stolowich, N. (1982). *Cancer Res.* **42,** 4072–4078.
Porter, C.W., Ozer, H., Cowens, W., Han, T., and Gavigan, M. (1983). *Proc. Biochem. Modulators Advisory Group Meet, June 13–14.*
Porter, C.W., Miller, J., and Bergeron, R.J. (1984a). *Cancer Res.* **44,** 126–128.
Porter, C.W., Sufrin, J.R., and Keith, D.D. (1984b). *Biochem. Biophys. Res. Commun.* **122,** 350–357.
Porter, C.W., Cavanaugh, P.F., Stolowich, N., Ganis, B., Kelly, E., and Bergeron, R.J. (1985). *Cancer Res.* **45,** 2050–2057.
Potter, V.R. (1951). *Proc. Soc. Exp. Biol. Med.* **76,** 41–46.
Prakash, N.J., and Sunkara, P.S. (1983). *Cancer Res.,* **43,** 3192–3196.
Prakash, N.J., Schechter, P.J., Mamont, P.S., Grove, J., Koch-Weser, J., and Sjoerdsma, A. (1980). *Life Sci.* **26,** 181–194.
Prakash, N.J., Sunkara, P.S., and Sjoerdsma, A. (1985). *Biochem. Pharmacol.* **34,** 1887–1890.
Roberts, J.J., and Pascoe, J.M. (1972). *Nature* (*London*) **235,** 282–284.
Rosenblum, M.G., and Gutterman, J.U. (1984). *Cancer Res.* **44,** 2339–2340.
Rupniak, H.T., and Paul, D. (1978). *J. Cell. Physiol.* **94,** 161–170.
Sano, Y., Deen, D.F., Oredsson, S.M., and Marton, L.J. (1984). *Cancer Res.* **44,** 577–581.
Santoro, A., Bonfante, V., and Bonadonna, G. (1982). *Ann. Intern. Med.* **96,** 139–143.
Sartorelli, A.C., and Creasy, W.A. (1982). *In* "Cancer Medicine" (J.F. Holland and E. Frei, III, eds.), pp. 720–730. Lea & Febiger, Philadelphia.
Schabel, F.M., Jr. (1969). *Cancer Res.* **29,** 2384–2389.
Schabel, F.M., Jr. (1975). *In* "Pharmacological Basis of Cancer Chemotherapy," pp. 595–621. Williams & Wilkins, Baltimore.
Seidenfeld, J., and Komar, K. (1985). *Cancer Res.* **45,** 2132–2138.
Seidenfeld, J., Deen, D.F., and Marton, L.J. (1980). *Int. J. Radiat. Biol.* **38,** *223–229.*
Sekar, V., Atmar, U.J., Joshi, A.R., Krim, M., and Kuehn, G.D. (1983). *Biochem. Biophys. Res. Commun.* **114,** 950–954.
Seppänen, P. (1981). *Acta Chem. Scand.* **35,** 731–736.
Seppänen, P., Alhonen-Hongisto, L., Pösö, H., and Jänne, J. (1980a). *FEBS Lett.* **111,** 99–103.
Seppänen, P., Alhonen-Hongisto, L., Siimes, M., and Jänne, J. (1980b). *Int. J. Cancer* **26,** 571–576.
Seppänen, P., Alhonen-Hongisto, L., and Jänne, J. (1981a). *Eur. J. Biochem.* **118,** 571–576.
Seppänen, P., Alhonen-Hongisto, L., and Jänne, J. (1981b). *Biochim. Biophys. Acta* **674,** 169–177.
Seppänen, P., Alhonen-Hongisto, L., and Jänne, J. (1983). *Cancer Lett.* **18,** 1–10.
Seppänen, P., Fagerstrom, R., Alhonen-Hongisto, L., Elo, H., Lumme, P., and Jänne, J. (1984a). *Biochem. J.* **221,** 483–488.
Seppänen, P., Ruohola, H., and Jänne, J. (1984b). *Biochim. Biophys. Acta* **803,** 331–337.
Seyfried, C.E., and Morris, D.R. (1979). *Cancer Res.* **39,** 4861–4867.
Shore, P.A., and Cohn, V.H. (1960). *Biochem. Pharmacol.* **5,** 91–95.
Siimes, M., Seppänen, P., Alhonen-Hongisto, L., and Jänne, J. (1981). *Int. J. Cancer* **28,** 567–570.

Sjoerdsma, A., and Schecter, P.J. (1984). *Clin. Pharmacol. Ther.* **35,** 287–300.
Skipper, H.E., Schabel, F.M., and Wilcox, W.S. (1964). *Cancer Chemother. Rep.* **35,** 1–11.
Smalley, R.V., Carpenter, J., Bartolucci, A., Vogel, C., and Krauss, S. (1977). *Cancer* **40,** 625–632.
Splinter, T.A.W., and Romijin, J.C. (1986). *Eur. J. Cancer Clin. Oncol.* **22,** *61–67.*
Sreevalsan, T., Taylor-Papadimitriou, J., and Rozengurt, E. (1979). *Biochem. Biophys. Res. Commun.* **87,** 679–685.
Sunkara, P.S., Fowler, S.K., Nishioka, K., and Rao, P.N. (1980). *Biochem. Biophys. Res. Commun.* **95,** 423–430.
Sunkara, P.S., Fowler, S.K., and Nishioka, K. (1981). *Cell Biol. Int. Rep.* **10,** 991–997.
Sunkara, P.S., Prakash, N.J., Chang, C.C., and Sjoerdsma, A. (1983a). *J. Natl. Cancer Inst.* **70,** 505–509.
Sunkara, P.S., Prakash, N.J., Mayer, G.D., and Sjoerdsma, A. (1983b). *Science* **219,** 851–853.
Sunkara, P.S., Prakash, N.J., and Nishioka, K. (1983c). *Ann. N.Y. Acad. Sci.* **419,** 268–272.
Sunkara, P.S., Prakash, N.J. Rosenberger, A.L., Hagan, A.C., Lachmann, P.J., and Mayer, G.D. (1984). *Cancer Res.* **44,** 2799–2802.
Sunkara, P.S., Chang, C.C., Prakash, N.J., and Lachman, P.J. (1985). *Cancer Res.* **45,** 4067–4070.
Talmadge, J.E., Benedict, K., Madsen, J., and Fidler, I.J. (1984). *Cancer Res.* **44,** 3801–3805.
Talpaz, M., Plager, C., Quesada, J., Benjamin, R., Kantarjian, H., and Gutterman, J. (1986). *Eur. J. Cancer* **22,** 685–689.
Tofilon, P.J., Oredsson, S.M., Deen, D.F., and Marton, L.J. (1982). *Science* **217,** 1044–1046.
Tofilon, P.J., Williams, M.E., and Deen, D.F. (1983). *Cancer Res.* **43,** 473–475.
Tsuruo, T., Iida, J., Nojiri, M., Tsukagoshi, S., and Sukurai, Y. (1983). *Cancer Res.* **43,** 2905–2910.
Tyrer, D.D., Kline, I., Venditti, J.M., and Goldin, A. (1967). *Cancer Res.* **27,** 873–879.
Wagner, J., Hirth, Y., Piriov, F., Zakett, D., Claverie, N., and Danzin, C. (1985). *Biochem. Biophys. Res. Commun.* **133,** 546–553.
Wallace, H.M., and Keir, H. (1986). *FEBS Lett.* **194,** 60–63.
Warrell, R.P., and Burchenal, J.H. (1983). *J. Clin. Oncol.* **1,** 52–65.
Warrell, R.P., Lee, B.J., Kempin, S.J., Lacher, M.J., Straus, D.J., and Young, C.W. (1981). *Blood* **57,** 1011–1014.
Warrell, R.P., Coonley, C.J., and Burchenal, J.H. (1983). *Cancer Chemother. Pharmacol.* **11,** 134–136.
Weil, M., Jacquillat, C., Boiron, M., and Bernard, J. (1969). *Eur. J. Cancer* **5,** 271–275.
Wheeler, G.P. (1976). *Am. Chem. Soc. Symp. Ser.* **30,** 87–119.
Wick, M.M. (1972). *J. Natl. Cancer Inst.* **63,** 1465–1467.
Williams-Ashman, H.G., and Schenone, A. (1972). *Biochem. Biophys. Res. Commun.* **46,** 288–295.
Wilson, C.B., and Gutin, P.H. (1984). *Cancer* **54,** 2702–2705.
Wiseman, A., Kramer, D.L., and Porter, C.W. (1983). *Cancer Res.* **43,** 5937–5942.
Wittes, R.E., and Goldin, A. (1986). *Cancer Treat. Rep.* **70,** 105–125.
Wolf, L.M., Leitzel, K.E., Pegg, A.E., Harvey, H.A., and Lipton, A. (1985). *J. Biol. Response. Mod.* **4,** 391–395.
Woo, P.W.K., Dion, H.W., Lange, S.M., Dahl, L.T., and Durham, L.J. (1974). *J. Heterocyclic Chem.* **11,** 641–643.
Zwelling, L.A., Kerrigan, D., and Marton, L.J. (1985). *Cancer Res.* **45,** 1122–1126.

11

Inhibition of Carcinogenesis by Inhibitors of Putrescine Biosynthesis

AJIT K. VERMA* AND R. K. BOUTWELL†

* *Department of Human Oncology*
Clinical Cancer Center

† *McArdle Laboratory for Cancer Research*
University of Wisconsin
Madison, Wisconsin 53706

I. INTRODUCTION

A rational approach in the choice of agents to prevent and/or treat cancer in humans must be based on knowledge of the biochemical mechanism of carcinogenesis. Multistep mouse skin carcinogenesis is an excellent system to identify and quantitate the sequence of biochemical alterations which are of mechanistic relevance to the initiation (first step) and the promotion (second step) of epithelial carcinogenesis (5, 43, 44, 53, 55).

Initiation can be accomplished by a single application of a carcinogen [e.g., 7,12-dimethylbenz[*a*]anthracene (DMBA)] at a sufficiently small dose so that it will not lead to the induction of visible tumors during the life-span of the animal. However, many tumors develop following repeated and prolonged application to the initiated skin of another chemical known as a tumor promoter. Applications of a tumor promoter alone rarely elicit tumors; it is only following initiation that promoters elicit tumors. Treatment with the initiator and the promoter in reverse order causes no more tumors than either agent alone. Initiation is essentially an irreversible process; a long time may elapse between the initiation and promotion steps without diminution of the tumor yield. In contrast, promotion is relatively reversible; if the period between applications of a promoter is extended too long, the ultimate tumor yield is decreased. Another important feature of initiation is that the action of the initiator is additive and there is no apparent threshold. In contrast, the effect of the promoter is not necessarily additive and there is a threshold in the dose of a promoter that is capable of eliciting effects of tumor promoters. Thus, initiation and promotion are qualitatively different components of the process of tumor formation (5, 6). Among a series of phorbol esters isolated from croton oil, 12-*O*-tetradecanoylphorbol-13-acetate (TPA) is a potent mouse skin tumor promoter (5, 17, 55).

The initiation and the promotion steps of tumor formation have also been demonstrated in organs other than skin (e.g., liver, urinary bladder, colon, and breast) (14, 19, 38–40, 42). The promotion stage of mouse skin carcinogenesis can be further dissected into two unique steps (15, 18, 43, 44). However, the specificity of stage 1 tumor promotion has been questioned by the findings of Fürstenberger *et al.* (15) and Ordman *et al.* (36). In their findings, stage 1 tumor promotion can occur before the initiation of the carcinogenic process. The significance of these results to the mechanism of carcinogenesis is not clear.

The step of initiation of tumors, which requires a noncarcinogenic exposure to a carcinogen, is irreversible. On the other hand, promotion for the autonomous tumor growth, which requires repeated and prolonged exposure to a promoter, is reversible at least in the early stages. Because of the properties of reversibility and the time required to accomplish the promotion stage of carcinogenesis, knowledge of the biochemical mechanisms for the phenomena of promotion is potentially important in the prevention of cancer. One of the properties ascribed to tumor promoters is the ability to induce ornithine decarboxylase (ODC) (27, 34, 35, 47). This has been shown in mouse skin (34, 35) and pulmonary tissues (10) following treatment with TPA, in rodent urinary bladder following treatment with tryptophan metabolites (27), in rat colon after treatment with bile salts (47), and in liver following exposure to phenobarbital (4).

ODC, which decarboxylates ornithine to form putrescine, is the first and the rate-limiting enzyme in the pathway of the biosynthesis of mammalian polyamines, putrescine, spermidine, and spermine. Polyamine metabolism plays a decisive role in nucleic acid synthesis, cell proliferation, and differentiation as well as in malignant transformation (3, 37, 48, 49).

Accumulating evidence indicates that the inhibition of the induction of ODC activity results in the inhibition of tumor promotion. This has been observed in the case of tumor promotion in various epithelial tissues including skin (50, 57, 58, 64), breast (13, 22, 51, 52), and colon (22). The association of the inhibition of ODC activity with the inhibition of carcinogenesis is the focus of this chapter.

II. EXPERIMENTS AND RESULTS

A. Essential Role of ODC Induction in Skin Tumor Promotion by TPA

O'Brien *et al.* (35) were the first to find that a single application of TPA to the shaved back of mouse skin leads to a dramatic induction of epidermal ODC activity between 4 and 6 hr following promoter treatment. The evidence in support of the suggestion that this phenotypic change may be essential, although not sufficient (29), for tumor promotion by TPA has been reviewed (34, 53, 54), but for the purpose of this chapter, the relevant information will be summarized. The magnitude of ODC induction is dependent on the dose of TPA and correlates with the dose in promoting skin tumor formation (35, 55). With the exception of mezerein (29), the degree of induction of ODC activity correlates with the tumor-promoting ability of a number of structurally unrelated tumor promoters (35, 46, 55). Furthermore, the ability of a series of phorbol esters to induce ODC activity correlates with their ability to promote tumor formation. Epidermal ODC activity is induced following treatment of mouse skin with tumor promoters and not with nonpromoting hyperplastic agents. In contrast, both tumor promoters and hyperplastic agents induce *S*-adenosylmethionine decarboxylase activity, an enzyme involved in the biosynthesis of spermidine (35). Furthermore, Clark-Lewis and Murray (9) have shown that wounding, which provides a promoting stimulus, induces epidermal ODC activity. In contrast, skin massage, sufficient to cause cell proliferation, induces neither ODC activity nor tumor formation. Also, the level of ODC is elevated in skin papillomas and carcinomas (34). In fact, the induction of mouse epidermal ODC can be used as a rapid screen for tumor promoters. Thus, Sugimura *et al.* (46), using ODC induction as a biochemical parameter, were able to discover, purify, and characterize new classes of naturally occurring tumor promoters such as teleocidins.

Further strong support for the role of ODC induction in skin tumor promotion by TPA was provided by the finding that α-difluoromethylornithine (DFMO), an enzyme-activated irreversible inhibitor of ODC (28), when given in the drinking water in conjunction with TPA, inhibits the formation of mouse skin tumors (50, 64). Furthermore, retinoic acid and certain of its analogs, applied 1 hr before each TPA treatment, inhibited the induction of ODC activity and the formation of skin tumors elicited by TPA; the retinoids, which did not inhibit ODC induction, failed to inhibit skin tumor formation promoted with TPA (57). Also, the inhibitors of arachidonic acid metabolism, such as indomethacin (58), nordihydroguaiaretic acid (NDGA) (30), and certain superoxide dismutase-mimicking agents (e.g., 3,5-diisopropylsalicylate) (12), which inhibit skin tumor promotion, also inhibit ODC induction by TPA. It is important to note that these agents do not inhibit ODC activity but rather inhibit the amount of enzymatically active ODC protein increased by TPA (54). These facts prove that ODC induction plays an essential role in skin tumor promotion by TPA. However, other evidence suggests that ODC induction is not a sufficient condition for tumor promotion (29). A specific example is our observation that mezerein, which is at least as potent as TPA on a molar basis for inducing ODC activity, is a weak promoter (29).

The exact molecular mechanism of ODC induction and skin tumor promotion by TPA is not clear (5, 65). Evidence indicates that TPA effects are transduced by its initial interaction with the plasma membrane (5, 65). A major breakthrough to unravel the mechanism of phorbol ester tumor promotion has been the identification of TPA receptors (2, 11, 21, 31). TPA receptors are ubiquitous in eucaryotes (61). Accumulating evidence indicates that protein kinase C (PKC) is a receptor of phorbol esters (2, 21, 31). Phorbol ester receptors from the brain have been purified to homogeneity. Data from Scatchard analysis indicate that one molecule of tumor promoter may bind to every molecule of PKC in the presence of calcium and phospholipid (21, 31). Castagna *et al.* (8) demonstrated that PKC is activated by TPA *in vivo* and *in vitro* in the presence of Ca^{2+} and phospholipid. PKC forms a part of the signal-transducing system involving the turnover of inositol phospholipid and is thought to be activated by diacylglycerol produced as a consequence of this turnover (32). Tumor-promoting phorbol esters appear to interact at the same site as diacylglycerols (24). We have shown, using surviving mouse skin pieces in culture, that the level of available Ca^{2+} determined the degree of ODC induction by TPA (56). We also found that diacylglycerol, an activator of PKC and phospholipase C, which releases diacylglycerol from membrane phospholipids, induced ODC activity in cultured newborn mouse epidermal cells. Phospholipase A_2 was ineffective. Phospholipase C- and TPA-induced ODC activity was preceded by an increased level of ODC mRNA quantitated by means of DNA–RNA hybridization analysis using a radiolabeled cDNA probe. Furthermore, palmitoylcarnithine (PC), an intermediate of long-chain fatty acid metabolism and an inhibitor of protein kinase C, inhibited epidermal ODC induction by TPA. Also, PC inhibited skin tumor promotion by TPA; application of 3 μmol of PC in conjunction with each promotion treatment with 10 nmol of TPA to female CD-1 mouse skin (initiated with DMBA) inhibited tumor formation completely (63). Taken together, we conclude that activation of protein kinase C may be an early event in ODC-gene transcription and skin tumor promotion by TPA. Furthermore, TPA-increased ODC activity is the result of increased synthesis of ODC protein apparently regulated at the transcriptional level (54).

Mouse skin tumors can be developed either by the initiation–promotion (DMBA–TPA) protocol or by the complete carcinogenesis regimen. However, the two protocols differ mechanistically (60). In the complete carcinogenesis process, mouse skin tumors are induced by a single application of a large dose, or by repeated applications of a small dose, of a carcinogen such as the polycyclic aromatic hydrocarbon DMBA. The biology of tumor formation by the initiation of tumors with a single application of DMBA followed by twice-weekly applications of TPA differs from that of tumor induction by once-weekly applications of DMBA only. In the latter case, papillomas developed more slowly and were less common (the tumor induction time was long; tumor yield was less), but carcinomas appeared much earlier. Retinoic acid, a potent inhibitor of tumor promotion by TPA, failed to inhibit, and under some experimental conditions enhanced, tumor formation by DMBA (60). These results indicate that tumor formation accomplished by the complete carcinogenesis process involves a promoting component with a mechanism different from that of TPA. This conclusion is compatible with the fact that DMBA does not compete for TPA receptors and may have different signal-transduction mechanisms from TPA for tumor promotion. Rather, DMBA, after being metabolized by the mixed-function oxidases or other systems to reactive electrophiles, reacts with bases in DNA and with protein components of the chromatin. The latter may result in a pleiotypic response that provides the promoting component of the complete carcinogen (60). Alternatively, ulceration and wounding, which are observed after DMBA application to skin, may lead to regenerative hyperplasia, which has been reported to be a promoting stimulus (1).

A single application of a complete carcinogenic dose of DMBA leads to dramatic induction of epidermal ODC activity; the peak ODC activity occurs at about 36 hr following DMBA application to skin (34, 59). The time of induction of ODC after DMBA is comparable to that of the maximum induction of ODC after partial hepatectomy (41) and is compatible with the idea that the induction is a component of skin regeneration resulting from a large cytotoxic dose of DMBA. The conclusion that DMBA-induced ODC activity may be an important component of the mechanism of DMBA carcinogenesis is supported by the fact that DFMO inhibits complete skin carcinogenesis by DMBA (Table I).

TABLE I

Inhibition by DFMO of DMBA-Induced Mouse Skin Carcinogenesis[a]

	At 22 weeks of DMBA treatment	
Treatment	Papilloma/mouse	Mice with papillomas (%)
Tap water	11.26	100
DFMO	3.05	64

[a] DMBA (10 nmol) in 0.2 ml acetone was applied twice weekly to the shaved backs of female SENCAR mice. DFMO (1%) in the drinking water was started 2 days before the first DMBA treatment and was continued for the duration of the experiment. There were 30 mice in each group and the tumor incidence was recorded weekly.

B. Essential Role of Polyamine Biosynthesis in Induction of Tumors in Organs other Than Skin

The facts presented for the mouse skin system indicate that the induction of the putrescine-synthesizing enzyme (ODC) is crucial to accomplish the process of carcinogenesis. These studies prompted others to explore the role of ODC induction and of putrescine in carcinogenesis in various organs in organ-specific models (13, 22, 51, 52). This was made possible because of the availability of DFMO, a specific suicide inhibitor of ODC (28). Administration of DFMO at 1% in the drinking water in conjunction with 1,2-dimethylhydrazine (DMH) reduces the colonic putrescine level and mucosal hyperplasia and decreases the incidence of DMH-induced colon tumors in mice (22). The development of mammary gland carcinoma in virgin Sprague–Dawley rats by 1-methyl-1-nitrosourea (MNU) is blocked by the administration of DFMO to rats in the drinking water (51, 52). Administration of putrescine alone to the diet of MNU-treated rats did not affect the number of induced cancers. However, when putrescine was combined with DFMO, it significantly blocked the inhibitory effect of DFMO (52). DFMO treatment has also been shown to prevent urinary bladder carcinogenesis in the heterotopically transplanted rat urinary bladder (HTB) model (20). In this model (HTB), whole rat urine or specific urinary fractions have been observed to enhance MNU-induced urinary bladder carcinogenesis. DFMO administered in conjunction with promotion treatments with normal rat urine inhibits bladder carcinogenesis in HTBs (33).

In these studies (13, 22, 51, 52), DFMO treatment was started after administration of the carcinogen, showing that DFMO appears to be effective in blocking the promotion stage of carcinogenesis. The fact that DFMO is not an effective inhibitor in the presence of putrescine implies that the inhibition of putrescine synthesis by DFMO is involved in mediating its inhibitory effect against carcinogenesis (33, 52).

Treatment with DFMO at high dosages has several side effects which may be attributable to polyamine depletion. Thus, DFMO suppresses early embryogenesis and arrests embryonic development in experimental animals (16), enhances ovulation in the rat (7), and impairs the development of the brain in preweanling rats (45).

III. PRACTICAL CONSIDERATION

Since ODC induction and the accumulation of putrescine are characteristic tissue responses to tumor promoters and higher than normal ODC activity is observed in established tumors, ODC activity and polyamine levels can be a potential marker for cancer prognosis. In this context, the work of Luk and Baylin (26) and Kingsnorth *et al.* (23) is noteworthy. Basal ODC levels are increased in the colon mucosa of patients with familial polyposis, an inherited disorder associated with a high incidence of colon carcinoma (26). Polyamine levels in breast cancer correlate with tumor aggressiveness (23). Based on the known facts, DFMO, an enzyme-activated irreversible inhibitor of ODC, may be a useful compound in the prevention and treatment of human cancer.

Because a number of chemopreventive agents, such as retinoids and DFMO, inhibit carcinogenesis by precluding increases in ODC activity, we developed a method for TPA induction of ODC activity *in vitro* in single 3-mm human punch biopsy samples (62). This methodology will aid the rational selection of agents and their doses and dose schedules for use in cancer prevention trials. For example, we have found that oral administration of the antitumor promoters, retinoic acid and DFMO, preclude the induction of ODC by TPA in mouse skin biopsy samples; similar data have also been obtained by us in human volunteers (25). Thus, TPA induction of ODC in human skin punch biopsy samples and its inhibition by orally administered agents may provide a simple model system for establishing a nontoxic dosage regimen that will be applicable to the prevention of cancer of other epithelial systems such as breast, colon, urinary bladder, and lung.

IV. SUMMARY AND CONCLUSIONS

TPA-induced ODC activity is one of the essential biochemical events of the mechanism of skin tumor promotion (34, 50, 57). Polyamine biosynthesis, especially putrescine formation, plays an essential role in the induction of cancers of breast, colon, and bladder in experimental animals; prevention of putrescine synthesis by the specific inhibitor, DFMO, prevents carcinogenesis in these models (13, 22, 51, 52). Evidence also indicates that DFMO may inhibit the growth of established tumors and may prevent metastasis, especially in combination with specific chemotherapeutic agents (see Chapter 10).

REFERENCES

1. Argyris, T.S. (1980). *J. Invest. Dermatol.* **75,** 360–362.
2. Ashendel, C.L., Staller, J.M., and Boutwell, R.K. (1983). *Cancer Res.* **43,** 4327–4332.
3. Bachrach, U. (1980). *In* "Polyamines in Biomedical Research" (J.M. Gaugas, ed.), pp. 81–107. Wiley, New York.
4. Bisschop, W., Van Rooijen, L.A., Derks, H.J., and Van Wijk, R. (1981). *Carcinogenesis* **2,** 1283–1287.
5. Boutwell, R.K. (1974). *CRC Crit. Rev. Toxicol.* **2,** 419–443.
6. Boutwell, R.K., Verma, A.K., and Ashendel, C.L. (1982). *In* "Carcinogenesis—A Comprehensive Survey" (E. Hecker, N.E. Fuseing, W. Kunz, F. Marks, and H.W. Thielmann, eds.), Vol. 7, pp. 1–12. Raven, New York.
7. Carpenter, J.R., and Fozard, J.R. (1982). *Eur. J. Pharmacol.* **80,** 263–266.
8. Castagna, M., Takai, Y., Kaibuchi, K., Sano, K., Kikkawa, U., and Nishizuki, Y. (1982). *J. Biol. Chem.* **257,** 7847–7851.
9. Clark-Lewis, I., and Murray, A.W. (1978). *Cancer Res.* **38,** 494–497.
10. Dinowitz, M., Nims, R., Bhooshan, B., Kouri, R.E., and Henry, C.J. (1980). *Proc. Am. Assoc. Cancer Res.* **21,** 31.
11. Driedger, P.E., and Blumberg, P.M. (1980). *Proc. Natl. Acad. Sci. U.S.A.* **77,** 567–571.
12. Egner, P.A., and Kensler, T.S. (1985). *Carcinogenesis* **6,** 1167–1172.
13. Fozard, J.R., and Prakash, N.J. (1982). *Arch. Pharmacol.* **28,** 4–6.
14. Fukushima, S., Friedell, G.H., Jacobs, J.B., and Cohen, S.M. (1981). *Cancer Res.* **41,** 3100–3103.
15. Fürstenberger, G., Kinzel, V., Schwarz, M., and Marks, F. (1985). *Science* **230,** 76–78.
16. Gupta, C., Sonawane, B.R., Yaffe, S.J., and Shapiro, B.H. (1980). *Science* **208,** 505–508.
17. Hecker, E., and Schmidt, R. (1974). *In* "Progress in the Chemistry of Organic Natural Products," Vol. 31, pp. 337–467. Springer-Verlag, Berlin.
18. Hennings, H., and Yuspa, S.H. (1985). *J. Natl. Cancer Inst.* **75,** 735–740.
19. Hicks, R.M., and Chowaniec, J. (1977). *Cancer Res.* **37,** 2943–2949.
20. Humma, Y., Ozono, S., Numata, I., Seidenfeld, J., and Oyasu, R. (1985). *Cancer Res.* **45,** 648–652.
21. Kikkawa, U., Takai, Y., Tanaka, Y., Miyake, R., and Nishizuka, Y. (1983). *J. Biol. Chem.* **258,** 11442–11445.

22. Kingsnorth, A.N., King, W.W.K., Diekema, K.A., McCann, P.P., Ross, J.S., and Malt, R.A. (1983). *Cancer Res.* **43,** 2245–2249.
23. Kingsnorth, A.N., Wallace, H.M., Bundred, N.J., and Dixon, J.M.J. (1984). *Br. J. Surg.* **71,** 352–356.
24. Leach, K.L., and Blumberg, P.M. (1985). *Cancer Res.* **45,** 1958–1963.
25. Loprinzi, C.L., Verma, A.K., Boutwell, R.K., and Carbone, P.P. (1985). *J. Clin. Oncol.* **3,** 751–757.
26. Luk, G.D., and Baylin, S.B. (1984). *N. Engl. J. Med.* **311,** 80–83.
27. Matsushima, M., Takano, S., Ertürk, E., and Bryan, G.T. (1982). *Cancer Res.* **42,** 3587–3591.
28. Metcalf, B.W., Bey, P., Danzin, C., Jung, M.J., Casara, P., and Vevert, J.P. (1978). *J. Am. Chem. Soc.* **100,** 2551–2553.
29. Mufson, R.A., Fischer, S.M., Verma, A.K., Gleason, G.L., Slaga, T.J., and Boutwell, R.K. (1979). *Cancer Res.* **39,** 4791–4795.
30. Nakadate, T., Yamamoto, S., Iseki, H., Sonoda, S., Takemura, S., Ura, A., Hosoda, Y., and Kato, R. (1982). *Gann* **73,** 841–843.
31. Neidel, J.E., Kuhn, L.J., and Vandenbark, G.R. (1983). *Proc. Natl. Acad. Sci. U.S.A.* **80,** 36–40.
32. Nishizuka, Y. (1984). *Nature (London)* **308,** 693–698.
33. Nowels, K., Homma, Y., Seidenfeld, J., and Oyasu, R. (1986). *Cancer Biochem. Biophys.* **8,** 257–263.
34. O'Brien, T.G. (1976). *Cancer Res.* **36,** 2644–2653.
35. O'Brien, T.G., Simsiman, R.C., and Boutwell, R.K. (1975). *Cancer Res.* **35,** 2426–2433.
36. Ordman, A.B., Cleaveland, J.S., and Boutwell, R.K. (1985). *Cancer Lett.* **29,** 79–84.
37. Pegg, A.E., and McCann, P.P. (1982). *Am. J. Physiol.* **243,** C212–C221.
38. Peraino, C., Fry, R.J.M., Staffeldt, E., and Kisieleski, W.E. (1973). *Cancer Res.* **33,** 2701–2705.
39. Pitot, H.C., and Sirica, A.E. (1980). *Biochim. Biophys. Acta* **605,** 191–215.
40. Reddy, B.S., Weisburger, J.H., and Wynder, E.L. (1978). *In* "Carcinogenesis—A Comprehensive Survey" (T.J. Slaga, A. Sivak, and R.K. Boutwell, eds.), Vol. 2, pp. 453–464. Raven, New York.
41. Russell, D.H., and McVicker, T.A. (1971). *Biochim. Biophys. Acta,* **244,** 85–93.
42. Schulte-Hermann, R. (1985). *Arch. Toxicol.* **57,** 147–158.
43. Slaga, T.J., Fischer, S.M., Nelson, K., and Gleason, G.L. (1980). *Proc. Natl. Acad. Sci. U.S.A.* **77,** 3659–3663.
44. Slaga, T.J., Klein-Szanto, A.J. P., Fischer, S.M., Weeks, C.E., Nelson, K., and Major, S. (1980). *Proc. Natl. Acad. Sci. U.S.A.* **77,** 2251–2254.
45. Slotkin, T.A., Grignolo, A., Whitmore, W.L., Lerea, L., Trepanier, P.A., Barnes, G.A., Weigel, S.J., Seidler, F.J., and Bartolome, J. (1982). *J. Pharmacol. Exp. Ther.* **222,** 746–751.
46. Sugimura, T., Fujiki, H., Mori, M., Nakayasu, M., Terada, M., Umezawa, K., and Moore, R.E. (1982). *In* "Carcinogenesis—A Comprehensive Survey" (E. Hecker, N.E. Fuseing, W. Kunz, F. Marks, and H.W. Thielmann, eds.), Vol. 7, pp. 69–73. Raven, New York.
47. Takano, S., Matsushima, M., Ertürk, E., and Bryan, G.T. (1981). *Cancer Res.* **41,** 624–628.
48. Takano, T., Takigawa, M., and Suzuki, F. (1981). *Med. Biol.* **59,** 423–427.
49. Takigawa, M., Takano, T., and Suzuki, F. (1981). *J. Cell. Physiol.* **106,** 259–268.
50. Takigawa, M., Verma, A.K., Simsiman, R.C., and Boutwell, R.K. (1983). *Cancer Res.* **43,** 3732–3738.

51. Thompson, H.J., Herbst, E.J., Meeker, L.D., Ronan, A.M., and Minocha, R. (1984). *Carcinogenesis* **5,** 1649–1651.
52. Thompson, H.J., Meeker, L.D., Herbst, E.J., Ronan, A.M., and Minocha, R. (1985). *Cancer Res.* **45,** 1170–1173.
53. Verma, A.K. (1983). *In* "Cancer: Etiology and Prevention" (R.G. Crispen, ed.), pp. 241–260. Elsevier, New York.
54. Verma, A.K. (1985). *Biochim. Biophys. Acta* **846,** 109–119.
55. Verma, A.K., and Boutwell, R.K. (1980). *Carcinogenesis* **1,** 271–276.
56. Verma, A.K., and Boutwell, R.K. (1981). *Biochem. Biophys. Res. Commun.* **110,** 375–383.
57. Verma, A.K., Shapas, B.G., Rice, H.M., and Boutwell, R.K. (1979). *Cancer Res.* **39,** 419–425.
58. Verma, A.K., Ashendel, C.L., and Boutwell, R.K. (1980). *Cancer Res.* **40,** 308–315.
59. Verma, A.K., Conrad, E.A., and Boutwell, R.K. (1980). *Carcinogenesis* **1,** 607–611.
60. Verma, A.K., Conrad, E.C., and Boutwell, R.K. (1982). *Cancer Res.* **42,** 3519–3525.
61. Verma, A.K., Bryan, G.T., and Reznikoff, C.A. (1985). *Carcinogenesis* **6,** 427–432.
62. Verma, A.K., Loprinzi, C.L., Boutwell, R.K., and Carbone, P.P. (1985). *J. Natl. Cancer Inst.* **75,** 85–90.
63. Verma, A.K., Pong, R.C., and Erickson, D. (1986). *Cancer Res.* **46,** 6149–6155.
64. Weeks, C.E., Herrmann, A.L., Nelson, F.R., and Slaga, T.J. (1982). *Proc. Natl. Sci. U.S.A.* **79,** 6028–6032.
65. Weinstein, I.B., Gattoni-Celli, S., Kirschmeier, P., Hsiao, W., Horowitz, A., and Jeffrey, A. (1984). *Fed. Proc., Fed. Am. Soc. Exp. Biol.* **43,** 2287–2294.

12

Inhibition of Polyamine Biosynthesis in Microorganisms

ALAN J. BITONTI AND PETER P. MCCANN

Merrell Dow Research Institute
2110 East Galbraith Road
Cincinnati, Ohio 45215

I. INTRODUCTION

It is our intention with this review to cover the work dealing with the inhibition of polyamine biosynthesis in selected microorganisms, including bacteria, fungi, and mycoplasma. Since the majority of work with inhibitors of polyamine biosynthesis has been done with bacteria, this will be our emphasis. Specific reference to the other microorganisms will be made when appropriate.

A. Polyamine Pathways

Bacteria generally contain putrescine and spermidine as their only polyamines, although *Acetobacteria* spp. are exceptional in that they also synthesize spermine (Paulin *et al.,* 1983). Yeast, other fungi, and some eucaryotic protozoans synthesize all three polyamines (Tabor and Tabor, 1985), while *Mycoplasma dispar* may contain cadaverine as its only polyamine (Poso *et al.,* 1984).

The biosynthetic pathways are detailed schematically in Fig. 1. Putrescine can be produced by bacteria in a single step by ornithine decarboxylase or in two steps by the sequential action of arginine decarboxylase and agmatinase (agmatine ureohydrolase; Morris and Pardee, 1966). It has

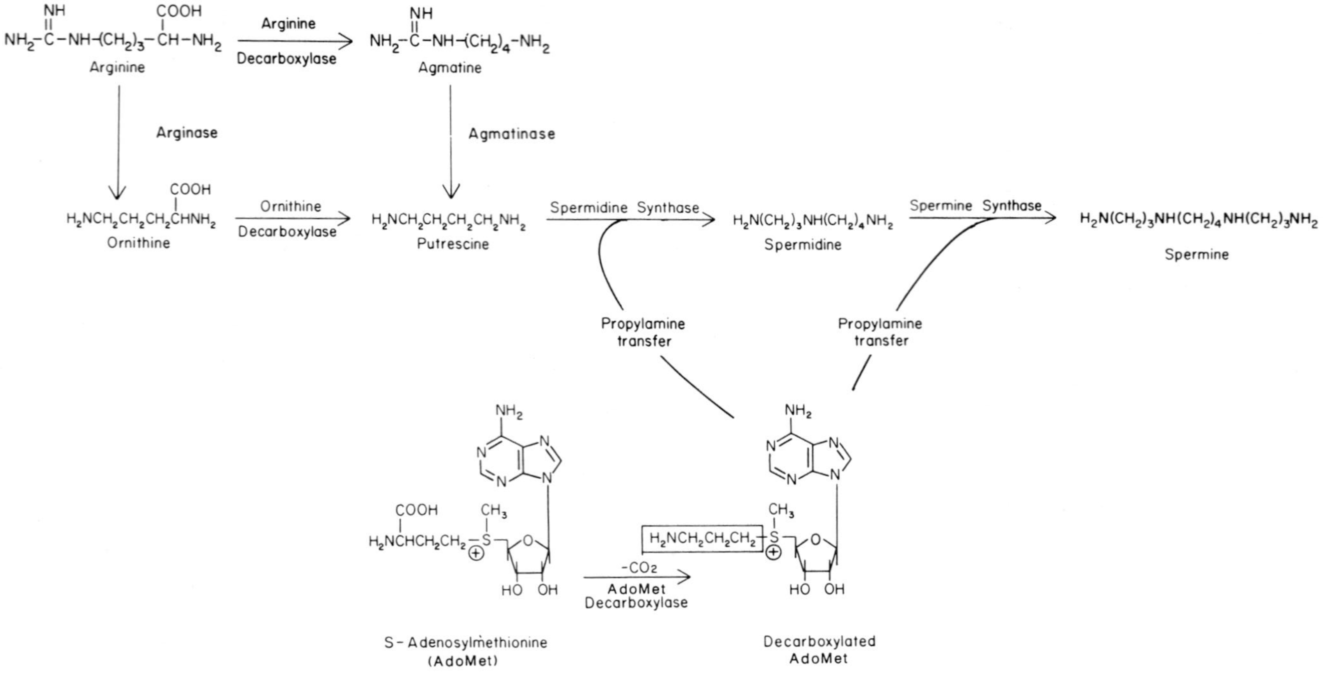

Fig. 1. Polyamine biosynthetic pathways.

been reported that putrescine may also be formed via *N*-carbamoylputrescine in *Streptococcus faecalis* (Roon and Barker, 1972) and in *Pseudomonas aeruginosa* (Haas *et al.*, 1984). There is also a single report of the decarboxylation of citrulline by *Escherichia coli* to form *N*-carbamoylputrescine (Akamatsu *et al.*, 1978). Putrescine could then be formed by the action of *N*-carbamoylputrescine aminohydrolase (Pegg and Williams-Ashman, 1981). We have not been able to confirm the presence of citrulline decarboxylase activity in *E. coli* using a direct assay, which measures the release of $^{14}CO_2$ from carboxyl-[^{14}C]citrulline (A.J. Bitonti, unpublished observations). Cataldi and Algranati (1986) have recently reported the possible existence of yet another pathway for the biosynthesis of putrescine in *E. coli* BGA8, a mutant bacteria which lacks ornithine decarboxylase and agmatinase but is still able to maintain nearly normal levels of putrescine and spermidine.

Fungi and eucaryotic protozoa apparently contain only ornithine decarboxylase as a source of putrescine (Tabor and Tabor, 1984; Pegg and Williams-Ashman, 1981). In bacteria both ornithine and arginine decarboxylase occur in two forms, the biosynthetic and biodegradative decarboxylases. The biodegradative enzymes originally described by Gale (1940) are inducible by growing bacteria in acidic media which are rich in the respective amino acids. Bacteria grown under normal conditions of neutral pH contain only the biosynthetic forms of the enzymes. All four decarboxylases have been purified to homogeneity and characterized (Morris and Boeker, 1983). All have in common dependence on pyridoxal phosphate as a cofactor, a point which is important in a discussion of inhibition of these enzymes since many of the reversible inhibitors as well as the enzyme-activated irreversible inhibitors are dependent on interaction with pyridoxal phosphate for their inhibitory effects.

A third decarboxylase, *S*-adenosylmethionine (AdoMet) decarboxylase, is involved in the synthesis of polyamines in microorganisms as it is in animal cells. This enzyme produces decarboxylated AdoMet which in turn becomes a donor of an aminopropyl moiety for the synthesis of spermidine and spermine from putrescine. The enzyme has been purified from *E. coli* and, unlike the ornithine and arginine decarboxylases, does not require pyridoxal phosphate but has a covalently attached pyruvoyl group (Wickner *et al.*, 1970) which forms a Schiff base intermediate with the substrate and is necessary for activity (Pankaskie and Abdel-Monem, 1980; Allen and Klinman, 1981; Markham *et al.*, 1982). AdoMet decarboxylase from bacteria is strongly activated by Mg^{2+} (Wickner *et al.*, 1970; Stevens *et al.*, 1978; Pösö *et al.*, 1976; Coppoc *et al.*, 1971), unlike

AdoMet decarboxylase from yeast (Jänne *et al.*, 1971; Pösö *et al.*, 1975a) or other fungi (Hart *et al.*, 1978), which is strongly activated by putrescine. Protozoal AdoMet decarboxylases appear to be both Mg^{2+} and putrescine insensitive (Pösö *et al.*, 1975a, 1976).

Spermidine is formed from putrescine by spermidine synthase (putrescine aminopropyltransferase), which transfers an aminopropyl moiety from decarboxylated AdoMet to a terminal amine of putrescine. Spermine is formed by spermine synthase (spermidine aminopropyltransferase) using decarboxylated AdoMet as an aminopropyl donor. Spermidine synthase has been purified from *E. coli* (Bowman *et al.*, 1973) while spermine synthase has not been isolated from those microorganisms in which it exists.

B. Polyamines and Rapid Growth

As in most other cells studied, polyamines appear to be essential for optimal growth of most microorganisms. In some microorganisms, such as the bacteria *Haemophilus influenzae* (Herbst and Snell, 1948) and mutants of the fungi *Aspergillus nidulans* (Sneath, 1955) and *Neurospora crassa* (Deters *et al.*, 1974), the polyamines are absolute requirements for cell proliferation.

In other microorganisms, such as *E. coli* or *Saccharomyces cerevisiae*, the polyamines are not absolutely required for growth but in the absence of polyamines growth is abnormally slow. this can be shown in studies utilizing mutants which lack measurable activities of one or more of the polyamine biosynthetic enzymes. Quadruple mutants of *E. coli* K12 have been constructed which lack arginine decarboxylase, agmatine ureohydrolase, ornithine decarboxylase, and *S*-adenosylmethionine decarboxylase. These bacteria do not synthesize putrescine or spermidine and grow at a rate only one-third that of wild-type *E. coli* (Hafner *et al.*, 1979). In a similar manner it was demonstrated that deficiencies in polyamine biosynthetic capability in yeast lead to a decreased growth rate (Cohn *et al.*, 1978; Whitney and Morris, 1978). It follows that use of specific enzyme inhibitors to interrupt polyamine biosynthesis in microorganisms should lead to a decrease in or a cessation of growth.

There are, as covered above, multiple enzyme sites suitable for intervention in the polyamine biosynthetic pathway; each specific enzyme has been attacked with numerous chemical agents in search of a potent inhibitor both of the enzyme activity and of polyamine biosynthesis *in vivo*. Below we will outline work done with inhibitors of each enzyme of the polyamine pathway.

II. INHIBITION OF ORNITHINE DECARBOXYLASE

As has been the case in animal cell studies, ornithine decarboxylase (ODC) of microorganisms has been the most popular target enzyme for the design of inhibitors. Both competitive and irreversible "suicide" inhibitors of these ornithine decarboxylases have been studied.

A. Competitive Reversible Inhibitors

The first of this type of inhibitor to be examined was α-hydrazinoornithine (Harik and Snyder, 1973). This compound inhibited *E. coli* ODC and was competitive with respect to the substrate ornithine with a $K_i = 5 \times 10^{-7}\ M$ at optimal pyridoxal phosphate concentration. Its inhibitory activity was diminished in the presence of 10- to 100-fold excess concentrations of pyridoxal phosphate, suggesting that α-hydrazinoornithine may interact with the carbonyl functionality of the cofactor. α-Hydrazinoornithine was more potent against *E. coli* ODC than are other hydrazino compounds, such as α-hydrazinohistidine or hydrazine, probably due to its resemblance to the substrate ornithine. The compound was fairly selective for ODC but did inhibit glutamic acid decarboxylase and DOPA decarboxylase at higher concentrations.

α-Methylornithine is another ornithine analog which competitively and reversibly inhibits mammalian (Abdel-Monem *et al.*, 1975) and *Lactobacillus* 30a (O'Leary and Herreid, 1978) ornithine decarboxylases. O'Leary and Herreid (1978) reported that α-methylornithine, while functioning as a competitive inhibitor, was also acting as a substrate for ODC, being decarboxylated by the enzyme from *Lactobacillus* 6000 times slower than the natural substrate ornithine. α-Methylornithine also has been used to inhibit ODC in *S. cerevisiae* and when added to growth medium the compound significantly slowed the growth of the yeast in parallel with depletion of polyamines (Brawley and Ferro, 1979).

Two putrescine analogs, 1,4-diaminobutanone (Stevens *et al.*, 1977) and 2-hydroxyputrescine (Stevens and Stevens, 1980), have been found to inhibit ODC in microbes. The first compound, 1,4-diaminobutanone, had a K_i of 0.9 μM against ODC in extracts from *A. nidulans*. The inhibition was competitive with respect to substrate and fully reversible by dialysis. When added to the growth medium, 1,4-diaminobutanone actually increased ODC activity when activity was measured in dialyzed extracts. This was probably due to stabilization of ODC as occurred with α-methylornithine in rat hepatoma cells (McCann *et al.*, 1977). 1,4-Diaminobutanone also caused a rise in putrescine concentration in *A. nidulans* but caused an 85 and 60% decrease in spermidine and spermine, respectively, when added at 0.6 m*M* to the growth medium. Germination was delayed by treatment with the drug. Against ODC from *E. coli* and *Dictyostelium discoideum,* 1,4-diaminobutanone was a weak competitive inhibitor (Stevens *et al.,* 1977). 2-Hydroxyputrescine also inhibited ODC from *A. nidulans* but was less potent (K_i = 60 μM) than 1,4-diaminobutanone (Stevens and Stevens, 1980).

More recently, 1-aminooxy-3-aminopropane was found to competitively inhibit *E. coli* ODC with a K_i = 1.0 n*M* (Paulin, 1986). This is by far the most potent inhibitor of bacterial ODC yet described. No data on the effects of this compound on *E. coli* polyamine concentrations or growth were available.

B. Irreversible Inhibitors

The most extensively tested compound in this category is DL-α-difluoromethylornithine (DFMO, eflornithine), which is an enzyme-activated irreversible inhibitor whose properties are described more fully elsewhere in this volume (Bey, Danzin and Jung, Chapter 1). DFMO was found to be an irreversible inhibitor of *P. aeruginosa* ODC (K_i = 1.4 mM) but not of the *E. coli* or *Klebsiella pneumoniae* enzymes (Kallio and McCann, 1981). The specific reason for the lack of effect of DFMO on the latter two ODC activities is not known. It is noteworthy, however, that DL-α-monofluoromethylornithine (MFMO) irreversibly inhibited ODC from both *P. aeruginosa* and *E. coli* (Bitonti *et al.,* 1982). Moreover, a putrescine analog, DL-α-monofluoromethylputrescine, was a potent irreversible inhibitor of *E. coli* ODC but had little or no effect against the enzyme from *P. aeruginosa* (Kallio *et al.,* 1982), in contrast to the findings with DFMO. Irreversible inhibition of bacterial ODCs by product and substrate analogs appears critically dependent on the number of fluorine atoms incorporated into the α-methyl group, since DL-α-difluoromethylputrescine was an inhibitor of neither *E. coli* nor *P. aeruginosa* ODC activities (A.J. Bitonti, unpublished observations). ODC from an extremely thermophilic bacteria, *Clostridium thermohydrosulfuricum,* was also somewhat resistant to inhibition by DFMO, with 50 m*M* DFMO giving only 64% inhibition after incubation with ODC for 30 min at 55°C (Pösö and Paulin, 1983).

The inhibition of bacterial ODCs produced by DFMO, MFMO, and DL-α-monofluoromethylputrescine was time-dependent, followed pseudo first-order kinetics, and was slowed by the natural substrate L-ornithine, suggesting that the inhibitors were all active-site directed. The inactivation of ODC by the three compounds was apparently irreversible since extensive dialysis of the inactivated enzymes failed to restore catalytic activity (Kallio and McCann, 1981; Kallio *et al.,* 1982; Bitonti *et al.,* 1982).

Recently DFMO has been shown to inhibit purified ODC from the cytosol and nucleus of the ciliated protozoan *Tetrahymena pyriformis* (Sklaviadis *et al.,* 1985). In *Physarum polycephalum* two distinct forms of ODC activity have been described, and in the presence of physiological concentrations of pyridoxal phosphate *in vitro* only the A-form of the enzyme was sensitive to inhibition by DFMO (Mitchell *et al.,* 1981). Only the A-form of the enzyme was inhibited by DFMO *in vivo,* but if ODC was assayed *in vitro* in the presence of high concentrations (20 μM or higher) of pyridoxal phosphate, the B-form of ODC was also susceptible to inactivation by DFMO. Yeast ODC is sensitive to DFMO (Tyagi *et al.,* 1982), and because of the specificity of DFMO for ODC, 5-[^{14}C]DFMO has been used to titrate the amount of enzyme present in *S. cerevisiae* and *Saccharomyces uvarum* (Pösö and Pegg, 1983).

There are several further particularly noteworthy reports on effects of DFMO in microorganisms. Rajam and Galston (1985) recently demonstrated that DFMO depletes polyamines and strongly inhibits the growth of four phytopathogenic fungi: *Botrytis* sp., *Botrytis cinerea, Rhizoctonia solani,* and *Monilinia fructiola.* The effects were seen at relatively low concentrations of DFMO (0.1, 0.5, and 1.0 m*M*) in the growth medium. In the same study α-difluoromethylarginine (DFMA) also inhibited the growth of the fungi but this was most likely due to intracellular conversion of DFMA to DFMO. In a second report Rajam *et al.* (1985) demonstrated that a spray containing 0.5% DFMO completely protected pinto bean (*Phaseolus vulgaris*) plants which were infected with the bean rust fungus, *Uromyces phaseoli.* The bean rust fungus would appear to be one of the most sensitive organisms yet described with regard to its response to DFMO. It was also recently demonstrated that DFMO completely inhibited ODC activity in *Helminthosporium maydis* and completely arrested sporulation of this plant fungus (Birecka *et al.,* 1986). A recent *in vitro* finding has been that DFMO will deplete polyamines in three yeast species that are human pathogens, *Candida albicans, Candida tropicalis,* and *Candida parapsilosis. Candida tropicalis* was significantly more susceptible to DFMO than the other two species. The action of DFMO was specifically reversed by exogenous polyamines added to the medium (Pfaller and Gerarden, 1987). The use of DFMO for the control of some phytopathogenic and human pathogenic fungi appears to be a promising area for future research.

While the irreversible inhibitors of ODC have been shown to inhibit ODC in bacteria *in vivo* (Table I), they have virtually no effect on intracellular polyamine concentrations since these organisms have the alternate

TABLE I

Effects of DL-α-Monofluoromethylputrescine (MFMP) and DL-α-Difluoromethylarginine (DFMA) on Ornithine Decarboxylase, Arginine Decarboxylase, and Polyamines in *E. coli*

Compound[a]	Ornithine decarboxylase (μmol CO_2 · min^{-1} · mg protein^{-1})	Arginine decarboxylase	Putrescine (nmol/10^8 cells)	Spermidine
None	0.97	0.64	1.58 ± 0.15	0.34 ± 0.06
MFMP	0.05	0.73	1.63 ± 0.13	0.33 ± 0.05
DFMA	1.28	0.03	1.45 ± 0.26	0.33 ± 0.04
MFMP + DFMA	0.06	0.05	0.54 ± 0.12	0.46 ± 0.06

[a] MFMP and DFMA were at a concentration of 1 m*M* when present.

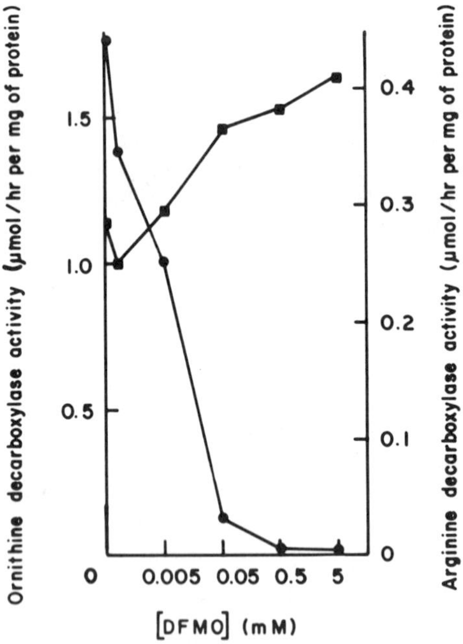

Fig. 2. Inhibition of ornithine decarboxylase and stimulation of arginine decarboxylase by DL-α-difluoromethylornithine in *P. aeruginosa*. Cells were grown in minimal medium in the presence or absence of various concentrations of DFMO, then sedimented, extracted, and assayed for either ODC (●) or ADC (■) activity.

arginine decarboxylase/agmatinase route for the biosynthesis of putrescine (see Fig. 1). In fact, as ODC activity is progressively inhibited in *P. aeruginosa* by DFMO, arginine decarboxylase activity is induced (Fig. 2; Kallio and McCann, 1981). One would guess from these data that inhibition of both ODC and arginine decarboxylase (ADC) would be necessary to lower putrescine and spermidine concentrations. In a mutant *E. coli* DR112 (speA, speB) which lacks arginine decarboxylase and agmatinase it is possible to deplete putrescine and spermidine and slow the growth rate with the addition of DL-α-monofluoromethylputrescine, an inhibitor of ODC activity, to the growth medium (Igarashi *et al.*, 1986).

III. INHIBITION OF ARGININE DECARBOXYLASE

Among the microorganisms, only bacteria have been found to contain arginine decarboxylase activity. This activity appears to be important for the maintenance of polyamine levels within the cell since putrescine remains unchanged when ODC is completely inhibited by a proper irreversible inhibitor (Kallio *et al.*, 1982).

One competitive reversible inhibitor of ADC is known, α-methylarginine. This compound inhibited both the biosynthetic (K_i = 0.27 mM) and biodegradative (K_i = 6 mM) ADCs from *E. coli* (Kallio *et al.*, 1981). DL-α-Difluoromethylarginine also inhibited both biosynthetic (K_i = 0.8 mM) and biodegradative (K_i = 0.14 mM) ADCs from *E. coli* as well as the biosynthetic ADCs from *P. aeruginosa* and *K. pneumoniae* (Kallio *et al.*, 1981). L-Arginine protected ADC against inhibition by DFMA, suggesting that the inhibitor was active-site directed. Furthermore, the loss of ADC activity due to DFMA was time and temperature dependent and could not be reversed by extensive dialysis, suggesting that the inhibition was enzyme catalyzed and irreversible. Five new irreversible inhibitors of *E. coli* ADC activity which are 10–100 times as potent as DFMA were described recently by Bitonti *et al.* (1987). The kinetic properties of α-monofluoromethylarginine, (E)-α-monofluoromethylagmatine, α-ethynylagmatine, and α-allenylagmatine are described in Chapter 1 (this volume).

DFMA was shown to inhibit ADC in *E. coli* (Table I) and *P. aeruginosa* when the cells were grown in the presence of 1 mM of the inhibitor, giving clear evidence that DFMA was taken up by the cells. However, the drug had no effect on the putrescine concentration of cells since the alternate route to putrescine through ODC was not inhibited. When *E. coli* were grown in the presence of DFMA, to inhibit ADC, and in the presence of α-monofluoromethylputrescine, to inhibit ODC, intracellular putrescine concentration did decrease by about 65%; however, the decrease in putrescine was offset by a 40% rise in spermidine (Table I) and growth was unaffected (Kallio *et al.*, 1982).

IV. INHIBITION OF *S*-ADENOSYLMETHIONINE DECARBOXYLASE

A. Diguanidines

S-Adenosylmethionine decarboxylase (AdoMetDC) varies widely among species in regard to its responsiveness to cations, as detailed above, but the enzymes from various sources are remarkably similar in their responses to inhibitors. Methylglyoxal bis(guanylhydrazone) (MGBG) is the most extensively studied inhibitor of AdoMetDC and was originally described as an inhibitor of putrescine-sensitive mammalian and yeast AdoMetDC (Williams-Ashman and Schenone, 1972). MGBG is competitive with respect to *S*-adenosylmethionine and its inhibition is reversible (Hölttä *et al.,* 1973; Corti *et al.,* 1974). The Mg^{2+}-stimulated AdoMetDC from *E. coli* is often regarded as being insensitive to MGBG since much higher concentrations of MGBG are needed to inhibit the *E. coli* enzyme (Williams-Ashman and Schenone, 1972; Pösö *et al.,* 1975b) than the mammalian enzyme. However, in more recent studies with purified *E. coli* AdoMetDC, a K_i for MGBG of 13 μM was obtained (Pegg and Jacobs, 1983; Markham *et al.,* 1982). Although this K_i is some 40 times higher than the K_i for MGBG against mammalian AdoMetDC, the compound is nonetheless quite a potent inhibitor of the bacterial enzyme. MGBG also inhibits AdoMetDC from *P. polycephalum* (Mitchell and Rusch, 1973) and from *T. pyriformis* (Pösö *et al.,* 1975b) to an extent similar to that seen with *E. coli* enzyme.

The ethyl analog of MGBG, ethylglyoxal bis(guanylhydrazone), or EGBG, is a potent inhibitor of AdoMetDC from microorganisms. The K_i values for EGBG against AdoMet DC from *S. cerevisiae, P. polycephalum,* and *E. coli* were 0.1, 2.6, and 1.6 μM, respectively (Pegg and Jacobs, 1983).

The addition of an amino group to each of the aminoguanidine moieties of MGBG to form 1,1′-[(methylethanediylidene)dinitrilo]bis(3-aminoguanidine), or MBAG, results in the formation of an irreversible inhibitor of AdoMetDC (Pegg and Conover, 1976). MBAG was very potent against AdoMetDC activity from yeast but less potent against the enzyme from *E. coli* (Pegg and Conover, 1976).

B. Analogs of S-Adenosylmethionine

A few analogs of the AdoMetDC substrate *S*-adenosylmethionine have been tested for inhibition of AdoMetDC. Of these analogs, two compounds were reasonably potent when tested against AdoMetDC from *S.*

cerevisiae, *P. polycephalum*, and *E. coli*. *S*-Adenosyl-3-methylthiopropylamine and 5′-(dimethylsulfonio)-5′-deoxyadenosine had K_i values in the range of 7–13 μM against the enzyme from the aforementioned sources (Pegg and Jacobs, 1983). *S*-(5′-Deoxy-5′-adenosyl)methylthiohydroxylamine was also reported to inhibit *E. coli* AdoMetDC (Paulin, 1986). Although kinetic constants were not reported, the compound was shown to inactivate AdoMetDC in a time-dependent fashion and the inhibited enzyme did not regain activity upon extensive dialysis. None of the compounds was tested for effects *in vivo*.

C. Other Inhibitors

Two structural analogs of MGBG, Berenil (4,4′-diaminidinodiazoaminobenzene) and pentamidine [*p*,*p*′-(pentamethylenedioxy)dibenzamidine], which are also antiprotozoal agents, have recently been shown to irreversibly inhibit AdoMetDC activity derived from both yeast and *P. aeruginosa* at concentrations as low as 10 μM (Karvonen *et al.*, 1985). The inhibition by these two compounds was not reversed by extensive dialysis.

V. INHIBITION OF SPERMIDINE SYNTHASE

A. Cyclohexylamine

Hibasami *et al.* (1980) first described the inhibition of mammalian spermidine synthase with so-called dicyclohexylamine sulfate. Dicyclohexylamine sulfate is a misnomer which has been propagated in the literature since that time. Elemental and mass spectral analyses carried out in our laboratories (A.J. Bitonti, unpublished observations) as well as the laboratory of Batchelor *et al.* (1986) with the compound labeled by Sigma Chemical Company as dicyclohexylammonium sulfate revealed that this compound is actually cyclohexylamine. Therefore, the actual inhibitory substance is cyclohexylamine and will be referred to as such in this chapter.

Cyclohexylamine is a potent inhibitor of bacterial spermidine synthase, 50% inhibition being obtained with 0.2 and 0.6 μM in *E. coli* and *P. aeruginosa*, respectively (Pegg *et al.*, 1983). The compound appeared to be competitive with respect to putrescine.

Notable is the finding that cyclohexylamine added to growing *P. aeruginosa* lowered intracellular spermidine concentration greater than 90% (Table II) and slowed the growth of the bacteria (Bitonti *et al.*, 1983,

TABLE II

Effect of DL-α-Monofluoromethylornithine (MFMO), DL-α-Difluoromethylarginine (DFMA), and Cyclohexylamine (CHA) on Putrescine and Spermidine Levels in *E. coli* and *P. aeruginosa*

	E. coli		*P. aeruginosa*	
Compound[a]	Putrescine (nmol/10[8] cells)	Spermidine	Putrescine (nmol/10[8] cells)	Spermidine
Control	1.70 (100)	0.26 (100)	3.11 (100)	0.64 (100)
MFMO + DFMA	0.24 (14)	0.37 (142)	0.21 (7)	0.76 (120)
CHA	1.68 (99)	0.22 (85)	2.44 (78)	0.05 (8)
MFMO + DFMA + CHA	0.57 (34)	0.13 (50)	0.63 (20)	0.03 (5)
MFMO + DFMA + CHA + putrescine + spermidine	0.72 (42)	0.27 (104)	1.51 (49)	0.66 (103)

[a] *Escherichia coli* and *Pseudomonas aeruginosa* were grown in the presence or absence of 2 m*M* MFMO, 2.5 m*M* DFMA, and 20 m*M* CHA for *E. coli* and 10 m*M* CHA for *P. aeruginosa*. Values in parentheses represent percentage of control.

1984). When combined with α-monofluoromethylornithine (MFMO) and α-difluoromethylarginine, cyclohexylamine depleted intracellular putrescine and spermidine further and growth of *E. coli* and *P. aeruginosa* was slowed more markedly than in the presence of cyclohexylamine alone (Fig. 3). That the effects of cyclohexylamine and the two decarboxylase inhibitors were specific for polyamine biosynthesis was demonstrated by a complete reversal of the inhibitory effects by the provision of 0.1 m*M* putrescine and 0.1 m*M* spermidine concurrently with the inhibitors. That cyclohexylamine's effects on the bacteria were specific for spermidine synthesis can be inferred from the finding that cyclohexylamine was more inhibitory to the growth of *P. aeruginosa* than to the growth of *E. coli*, probably because the drug lowered *P. aeruginosa* spermidine concentration by greater than 90%, whereas *E. coli* spermidine concentration was only lowered 15% by cyclohexylamine alone or was reduced 50% when combined with DFMO and DFMA (Table II). From these experiments we would also conclude that a decrease in spermidine appears to be more critical for limiting the growth of bacteria than a decrease in putrescine. The combination of polyamine biosynthesis inhibitors not only inhibited growth, but also was apparently lethal to *Serratia marcescens* (Fig. 4) (Bitonti *et al.*, 1983). The growth inhibitory effects of the three compounds on *S. marcescens* were completely reversed by the addition of

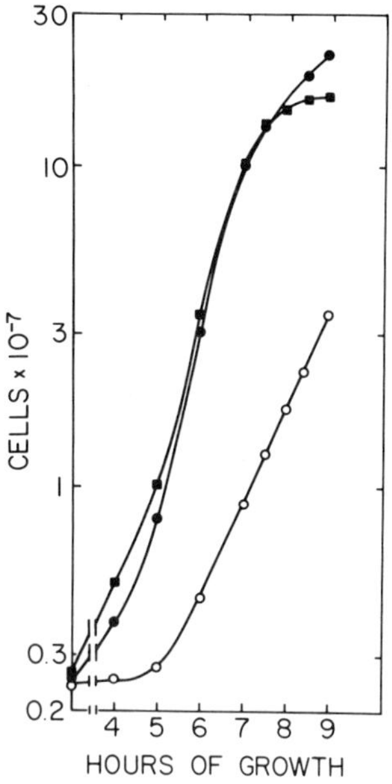

Fig. 3. Inhibition of *P. aeruginosa* growth with DL-α-monofluoromethylornithine, DL-α-difluoromethylarginine, and cyclohexylamine and reversal of inhibition by polyamines. Cultures of *P. aeruginosa* were grown without additions (●) or in the presence of 2 m*M* MFMO, 2.5 m*M* DFMA, and 10 m*M* CHA with (■) or without (○) 0.1 m*M* spermidine and 0.1 m*M* putrescine.

putrescine and spermidine to the growth medium. Cyclohexylamine alone was bactericidal to *Streptococcus uberis, Staphylococcus aureus, E. coli,* and *P. aeruginosa* when incubations of the bacteria with cyclohexylamine were continued for relatively long periods (Mattila *et al.*, 1984).

B. S-Adenosyl-1,8-diamino-3-thiooctane

S-Adenosyl-1,8-diamino-3-thiooctane (AdoDato) was originally described as an inhibitor of mammalian spermidine synthase (Tang *et al.*, 1980). The compound theoretically resembles the transition-state structure of the aminopropyltransferase (spermidine synthase) reaction and is quite specific, therefore, for spermidine synthase, having little effect on spermine synthase. AdoDato inhibits bacterial spermidine synthase, with 50% inhibition of the enzyme from *E. coli, P. aeruginosa,* and *S. marcescens* occurring at approximately 20 n*M* AdoDato (Table II) (Pegg *et al.*, 1983). Although more potent than cyclohexylamine as an inhibitor of

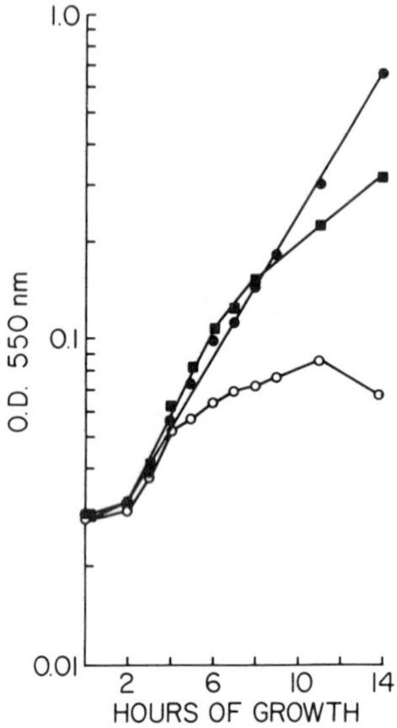

Fig. 4. Inhibition of *S. marcescens* growth with DL-α-monofluoromethylornithine, DL-α-difluoromethylarginine, and cyclohexylamine. Cultures of *S. marcescens* were grown without additions (●) or with 2 m*M* MFMO, 2.5 m*M* DFMA, and 10 m*M* CHA with (■) or without (○) 0.1 m*M* spermidine and 0.1 m*M* putrescine.

bacterial spermidine synthase *in vitro,* AdoDato had little if any effect on polyamine biosynthesis in growing bacteria and was not inhibitory to bacterial growth either alone or in combination with inhibitors of ornithine and arginine decarboxylases (Pegg *et al.,* 1983). The lack of an effect of AdoDato *in vivo* is probably due to lack of uptake of the compound.

C. Other Spermidine Synthase Inhibitors

It has recently been discovered that the dextro isomer of ethambutol, a drug useful in the treatment of tuberculosis, is a very potent inhibitor of spermidine synthase (50% inhibition at 10 μM) from *Mycobacteria* spp. but not from other bacteria, which are not killed by the drug (Paulin *et al.,* 1985). Ethambutol was also effective at limiting spermidine synthesis in growing *Mycobacteria* spp. at concentrations of the drug known to be bactericidal.

There has also been a report that 1,4-diaminobutanone is a weak competitive inhibitor of spermidine synthase from *A. nidulans* (Stevens and Stevens, 1980). There were no data on the effects of the latter compound on polyamine biosynthesis in growing *A. nidulans*.

VI. SPERMINE SYNTHASE

Little work has been done with spermine synthase in microorganisms; this is quite understandable since the most extensively studied microorganisms (with respect to polyamine biosynthesis), the bacteria, for the most part do not contain spermine synthase activity. There has been one report, however, of *Acetobacteria* which do contain spermine synthase activity that is inhibited by cyclohexylamine (Paulin *et al.,* 1983).

VII. MYCOPLASMA

Mycoplasma has been granted a separate section in this review because certain strains seem to be unique in their ability to synthesize cadaverine (diaminopentane) as their only polyamine. In these particular microorganisms, lysine decarboxylase activity is present while ODC is lacking (Pösö *et al.,* 1984). The lysine decarboxylase activity was completely inhibited by DL-α-difluoromethyllysine (DFML). The compound completely inhibited the growth of *Mycoplasma dispar* when added at a concentration of 1 m*M* at the beginning of the growth period. The growth inhibition was fully reversed by the concurrent provision of 1 m*M* cadaverine in the growth medium.

VIII. SUMMARY

All the enzymes for polyamine biosynthesis in microorganisms, with the exception of agmatinase, have been targets for inhibition with various chemical agents. In many cases the inhibitors have been rationally designed (for example the irreversible inhibitors of ODC such as α-difluoromethylornithine or the transition-state analog AdoDato for the inhibition of spermidine synthase) while others have been discovered through serendipity (as in the case of cyclohexylamine, an inhibitor of spermidine synthase). In one case a compound's effects on polyamine biosynthesis were discovered long after the compound became a therapeutic tool. A widely used tuberculostatic agent, ethambutol, was found recently to inhibit polyamine biosynthesis in sensitive *Mycobacteria,* suggesting that this effect might contribute to the drug's therapeutic efficacy. The manner in which these agents have come to be known as inhibitors of polyamine biosynthesis is not important; they have all contributed to understanding not only the routes of polyamine biosynthesis in microorganisms but also the role of polyamines in microbial physiology. And while there is promise that some of the inhibitors (e.g., DFMO) would, in fact, become useful agents for the treatment of diseases such as bean rust caused by phytopathogenic fungi, the development of specific enzyme inhibitors as research tools also remains a goal of high importance.

REFERENCES

Abdel-Monem, M.M., Newton, N.E., Ho, B.C., and Weeks, C.E. (1975). *J. Med. Chem.* **18,** 600–604.

Akamatsu, N., Oguchi, M., Yajima, Y., and Ohno, M. (1978). *J. Bacteriol.* **133,** 409–410.

Allen, R.R., and Klinman, J.P. (1981). *J. Biol. Chem.* **256,** 3233–3239.

Batchelor, K.W., Smith, R.A., and Watson, N.S. (1986). *Biochem. J.* **233,** 307–308.

Birecka, H., Garraway, M.O., Baumann, R.J., and McCann, P.P. (1986). *Plant Physiol.* **80,** 798–800.

Bitonti, A.J., McCann, P.P., and Sjoerdsma, A. (1982). *Biochem. J.* **208,** 435–441.

Bitonti, A.J., Kallio, A., McCann, P.P., and Sjoerdsma, A. (1983). *Adv. Polyamine Res.* **4,** 495–506.

Bitonti, A.J., Kelly, S.E., and McCann, P.P. (1984). *Life Sci.* **34,** 1513–1520.

Bitonti, A.J., Casara, P.J., McCann, P.P., and Bey, P. (1987). *Biochem. J.* **242,** 69–74.

Bowman, W.H., Tabor, C.W., and Tabor, H. (1973). *J. Biol. Chem.* **248,** 2480–2486.

Brawley, J.V., and Ferro, A.J. (1979). *J. Bacteriol.* **140,** 214–220.

Cataldi, A.A., and Algranati, I.D. (1986). *Biochem. J.* **234,** 617–622.

Cohn, M.S., Tabor, C.W., and Tabor, H. (1978). *J. Bacteriol.* **134,** 208–213.

Coppoc, G.L., Kallio, P., and Williams-Ashman, H.G. (1971). *Int. J. Biochem.* **2,** 673–681.

Corti, A., Dave, C., Williams-Ashman, H.G., Mihich, E., and Schenone, A. (1974). *Biochem. J.* **139,** 351–357.

Deters, J., Miskimen, J., and McDongall, K.J. (1974). *Genetics* **77,** S16–S17.

Gale, E.F. (1940). *Biochem. J.* **34,** 392–413.

Haas, D., Matsumoto, H., Moretti, T., Stalon, D., and Mercenier, A. (1984). *Mol. Gen. Genet.* **193,** 437–444.

Hafner, E.W., Tabor, C.W., and Tabor, H. (1979). *J. Biol. Chem.* **254,** 12419–12426.

Harik, S.I., and Snyder, S.H. (1973). *Biochim. Biophys. Acta* **327,** 501–509.

Hart, D., Winther, M., and Stevens, L. (1978). *FEMS Microbiol. Lett.* **3,** 173–175.

Herbst, E.J., and Snell, E.E. (1948). *J. Biol. Chem.* **176,** 989–990.

Hibasami, H., Tanaka, M., Nagai, J., and Ikeda, T. (1980). *FEBS Lett.* **116,** 99–101.

Hölttä, E., Hanonnen, P., Pispa, J., and Jänne, J. (1973). *Biochem. J.* **136,** 669–676.

Igarashi, K., Kashiwagi, K., Hamasaki, H., Miura, A., Kakegawa, T., Hirose, J., and Matsuzaki, S. (1986). *J. Bacteriol.* **166,** 128–134.

Jänne, J., Williams-Ashman, H.G., and Schenone, A. (1971). *Biochem. Biophys. Res. Commun.* **43,** 1362–1368.

Kallio, A., and McCann, P.P. (1981). *Biochem. J.* **200,** 69–75.

Kallio, A., McCann, P.P., and Bey, P. (1981). *Biochemistry* **20,** 3163–3166.

Kallio, A., McCann, P.P., and Bey, P. (1982). *Biochem. J.* **204,** 771–775.

Karvonen, E., Kauppinen, L., Partanen, T., and Pösö, H. (1985). *Biochem. J.* **231,** 165–169.

McCann, P.P., Tardif, C., Duchesne, M.-C., and Mamont, P.S. (1977). *Biochem. Biophys. Res. Commun.* **76,** 893–899.

Markham, G.D., Tabor, C.W., and Tabor, H. (1982). *J. Biol. Chem.* **257,** 12063–12068.

Mattila, T., Honkanen-Buzalski, T., and Pösö, H. (1984). *Biochem. J.* **223,** 823–830.

Mitchell, J.L.A., and Rusch, H.P. (1973). *Biochim. Biophys. Acta* **297,** 503–516.

Mitchell, J.L.A., Yingling, R.A., and Mitchell, G.K. (1981). *FEBS Lett.* **131,** 305–309.

Morris, D.R., and Boeker, E.A. (1983). *In* "Methods in Enzymology" (H. Tabor and C.W. Tabor, eds.), Vol. 94, pp. 125–134. Academic Press, New York.

Morris, D.R., and Pardee, A.B. (1966). *J. Biol. Chem.* **241,** 3129–3135.

O'Leary, M.H., and Herreid, R.M. (1978). *Biochemistry* **17,** 1010–1014.

Pankaskie, M., and Abdel-Monem, M.M. (1980). *J. Med. Chem.* **23,** 121–127.

Paulin, L.G. (1986). *FEBS Lett.* **202,** 323–326.

Paulin, L., Vehmaanpera, J., Nykanen, I., and Pösö, H. (1983). *Biochem. Biophys. Res. Commun.* **114,** 779–784.

Paulin, L.G., Brander, E.E., and Pösö, H. (1985). *Antimicrob. Agents Act.* **28,** 157–159.

Pegg, A.E., and Conover, C. (1976). *Biochem. Biophys. Res. Commun.* **69,** 766–774.

Pegg, A.E., and Jacobs, G. (1983). *Biochem. J.* **213,** 495–502.

Pegg, A.E., and Williams-Ashman, H.G. (1981). *In* "Polyamines in Biology and Medicine" (D.R. Morris and L.J. Marton, eds.) pp. 3–42. Dekker, New York.

Pegg, A.E., Bitonti, A.J., McCann, P.P., and Coward, J.K. (1983). *FEBS Lett.* **155,** 192–196.

Pfaller, M.A., and Gerarden, T. (1987). *Mycopathology* **98,** 3–8.

Pösö, H., and Paulin, L. (1983). *Adv. Polyamine Res.* **4,** 489–494.

Pösö, H., and Pegg, A.E. (1983). *Biochim. Biophys. Acta* **747,** 209–214.

Pösö, H., Sinervirta, R., and Jänne, J. (1975a). *Biochem. J.* **151,** 73–76.

Pösö, H., Sinvervirta, R., Himberg, J.-J., and Jänne, J. (1975b). *Acta Chem. Scand.* **B29,** 932–936.

Pösö, H., Hannonen, P., Himberg, J.-J., and Jänne, J. (1976). *Biochem. Biophys. Res. Commun.* **68,** 227–234.

Pösö, H., McCann, P.P., Tanskanen, R., Bey, P., and Sjoerdsma, A. (1984). *Biochem. Biophys. Res. Commun.* **125,** 205–210.

Rajam, M.V., and Galston, A.W. (1985). *Plant Cell Physiol.* **26,** 683–692.

Rajam, M.V., Weinstein, L.H., and Galston, A.W. (1985). *Proc. Natl. Acad. Sci. U.S.A.* **82,** *6874–6878.*

Roon, R.J., and Barker, H.A. (1972). *J. Bacteriol.* **109,** 44–50.

Sklaviadis, T.K., Georgatsos, J.G., and Kyriakidis, D.A. (1985). *Biochim. Biophys. Acta* **831,** 288–296.

Sneath, P.H.A. (1955) *Nature (London)* **175,** 818.

Stevens, L., and Stevens, E. (1980). *In* "Polyamines in Biomedical Research" (J.M. Gaugas, ed.), pp. 167–183. Wiley, New York.

Stevens, L., McKinnon, I.M., and Winther, M. (1977). *FEBS Lett.* **75,** 180–182.

Stevens, L., McKinnon, I.M., Turner, R.M., and North, M.J. (1978). *Biochem. Soc. Trans.* **6,** 407–409.

Tabor, C.W., and Tabor, H. (1984). *Annu. Rev. Biochem.* **53,** 749–790.

Tabor, C.W., and Tabor, H. (1985). *Microbiol. Rev.* **49,** 81–99.

Tang, K.-C., Pegg, A.E., and Coward, J.K. (1980). *Biochem. Biophys. Res. Commun.* **96,** 1371–1377.

Tyagi, A.K., Tabor, H., and Tabor, C.W. (1982). *Biochem. Biophys. Res. Commun.* **109,** 533–540.

Whitney, P.A., and Morris, D.R. (1978). *J. Bacteriol.* **134,** 214–220.

Wickner, R.B., Tabor, C.W., and Tabor, H. (1970). *J. Biol. Chem.* **245,** 2132–2139.

Williams-Ashman, H.G., and Schenone, A. (1972). *Biochem. Biophys. Res. Commun.* **46,** 288–295.

13

Inhibition of Viral Polyamine Biosynthesis

A. STANLEY TYMS AND JOHN D. WILLIAMSON

Division of Virology, Department of Medical Microbiology
St. Mary's Hospital Medical School
Paddington, London W2 1PG, England

I. INTRODUCTION

A rational approach to the development of new therapies for the treatment of virus infections must begin with a clear recognition of the problems to be faced. Unlike many other pathogens, viruses are obligate intracellular parasites and their replication depends almost completely on the metabolic pathways available in the host cell. Outside the cell, a virus particle consists essentially of a genome, which may be either DNA or RNA, protected by a protein coat. This nucleocapsid may be naked or, in the case of enveloped viruses, surrounded by a lipoprotein membrane. All viruses are too small to be visualized by light microscopy and their size also precludes any viral structures analogous to the subcellular organelles such as mitochondria or ribosomes found in procaryotic and eucaryotic cells. Consequently, the extracellular virion is metabolically inert and it is not until the viral genome has been released within the infected cell that the specific metabolic processes associated with virus replication are initiated. These intracellular events, which will result in the production of infectious progeny virus, require the synthesis of viral nucleic acid and proteins. Cellular enzymes may be utilized for viral DNA replication, but unique enzymes not normally found in the host cell are needed for viral RNA replication and such enzymes are, in fact, encoded within the genomes of RNA viruses. Two families of DNA viruses, the herpesviruses and poxviruses, also effect genome replication using virus-specific enzymes. Apart from the poxviruses, which have a complete processing system for the production of virus-specific messenger RNA within the virion, host cell enzymes transcribe the genomes of all other DNA viruses. Sophisticated mechanisms resulting in the temporal control of transcription have been demonstrated with different families of DNA viruses, particularly the division of the virus growth cycle into early, prereplicative and late, postreplicative events. The genome of RNA viruses may be translated directly or only after its transcription, as in the case of negative-strand RNA viruses and the retroviruses. With all viruses the translation of virus-specific mRNA occurs in association with host cell ribosomes and these viral polysomes may be identified by their characteristic buoyant density. Posttranslational modification of viral proteins has been shown to take place during the replication of both RNA and DNA viruses. The production of virus nucleic acid and capsid proteins is followed by their assembly into progeny virions, which are released from the host cell either by cytolysis or by extrusion from the cell surface in noncytolytic infections. The latter process often occurs with enveloped viruses because the viral envelope is derived from host cell membranes after they have been modified by the incorporation of virus-specific glycoproteins. The herpesviruses, which may be released by cytolytic or noncytolytic mechanisms, acquire their envelope from the nuclear membrane of the infected cell. Cell death or other noncytocidal changes accompanying virus replication together with humoral and cellular immune responses to virus antigens in infected tissues and organs in the whole animal can account for the pathogenesis of viral diseases. Antiviral agents to be used therapeutically must be designed, therefore, with regard to the virion's essential lack of metabolic activity and the intimate dependence of virus replication on the metabolism of its host cell. It must also be noted that many of these cellular processes are dependent upon the availability of polyamines (see Chapter 4).

The first indication that polyamines may have a role in virus replication was provided by chemical analyses of purified virus particles. In 1957 Hershey reported the presence in T2 bacteriophage of acid-soluble, dialyzable compounds derived from arginine, which, in the following year, were identified as putrescine and spermidine (Ames *et al.*, 1958). The polyamines were shown to be specifically associated with T-even bacteriophages in amounts sufficient to neutralize about 50% of the negatively charged phosphate groups in the bacterial virus DNA. Interestingly, similar quantities of spermine together with its acetylated derivative were recovered with the phage nucleic acid if the host bacterium was grown in the presence of spermine, although this polyamine is not normally found in prokaryotic cells. Gibson and Roizman (1971) provided the earliest evidence for polyamines in an animal virus when they reported the presence of spermidine and spermine in herpes simplex virus type 1. Spermidine is present in the virus envelope or tegument region but spermine is found in the nucleocapsid in quantities sufficient to neutralize over 40% of the virus DNA phosphates. Other animal DNA viruses known to contain polyamines include another herpesvirus, human cytomegalovirus (Gibson *et al.*, 1984), and a poxvirus, vaccinia virus (Lanzer and Holowczak, 1975), although quantitative data are not available. Spermidine and spermine have been reported to be present in human type 5 adenovirus (Shortridge and Stevens, 1973) but this claim has been disputed (Pett and Ginsberg, 1975). Animal RNA viruses containing polyamines include influenza virus and Newcastle disease virus, a pathogen of veterinary importance, with amounts sufficient to neutralize 38 and 31% of the RNA phosphates, respectively (Bachrach *et al.*, 1974). Two picornaviruses, encephalomycarditis virus and human rhinovirus 14, contain putrescine, spermidine, and spermine (Sheppard *et al.*, 1980; Fout *et al.*, 1984), but quantitative analysis of purified poliovirus particles has failed to detect any polyamine (Ames and Dubin, 1960; Fout *et al.*, 1984). Polyamines do not appear to be significant structural components of Semliki Forest virus, a togavirus (Raina *et al.*, 1981).

The fact that they are components of several animal viruses, including some important human pathogens, does not directly indicate a role for polyamines in virus replication. Such evidence had been provided by Ferrari and colleagues in 1964 when they demonstrated the antiviral effects of methylglyoxal bis(guanylhydrazone) (MGBG). This compound, a known antitumor agent, was shown to inhibit the replication of poliovirus type 1, Coxsackie B3 virus, and echovirus 6, although a less marked effect was obtained against vaccinia virus. This antiviral effect was not directed against the virus particle but against the virus-infected cell. Such selective action was confirmed in a later study which showed MGBG treatment prolonged the mean survival time of mice infected with influenza virus (Kuchler *et al.*, 1968). It was not until several years later, however, that MGBG was shown to be an inhibitor of *S*-adenosylmethionine decarboxylase (Williams-Ashman and Schenone, 1972). Thus, the antiviral properties of one polyamine inhibitor was demonstrated more than 20 years ago but the toxic properties of MGBG discouraged further assessment of its chemotherapeutic properties. During the past decade other inhibitors with specific action against polyamine metabolism have been developed which also have antiviral properties. They have been used to identify the nature of the essential requirement for polyamine metabolism in the replication of certain animal viruses, and such studies also permit an evaluation of the therapeutic potential of polyamine inhibitors in the clinical treatment of virus infections.

II. POLYAMINE METABOLISM AND THE REPLICATION OF ANIMAL VIRUSES

A. Poxviruses

The replication of vaccinia virus, in common with other DNA viruses, is dependent on the availability of arginine (Archard and Williamson, 1971). In fact, vaccinia-infected cells have been shown to contain virus-specific enzymes which affect the biosynthesis of arginine from its metabolic precursor, citrulline (Cooke and Williamson, 1973; Williamson and Cooke, 1973). In addition to an essential requirement in the synthesis of viral proteins containing arginyl residues, arginine through the action of arginase is the source of ornithine for polyamine biosynthesis in vaccinia-infected cells (Obert *et al.,* 1980). Radiotracer studies have shown the utilization of ornithine for the biosynthesis of spermidine and spermine recovered from purified virions even if [^{3}H]ornithine was made available at a late stage in the vaccinia replication cycle (Lanzer and Holowczak, 1975). This observation was especially remarkable in view of the rapid inhibition of host cell protein synthesis following vaccinia infection and the very short half-life of ornithine decarboxylase in animal cells. Consequently, the demonstration that ornithine decarboxylase activity does increase markedly in vaccinia-infected cells, to reach peak levels at about 6 hr postinfection, was consistent with the results of further experiments which provided evidence for such activity being due to a new enzyme encoded by the vaccinia virus genome (Hodgson and Williamson, 1975). Expression of this virus-specific ornithine decarboxylase requires both RNA and protein synthesis and temporal control is influenced by viral DNA synthesis. Treatment of vaccinia-infected cells with inhibitors of DNA synthesis resulted in prolonged expression of ornithine decarboxylase at times when the enzyme activity had fallen to basal levels in control, infected cells and in uninfected cells treated similarly. A switch-off mechanism affected by the synthesis of viral DNA controls the expression of other enzymes known to be encoded within the vaccinia virus genome, such as thymidine kinase or DNA polymerase, and their activities also continue undiminished in vaccinia-infected cells exposed to DNA inhibitors. The K_m value of ornithine decarboxylase from uninfected HeLa cells was calculated to be 0.12×10^{-3} M but 0.05×10^{-3} M for the enzyme from vaccinia-infected cells (Hodgson and Williamson, 1975). These properties argue strongly that the production of putrescine for polyamine biosynthesis is essential during vaccinia virus replication, a conclusion supported by the further demonstration of *S*-adenosylmethionine decarboxylase activity in vaccinia-infected HeLa cells (Hodgson, 1976).

The expression of these enzyme activities, which catalyze the formation of putrescine, spermidine, and spermine, can be related to polyamine-dependent events in the vaccinia replication cycle. The formation of virus-specific, DNA-containing inclusions detected by cytochemical staining techniques was inhibited by MGBG, an effect reversible by spermidine, although viral DNA synthesis measured by incorporation of radiolabeled thymidine was unaffected by the polyamine inhibitor (Williamson, 1976). Both spermidine and spermine have been shown to be constituents of the core of vaccinia virus particles (Lanzer and Holowczak, 1975), and polyamines stimulate the transcription of viral DNA by the vaccinia core-associated RNA polymerase (Moussatche, 1985). The condensation of electron-dense material, possibly viral DNA, into immature core structures during the morphogenesis of vaccinia virus was also dependent on the availability of polyamines or their metabolic precursors (Obert *et al.*, 1980). The latter authors also showed that the transcription of late, virus-specific mRNA can be initiated by the provision of ornithine, putrescine, spermidine, or spermine to vaccinia-infected cells deprived of arginine.

Complementary investigations of the antiviral properties of specific inhibitors have confirmed the essential requirement for continued polyamine biosynthesis in the replication of vaccinia virus. Both α-methylornithine, a competitive inhibitor, and α-difluoromethylornithine (DFMO), a catalytically activated, irreversible inhibitor of ornithine decarboxylase, have been shown to inhibit the growth of vaccinia virus when added to the medium supplied to cell cultures immediately after infection. The concentration of DFMO required to obtain 50% reduction in virus yield (ED_{50} value) was 5.9 and 4.4 m*M* using human epithelial carcinoma (HeLa) cells or human embryo fibroblasts (MRC-5), respectively, as the host cell system (Tyms *et al.*, 1983). Although high extracellular concentrations are required to exert an antiviral effect, 10 μM DFMO completely inhibits the activity of ornithine decarboxylase prepared from extracts of vaccinia-infected HeLa cells (J.D. Williamson, unpublished results). Such comparisons indicate the relative inefficiency of this drug's uptake into mammalian cells infected with vaccinia virus. Earlier reference has been made to the antiviral effects of MGBG, an inhibitor of mammalian *S*-adenosylmethionine decarboxylase, but the drug is known to exert other effects unrelated to polyamine metabolism. It is pertinent, however, to note the widely different ED_{50} values of 300 and 3.5 μM MGBG obtained in determinations of antiviral action against the replication of vaccinia virus in HeLa or MRC-5 cells, respectively (Tyms *et al.,* 1983). These differences can be related to cellular transport mechanisms, which, as will be discussed later, may be utilized to enhance the selective action of polyamine inhibitors used as antiviral chemotherapeutic agents against virus infections.

B. Herpesviruses

The five herpesviruses that cause infection and disease in man are herpes simplex virus (HSV) type 1 and type 2, human cytomegalovirus (CMV), varicella–zoster virus (VZV), and Epstein–Barr virus (EBV). A major feature of herpesviruses is the ability to establish latent or chronic long-term infections. During latency the virus genome is considered to remain in an integrated or episomal form but inhibitors of DNA synthesis have little effect on this nonreplicate state.

The presence of spermidine and spermine within the intact herpes simplex virion (Gibson and Roizman, 1971) was evidence that polyamines have an important role in herpesvirus replication. The nucleocapsids of HSV, as in all herpesviruses, have an icosahedral structure surrounded by an outer membrane-like envelope with an intermediate tegument region between the nucleocapsid and envelope (see Fig. 1). Spermine was associated with nucleocapsids and spermidine with the outer components when purified virions were fractionated. This apparent structural requirement for polyamines within the virion was not related to their *de novo* synthesis. Ornithine decarboxylase activity is not induced during HSV infection (Gibson and Roizman, 1971; Isom, 1979), and this is evident by a lack of spermidine and spermine biosynthesis (McCormick and Newton, 1975; Tyms *et al.,* 1979). Most pertinent to this review, HSV growth was consequently unresponsive to treatment with polyamine inhibitors (Tyms *et al.,* 1979) unless depletion of polyamines was achieved by prior treatment of cultured cells (Tuomi *et al.,* 1980; McCormick and Newton, 1975; Tyms *et al.,* 1983; Gibson *et al.,* 1984). What was apparent from the earlier studies was the association of the lytic-type infection of HSV, which resulted in a rapid inhibition of many host cell functions, with the nonrequirement for *de novo* polyamine biosynthesis. Polyamines extant within host cells at the time of infection were considered sufficient to fulfill the needs of HSV growth. The association between leaky cells and poor HSV virus yield (McCormick, 1978) provided support for this hypothesis.

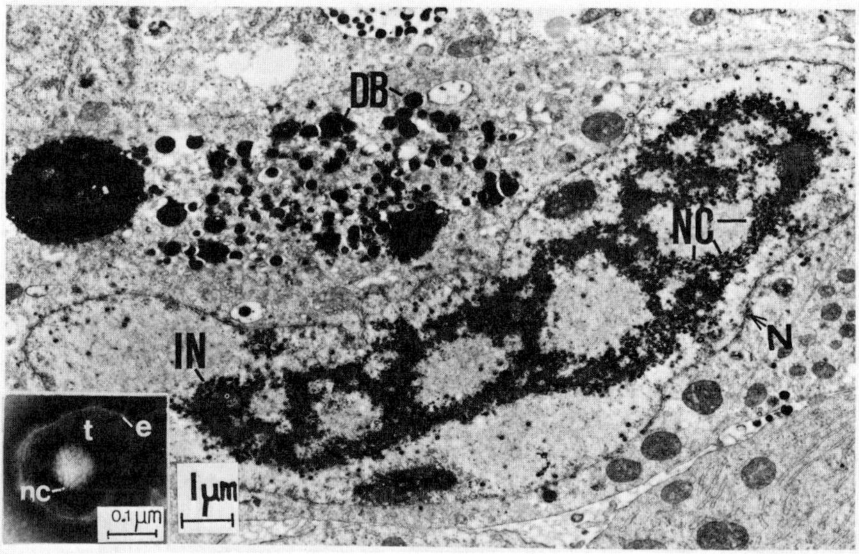

Fig. 1. An electron micrograph of an ultrathin section of human embryo fibroblast cells infected for 10 days with cytomegalovirus (strain AD1-69). Cell pellets were fixed and embedded, and ultrathin sections cut and stained as described by Tyms *et al.* (1984). Sections were viewed under a Hitachi HU 12A transmission electron microscope. Nuclei (N) of infected cells contained large inclusion bodies (IN), which are the site of viral DNA synthesis and nucleocapsid (NC) assembly. Mature virions in the enveloped form collect in the Golgi region along with the virus-induced dense bodies (DB), which are composed of viral proteins but no DNA. The inset shows a negatively stained preparation of a mature virion with fine details of the nucleocapsid (nc) surrounded by the envelope (e) and intervening tegument region (t). (Presented by kind permission of Tim Ryder and Margaret Mobberley, Electron Microscope Unit, Queen Charlotte's Hospital for Women, London.)

Human CMV causes serious disease in the immunocompromised patient (Meyers *et al.*, 1982; Macher *et al.*, 1983) and historically is known for its ability to induce intrauterine infection at relatively high frequency. After congenital infection, the newborn may present with symptoms of disease, sometimes severe and including neurological damage, although up to 90% of cases are asymptomatic. The continuous excretion of infective virus is often for months or years of early life and is associated with more subtle sequelae in later childhood such as sensorineural deafness (Stagno *et al.*, 1975). This is particularly disturbing because disease can occur in children who are initially symptom-free. Infections with human CMV are known to transform resting cells in culture to a highly energetic state involving the synthesis of major cell macromolecules and accompanying enzymatic activities. The stimulation of host cell functions, in particular the induction of ODC (Isom, 1979) and abundant synthesis of spermidine and spermine (Tyms and Williamson, 1980, 1982; Tyms *et al.*, 1983), is in total contrast to infection with HSV. A metabolically active host cell seems essential to maintain cell viability in order to provide for long-term CMV infection *in vitro* (Tyms *et al.*, 1987), and this probably mimics the persistent excretion of infective virus in clinical infections (Stagno *et al.*, 1975).

Earlier studies showed that the induction of polyamine biosynthesis in CMV-infected cells in culture was obligatory for the successful replication of human CMV. The antiviral action of MGBG, assumed to be through inhibition of *S*-adenosylmethionine decarboxylase (Williams-Ashman and Schenone, 1972), provided the first indication of the dependence of human CMV on polyamine biosynthesis (Tyms *et al.*, 1979). Human embryo fibroblasts treated with MGBG from the time of infection failed to support the replication of a laboratory strain of human CMV, but the compound had little or no effect on parallel cultures infected with HSV type 1 or 2 (Tyms *et al.*, 1979). The antiviral action of MGBG was later confirmed in a study of clinical isolates of human CMV (Tyms and Williamson, 1982). MGBG had potent activity against murine CMV grown in whole embryo mouse (WEM) cells at drug concentrations that had no effect on HSV-infected WEM cells (Tyms *et al.*, 1983). Interestingly, experiments with the mouse virus used a plaque reduction assay, a method based on low virus-to-cell infection ratio; this subjects cells to drug treatment prior to the spread of infection. Under these conditions HSV showed moderate response to treatment with MGBG, presumably through polyamine depletion, and this agreed with other published work (Tuomi *et al.*, 1980; Gibson *et al.*, 1984). MGBG is known to have other antiproliferative effects on animal cells in culture metered at the mitochondrial level (Pathak *et al.*, 1977). Although this compound has a history of clinical application in the treatment of malignant conditions, its well-documented clinical side effects and ill-defined mechanism of action argue against its being used to treat CMV or other infectious disease.

DFMO inhibits the replication of human CMV *in vitro* (Tyms and Williamson, 1982; Gibson *et al.*, 1984). Extracellular concentrations of this compound in the order of 5–10 m*M* were essential to produce observable effects. In our cell culture systems, the potency of DMFO when added at the time of infection varied with type and individual strain of cell used (Tyms *et al.*, 1983; A.S. Tyms, unpublished observations). In human embryo fibroblast cells the inhibition of CMV growth appeared to be mediated through a block in viral DNA synthesis (Tyms and Williamson, 1982). Consistent inhibition was seen in different fibroblast cultures when DFMO was tested by plaque reduction assay (see Fig. 4a). As described earlier, the exposure of cells prior to the spread of virus (assay time 7–10 days) resulted in drug priming, which was akin to the protocols used by Gibson *et al.*, (1984), who pretreated cells for 3 days before infection. The latter studies indicated that a late event in CMV growth was affected by DFMO treatment and was probably concerned with virion assembly. This related to the proven association of spermidine and spermine with intact CMV particles (Gibson *et al.*, 1984).

Other analogs of ornithine and putrescine which inhibit ODC in a time-dependent manner have been synthesized (Sjoerdsma and Schechter, 1984). The methyl ester of α-monofluoromethyldehydroornithine (Δ-MFMO methyl ester) has 10–12 times greater inhibitory activity against ODC than the parent compound Δ-MFMO or DFMO (Sjoerdsma and Schechter, 1984). Δ-MFMO methyl ester (1 m*M*) completely blocked the replication of CMV in human embryo fibroblast cells when infected at high multiplicity. Polyamine metabolism under these conditions was significantly altered (Clarke and Tyms, 1986b) and this is illustrated in Fig. 2. A major reduction in the level of spermidine was observed in the infected cells treated with Δ-MFMO methyl ester when compared with infected, untreated cells or uninfected control cells. In all cultures, the level of spermine was maintained at or near that associated with quiescent cells except for a marked increase seen in the infected, untreated cells as reported previously (Tyms and Williamson, 1982; Tyms *et al.*, 1983; Clarke and Tyms, 1986a). Levels of spermine in HTC cells treated with DFMO, Δ-MFMO, or Δ-MFMO methyl ester also remained relatively unchanged (Mamont *et al.*, 1986). The molar ratio of spermidine to spermine shown in Fig. 2c summarizes the overall effects of Δ-MFMO methyl ester on polyamine metabolism, in CMV infected and sham infected cells. The ratio was significantly lower in the drug-treated infected cells (ratio 0.12) than in drug-treated uninfected cells (ratio 0.71) or drug-treated actively growing cells (ratio 0.78—results not shown). The much lower spermidine:spermine ratio observed in the infected series when compared to either resting or actively growing cells may be related to the two- to fourfold increase in spermidine and spermine synthase activities recorded in cells infected with either of two strains of CMV (A.S. Tyms and T. Eloranta, unpublished data).

The apparent increase in polyamine synthase activity induced by CMV infection coupled with the stimulation of *S*-adenosylmethionine decarboxylase by the time-dependent inhibitors of ODC activity (Mamont *et al.*, 1986) might explain the very low spermidine:spermine ratio seen after Δ-MFMO methyl ester treatment. These observations are important from two points of view. First, the spermidine and spermine synthase activities may constitute an important target for potential use in combined drug

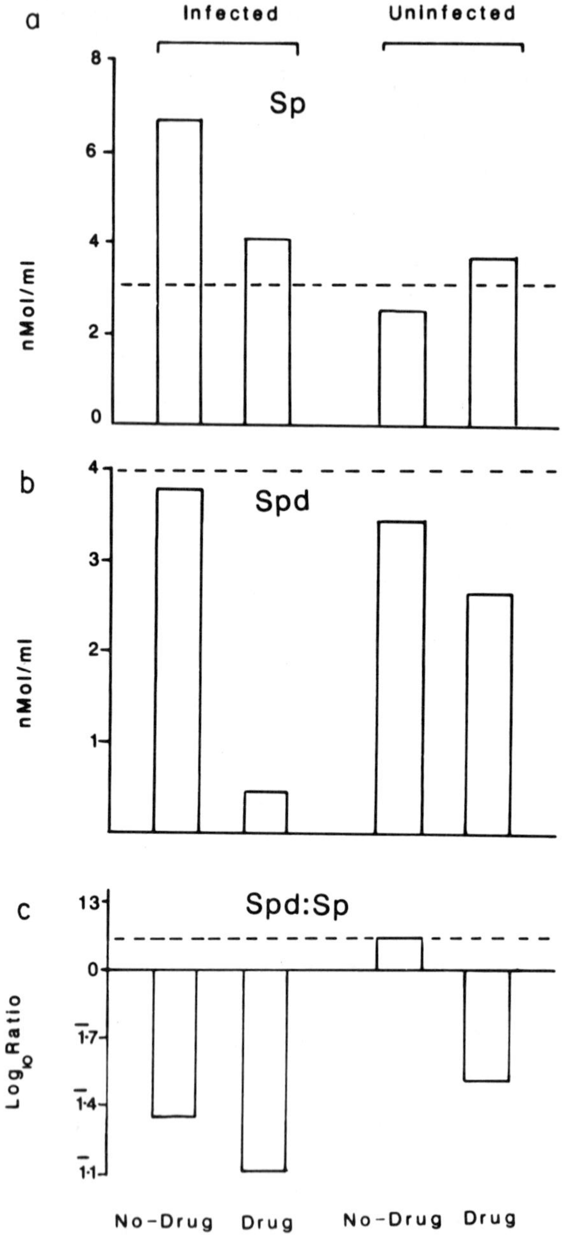

Fig. 2. Human embryo fibroblasts were infected with CMV strain AD-169 or were sham infected. Cells were treated with Δ-MFMO methyl ester (1 m*M*) from the time of infection or left untreated. Cells were harvested at 4 days postinfection, processed, and levels of polyamines measured by reverse-phase HPLC as described by Clarke and Tyms (1986a). Con-

therapy. Second, the high level of spermine and low level of spermidine in virus-infected cells in conditions where Δ-MFMO methyl ester completely blocks CMV growth seem to suggest that spermidine levels may be a critical factor in the replication of this virus. The putrescine analog (2*R*, 5*R*)-6-heptyne-2,5-diamine (2*R*,5*R* methyl acetylenic putrescine; MAP), a time-dependent inhibitor of mammalian ODC, depleted both spermidine and spermine levels in cultures of rat hepatoma cells (Mamont *et al.*, 1984). This inhibitor had some activity against CMV growth and also reduced the levels of spermine as well as spermidine after treatment of infected cells (Clarke and Tyms, 1986b). The different effects of 2*R*,5*R* MAP and Δ-MFMO methyl ester on polyamine metabolism in CMV-infected cells are currently under investigation and may enhance our understanding of the role of polyamines in virus replication.

Our previous studies have shown that viral DNA synthesis was particularly susceptible to depletion of polyamines when MGBG, α-methylornithine (Tyms *et al.*, 1979), or α-difluoromethylornithine (Tyms and Williamson, 1982) was used to treat cells infected with human CMV. This may have involved genome synthesis per se due to a direct effect on virus-specific enzyme activities or possibly was related to the selective expression of viral genes. The sequential transcription of the immediate-early, early, and then late genes, the latter restricted to the period after the initiation of viral DNA synthesis, provides a number of novel targets for regulation by polyamine depletion. In the study by Gibson *et al.* (1984), depletion of polyamines by DFMO treatment had no apparent effect on early or on late CMV-coded proteins or on viral DNA synthesis. Even so, infective virus production was reduced by greater than 4 $\log_{10}$ units, and in this case DFMO treatment was considered to exert an effect on the assembly of nucleocapsids or the final stages of virion maturation involving its envelopment. The poor penetration of this ODC inhibitor into mammalian cells may result in polyamine depletion to a varying degree which could produce differential effects on virus growth.

Recent data from our studies (Tyms and Davis, 1986) have shown that in conditions where Δ-MFMO methyl ester completely inhibited CMV growth, the synthesis of late viral proteins was consistently reduced (Fig. 3), even though viral DNA synthesis was not always completely blocked. As indicated earlier, a major constraint on the synthesis of the late viral proteins early on in CMV infection is the lack of viral DNA synthesis. Our

---

fluent cell monolayers not exposed to virus or serum stimulation were harvested at the beginning and end of the experimental phase (broken line). Infected cultures treated with Δ-MFMO methyl ester were completely devoid of infectivity at 4 days postinfection in contrast to the virus controls. The spermine (Sp) and spermidine (Spd) levels shown were consistent for duplicate samples.

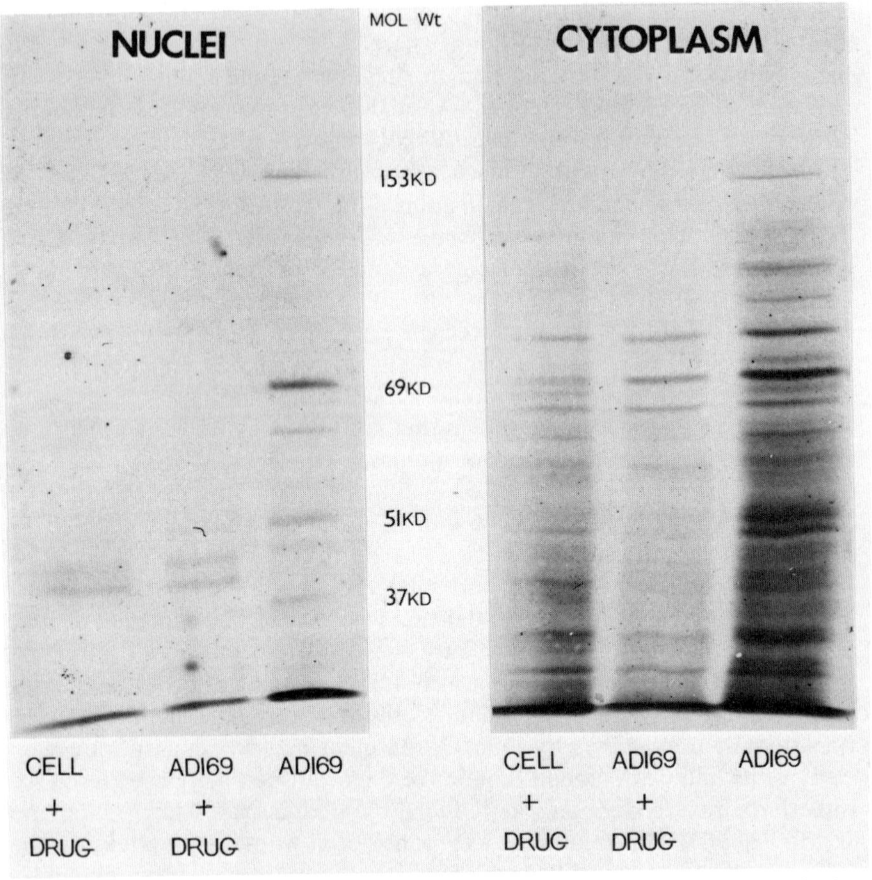

Fig. 3. Cells were infected and treated with Δ-MFMO methyl ester (1 m*M*) as in the legend for Fig. 2. At 4 days postinfection the nuclear and cytoplasmic fractions were separated by treatment with nonionic detergent (Gibson, 1983) and subjected to SDS–PAGE electrophoresis as described by Tyms *et al.* (1984). Proteins were electroblotted onto nitrocellulose and stained by 0.1% amido black. The Δ-MFMO methyl ester-treated, uninfected control was similar to its untreated counterpart and is shown to facilitate the identification of virus-specific proteins (see text).

results indicate that polyamine inhibitors may exert antiviral activity prior to viral DNA synthesis but also may have a major effect on the expression of late proteins. The predominant intracellular viral proteins illustrated in Fig. 3 correspond to those characterized by Gibson (1983). They consist of the major capsid protein (153 kDa), the main component of the virion shell, and abundant matrix protein (69 kDa), found in the tegument region (Fig. 1). The latter protein along with the abundant DNA-binding (51 kDa) and virion assembly proteins (37 kDa) undergo phosphorylation (Gibson, 1983; Gibson and Irmiere, 1984). It is well known that polyamines have the potential to neutralize nucleotide phosphate groups, which has been related to genome packaging in the virion. However, as suggested by Gibson and Irmiere (1984), their interaction with phospholipid and phosphoproteins may be equally important during the maturation of infective virions.

C. Other DNA Viruses

The replication of human adenovirus type 5 in oral epidermoid carcinoma (KB cells) inhibits both polyamine biosynthesis and host cell DNA synthesis (Pett and Ginsberg, 1975). In quiescent rat cells, however, ornithine decarboxylase activity is not induced but infection results in a progression of the cell cycle from the G_1 phase into the S phase (Cheetham and Bellett, 1982). An increase in thymidine kinase activity, which accompanied the induction of host cell DNA synthesis in adenovirus-infected rat cells, could be inhibited by MGBG but not by α-methylornithine, although virus replication was unaffected.

Infection of primary mouse embryo cells with polyoma virus, a papovavirus, also causes a progression into the S phase but this is accompanied by a biphasic stimulation of both ornithine decarboxylase and *S*-adenosylmethionine decarboxylase activities (Goldstein *et al.*, 1976). The initial peak occurs during early viral mRNA synthesis and the second, major peak coincides with the maximum rate of virus-induced cellular DNA synthesis. Inhibition of viral and cellular DNA synthesis by 5-fluoro-2′-deoxyuridine did not affect the biphasic increases in polyamine biosynthesis, but the second peak of ornithine decarboxylase activity and each peak of *S*-adenosylmethionine decarboxylase activity required both transcriptional and translational events.

D. RNA Viruses

Unlike the DNA viruses, there have not been any studies of polyamine metabolism in cell cultures infected with RNA viruses which are also important human pathogens. The late Aarne Raina and his colleagues, however, have made a series of investigations into the effect of polyamine depletion on the replication of a togavirus, Semliki Forest virus. The growth of this virus in baby hamster kidney (BHK-21) cells was shown to be inhibited by 5 m*M* DFMO added to cell cultures 72 hr before infection; pretreatment for shorter periods even increased virus yields. The specificity of the antiviral effect was indicated by its reversal if putrescine or spermidine was added 24 hr before infection (Tuomi *et al.*, 1980). Purified preparations of Semliki Forest virus were found to contain only small amounts of spermidine and spermine in quantities sufficient to neutralize about 3% of the viral RNA phosphate. This indicated that polyamines are not signficant structural components of the virion and no increase in ornithine decarboxylase or *S*-adenosylmethionine decarboxylase activity occurred in BHK-21 cells after infection. In fact, there was a rapid decline in both enzyme activities, indicating virus growth was not dependent on polyamines produced after infection and the reduction in virus yield in cell cultures pretreated with DFMO was due to changes in macromolecular synthesis concomitant with polyamine deficiency (Raina *et al.*, 1981). These conclusions were supported by further studies showing polyamine depletion prior to infection markedly reduced viral RNA polymerase activity. The effect was reversed by the addition of spermidine to the culture medium supplied to infected cultures, although the polyamine did not affect enzyme activity assayed *in vitro*. It appears the inhibition of Semliki Forest virus replication caused by inhibition of polyamine biosynthesis is due to malfunctions of protein-synthetic mechanisms of the host cell (Tuomi *et al.*, 1982). The synthesis of mRNA for viral structural proteins was relatively unaffected, however, in infected cells treated with DFMO, whereas the synthesis of viral genomic RNA was markedly inhibited (Tuomi, 1983).

A picornavirus, encephalomyocarditis virus, has also been shown to contain polyamines, but the concentrations of putrescine, spermidine, and spermine present in purified virions differed from those found in the host cell (Sheppard *et al.*, 1980). This preferential uptake of polyamines from the host cell pool for incorporation into virions suggests these molecules have an important, but unknown, function, although they are present in amounts which neutralize only 9–13% of the phosphate charges on the viral genome.

The oncornaviruses (oncoviruses) are so-called because these RNA virus-infected cell (Sissons, 1986). Additionally, exogenous putrescine transformation of their host cells. Such changes following infection of chick embryo fibroblasts with Rous sarcoma virus (Don and Bachrach, 1975) or mouse embryo BALB/3T3 cells with murine sarcoma virus (Gazdar *et al.*, 1976) are accompanied by a marked increase in ornithine decarboxylase activity and putrescine production followed by an accumulation of polyamines. A more recent study has shown treatment with 3 m*M* DFMO reduced the concentrations of putrescine and spermidine but did not affect the induction or the maintenance of morphological transformation of Rous sarcoma virus-infected cells (Hölttä *et al.*, 1981).

III. QUALITATIVE AND QUANTITATIVE ASPECTS OF ANTIVIRAL CHEMOTHERAPY

The ideal chemotherapeutic agent must show absolute discrimination between the pathogen and its host, a concept epitomized in Paul Erhlich's "magic bullet." Unfortunately, it is particularly difficult to achieve such selective action in the treatment of virus infections due to the pathogen's intimate dependence on the host cell for its replication. Only recently has it been possible to develop antiviral compounds based on nucleoside analogs, which have unique action because they are activated by enzymes in the infected cell which are encoded by the viral genome. Thus, 9-(2-hydroxyethoxymethyl)guanine and (*E*)-5-(bromovinyl-2′-deoxyuridine) are substrates for the virus-specific thymidine kinase found in cells infected with certain herpesviruses and which produces phosphorylated derivatives with inhibitory action against DNA polymerase, another virally encoded enzyme. Virus replication is blocked selectively because both of the equivalent cellular enzymes have very low affinity for the respective substrates.

Similar qualitative distinctions have not emerged, to date, between the enzymes associated with polyamine metabolism in uninfected and in virus-infected cells. Although ornithine decarboxylase in vaccinia-infected HeLa cells is distinguishable from the host cell's enzyme by biological and physiocochemical criteria, both enzymes are inhibited similarly by DFMO, α-monofluoromethylornithine, and α-monofluoromethylputrescine (J.D. Williamson, unpublished results). These results were disappointing in view of the selective effects these polyamine inhibitors have had against procaryotic ornithine decarboxylases (Kallio and McCann, 1981; Kallio *et al.,* 1982). Similarly, the regulation of ornithine decarboxylase activity induced by infection with human cytomegalovirus was altered such that the enzyme activity was no longer controlled by putrescine in the culture medium but the inhibitory effect of DFMO failed to distinguish between the host and virus-induced enzyme (Isom and Pegg, 1979).

Although the ability to synthesize polyamines *de novo* appears to be an innate property of all eucaryotes, certain animal cells and tissues have been shown to accumulate polyamines by an energy-dependent transport system. This uptake mechanism has been demonstrated in mouse, rat, and rabbit brain; in mouse, rat, and human lung; in human leukocytes; and in various continuous cell lines. Although its physiological significance is not known, the accumulation system has a high degree of specificity. Putrescine, spermidine, and spermine are accumulated into slices of rat lung by a process which obeys saturation kinetics; this system has been used to determine the structural requirements of compounds to inhibit pulmonary diamine accumulation (Gordon-Smith *et al.*, 1983). Effective inhibitors were α,ω-dialkenes with a distance of approximately four methylene units (6.6 Å) between the nitrogen atoms. MGBG and its congeners proved to be very potent inhibitors and the uptake of MGBG into rat lung slices is a process identical to the process of the putrescine and polyamine accumulation system (Gordon-Smith *et al.*, 1985). Lung, seminal vesicle, and brain cortex slices also accumulated both spermidine and MGBG by a temperature- and energy-dependent process but this mechanism was not demonstrable in slices of heart, liver, spleen, kidney cortex, or submaxillary gland.

Depletion of intracellular polyamines by DFMO activates homeostatic mechanisms which can increase the uptake of exogenous polyamines, and because MGBG is transported through the polyamine accumulation system, its uptake is greatly enhanced by polyamine deprivation (Alhonen-Hongisto *et al.*, 1980). This "priming" effect by DFMO increased the antiproliferative activity of MGBG against cultured Ehrlich ascites tumor cells, a phenomenon which was critically dependent on the growth rate of the cells and the extent of intracellular polyamine depletion (Seppänen *et al.*, 1981). Although DFMO–MGBG combination therapy has been reported to have clinical efficacy in childhood leukemia (Siimes *et al.*, 1981), this has not been confirmed subsequently. Further, DFMO pretreatment failed to enhance the selective uptake of MGBG into tumor tissue in L1210 leukemic mice but resulted in increased accumulation of the drug in normal tissues (Kramer *et al.*, 1985).

Despite the disappointing results from studies *in vivo,* other properties of MGBG point toward the development of alternative polyamine-related compounds with antiviral properties which can be expected to have selective action against virus-infected cells. CMV causes severe, often fatal, pneumonitis, and lung tissue is known to have an active polyamine transport system. Autoradiographic studies of rat lung slices supplied with radiolabeled putrescine and spermidine have shown that polyamine accumulation is restricted to type I and type II pneumocytes and to Clara cells, although in similar studies *in vivo* the distribution was limited to the alveolar epithelium of the lung but without Clara cell involvement (Smith, 1985). The uptake of radiolabeled putrescine into cell cultures derived from human embryonic lung is markedly stimulated following CMV infection (Tyms and Williamson, 1980). It may be possible, therefore, to obtain enhanced chemotherapeutic action against CMV-infected lung tissue by the design of antiviral compounds based on inhibitors of polyamine metabolism which are taken selectively into infected cells through the polyamine transport system.

IV. POLYAMINES AND VIRAL DISEASE

In the preceding section, the replication of certain animal viruses has been shown to be accompanied by altered polyamine metabolism in the host cell. A requirement for polyamines synthesized *de novo* has been linked to their essential role at various stages during the growth cycle of vaccinia virus. In contrast, inhibition of anabolic polyamine metabolism has been demonstrated in cell cultures infected with herpes simplex viruses. It is interesting to note such differences because both vaccinia virus and the herpesviruses cause cytocidal infections *in vitro* which are characterized by rapid inhibition of host-specific macromolecular synthesis. The contrary effect is observed in human CMV infections in which the virus-mediated stimulation of host cell DNA, RNA, and protein synthesis is accompanied by markedly enhanced polyamine metabolism. Such perturbations in polyamine metabolism may be related through other studies to the cytopathology and pathogenesis of virus infections.

The polyamine oxidases are widespread in the plant and animal kingdoms, although their sites of enzymatic action vary according to the source (Morgan, 1985). In animal cells it is now recognized that the enzyme has a key role in the conversion of *N*-acetylspermine or *N*-acetylspermidine to spermine or putrescine accompanied by the formation of monoacetylpropionaldehyde (Tabor and Tabor, 1984). It would appear, however, that a "putrescine cycle" exists in animal cells, the anabolic route forming polyamines which stimulate cell growth and proliferation and the catabolic route producing their oxidation products which can act as growth inhibitory substances. Although exogenous spermidine and spermine are able to stimulate the growth of animal cells in culture, cytotoxic products can be formed if the culture medium also contains bovine serum (Alarcon *et al.*, 1961). This effect has been ascribed to polyamine oxidase in ruminant sera and the production of oxidized polyamines or their derivatives. During the past decade some attention has been paid to the ability of low concentrations of spermidine or spermine to inhibit lymphocyte proliferative responses *in vitro*, an effect also dependent on the presence of bovine serum in the culture medium (Byrd *et al.*, 1977). Another, contemporary, report drew attention to the association of spermine with a thymus-derived inhibitor of lymphocyte transformation indicating the existence of similar immunoregulatory mechanisms *in vivo* (Allen *et al.*, 1977). Subsequent studies have shown such inhibitory effects are reversible and are probably mediated by the formation of aminoaldehydes rather than acrolein, both potential products from the oxidative degradation of spermine or spermidine (Smith *et al.*, 1985).

Some physiological significance of the polyamine–polyamine oxidase interaction may be perceived from studies of the protection of the conceptus against immune rejection and the immunosuppressive effects of seminal plasma. The human placenta is rich in spermine and, although its activity is low in normal human sera, polyamine oxidase activity increases during pregnancy. Human syncytiotrophoblast cells produce immunosuppressive factors which appear to consist of polyamines associated with nucleic acids, and trophoblast cell extracts also contain polyamine oxidase (Remacle-Bonnet *et al.*, 1985). The high concentration of spermine together with polyamine oxidase in human seminal plasma may also provide the immunosuppressive effects which protect sperm from immunological damage and prevent the sensitization of the female by sperm antigens following coitus (James and Hargreave, 1984).

Although some animal viruses are inactivated by exposure to oxidized polyamines *in vitro* (Bachrach and Don, 1971), there is no evidence for such effects *in vivo* during virus replication. The immunomodulatory effects of polyamine oxidation products may, however, be related to the pathogenesis of certain viral diseases. The accumulation of polyamines in cell cultures infected with human CMV achieves concentrations which can be calculated to be sufficient to inhibit lymphocyte transformation *in vitro*. Monocytes from patients with primary human CMV infection have been shown to have reduced responses to mitogenic stimulation during the acute phase of the infection, although normal responses were obtained in the convalescent stage (Carney *et al.*, 1981). Monocytes from uninfected control donors after infection with human CMV *in vitro* were able to suppress the mitogenic responses of autologous lymphocytes (Carney and Hirsch, 1981). The molecular mechanism of the effect remains uncertain, although an inhibitor of interleukin-1 appears to be released from the virus-infected cell (Sissons, 1986). Additionally, exogenous putrescine can affect suppressor cell activity associated with the regulation of hematopoiesis in mice (Sharkis *et al.*, 1983). It is interesting that in human CMV mononucleosis there is an increase in the population of T-suppressor cells (Carney *et al.*, 1981). Seroepidemiological studies have shown a high frequency of human CMV infections in male patients with the acquired immune deficiency syndrome (AIDS) and it has been suggested that the pathogenesis of AIDS may be exacerbated by the presence of spermine and human CMV in semen (Williamson, 1984).

Margination of chromatin is characteristic of cells infected with HSV, and chromosomal damage with release of deoxynucleotides has been documented (Cheng *et al.*, 1975). DNase activities associated with HSV type 1 and type 2 infections have similar properties and are unusual in that both enzymes contain endo- and exonuclease activity (Hoffmann and Cheng, 1979). The susceptibility of viral DNA to nuclease activity when tested *in vitro* was dependent on spermine levels; the endonuclease activity was much more sensitive than the exonuclease activity to the concentrations of the polyamine. As previously discussed by Cheng *et al.* (1979), the protection of viral DNA from its own coded nuclease activity by polyamines may constitute an antiviral target. Inhibition of polyamine biosynthesis had little effect on HSV growth unless preexisting, intracellular levels were reduced by prior treatment of infected cells. This may be influenced by the inability of cells from some species to retain polyamines after HSV infection (McCormick, 1978; Tyms *et al.*, 1983). It has been postulated that polyamine depletion is further complicated by the role of polyamine oxidase activity in the regeneration of putrescine and spermidine from spermine by a catabolic pathway (see Williamson and Tyms, 1984).

There is evidence that the restriction of polyamine biosynthesis induced by adenovirus infection does affect the molecular integrity of other cellular macromolecules. Cheetham and colleagues (1982) were able to prevent chromosomal damage, also a feature of adenovirus type 5 infections, by supplementing the cellular levels of spermine by exogenous addition. Evidence that polyamine oxidase activity was involved in the degradation of spermine extant within the cells was provided by treatment with the polyamine oxidase inhibitor, aminoguanidine (Cheetham *et al.*, 1982).

The modulation of ODC activity in cells infected with different viruses has interesting implications, particularly the absence of ODC activity after infection with HSV or human adenovirus strains. In the latter situation the unusual separation of ODC from another periodic enzyme activity, thymidine kinase, presents the possibility of selective degradation of ODC protein or perhaps a novel means of down-regulation of this enzyme activity. Either way it demonstrates the potential of studying a number of different DNA viruses with respect to one enzyme activity in a defined cell culture system.

The exploitation of molecular degradation as a means for restricting virus growth may be perceived in relation to the formation of 5-deoxy-5-methylthioadenosine (MTA). This compound is formed in stoichiometric amounts to spermidine and spermine during the synthase reactions. The enzyme 5-deoxy-5-methylthioadenosine phosphorylase cleaves MTA to 5-methylthioribose-1-phosphate with the recovery of the purine base.

MTA has allosteric properties with respect to the regulation of polyamine synthase reactions. The rapid removal of MTA by 5-methylthioadenosine phosphorylase suggests that feedback inhibition by the nucleoside may not be an important control factor in mammalian cells. Modulation of this enzyme activity by the use of specific inhibitors could induce important changes in the synthesis of polyamines during virus infection. For example, the long-term growth of cytomegalovirus *in vitro* is accompanied by the continuous synthesis of spermidine and spermine (Tyms *et al.*, 1987), thus maintaining high levels of both polyamines in the cell late in virus growth. In view of this high metabolic activity, a restriction on MTA degradation might prove effective as an antiviral mechanism either singularly or in combination with other established inhibitors. The possibility of regulating polyamine biosynthesis by MTA accumulation induced by inhibition of MTA phosphorylase has been examined by Williams-Ashman *et al.* (1982).

The importance of degradative products of the polyamine pathway in relation to immunological mechanisms of virus control has been covered in the previous section. In all, this discussion illustrates the importance of catabolic as well as anabolic pathways in the control of polyamine metabolism and perhaps its subsequent influence on virus replication.

V. DRUG COMBINATIONS AND ANTIVIRAL CHEMOTHERAPY

A major finding in the recent development and clinical application of polyamine inhibitors has concerned the potential of drug combinations in the treatment of parasitic infections and neoplastic growth. Two studies were of particular interest with regard to the mechanisms of action. The first concerned the treatment of leukemia with MGBG after polyamine levels in patients' tissues were first reduced by prior treatment with DFMO. This was discussed in a previous section. A second study demonstrated synergism between DFMO and the bleomycins A5 and A6 in curing semichronic infections with *Trypanosoma brucei brucei* in mice (Clarkson *et al.*, 1983). The presence of spermidine or spermine as terminal amines on these antitumor antibiotics enabled binding of the molecules to DNA, resulting in double-stranded scission (Umezawa, 1983). Depletion of polyamines by DFMO facilitated this reaction. These studies clearly indicate the need for further exploitation of the well-defined principles involved. The potential application of DFMO in conjunction with interferons (Sunkara *et al.*, 1982) or 3,4-dihydroxybenzylamine (Prakash *et al.*, 1985) against melanoma could also be important. Similar protocols may have a place in the treatment of virus infections. In our studies, DFMO has been used against herpesvirus infections *in vitro* in combination with MGBG or acyclovir.

Human CMV responds readily to DFMO treatment in the plaque reduction assay (Fig. 4a); the 50% effective dose (ED_{50}) for this experiment was 4 m*M* DFMO. The dose response against human CMV for MGBG is also shown in this figure. The ED_{50} for this drug was 1.7 μM, which is within the range of values for CMV recorded previously (Tyms and Williamson, 1982). In the presence of 2 m*M* DFMO, a concentration of the drug predicted to have no effect on CMV replication alone, the ED_{50} for MGBG was reduced to 0.98 μM. Similar changes in the sensitivity of murine CMV and HSV type 2 were observed when mouse cells (WEM) were infected and treated with this drug combination (Fig. 4b). The small but consistent shift in dose–response curves in both cell systems suggested that priming of cells with the ODC inhibitor added to the inhibitory effect of MGBG. DFMO treatment alone was shown to reduce levels of putrescine and spermidine in human and mouse cells, although the efficiency of depletion was not nearly as great as that achieved with the newer compounds, MFMO methyl ester or *R,R* MAP (A.S. Tyms, unpublished observations). It will be expedient to investigate these more potent inhibitors in combination with MGBG against herpesviruses.

The second approach to modeling combined-treatment protocols involved the use of antivirals with activity against herpesviruses but not targeted directly at enzymes in the polyamine pathway. Acyclovir is a potent inhibitor of HSV type 1, HSV type 2, and murine CMV, as stated earlier, by mechanisms involving virus-coded enzymes. Phosphorylation of this compound is initiated by a virus-coded thymidine kinase, and the triphosphate derivative specifically inhibits viral DNA synthesis through interaction with the viral DNA polymerase. The DNA polymerases coded for by HSV type 1 and type 2 viruses have been purified (Ostrander and Cheng, 1977) and polyamines are known to exert regulatory effects on these enzyme activities at concentrations within the physiological range (Ostrander and Cheng, 1980; Cheng *et al.*, 1979; Wallace *et al.*, 1980). The synthesis of HSV DNA usually occurs within 12 hr of infection, and modulation in cellular polyamine content during the early phase of virus growth could infuence DNA polymerase function. It was considered possible that DFMO treatment might directly effect the activity of drugs acting at the polymerase level even though the ODC inhibitor alone had no effect without prior treatment of cells (Tuomi *et al.*, 1980; Gibson *et al.*, 1984). Plaque reduction assays showed that concomitant treatment with DFMO had little effect on the dose–response curves for acyclovir against murine CMV (Fig. 4c) or HSV type 1 (results not shown). Small

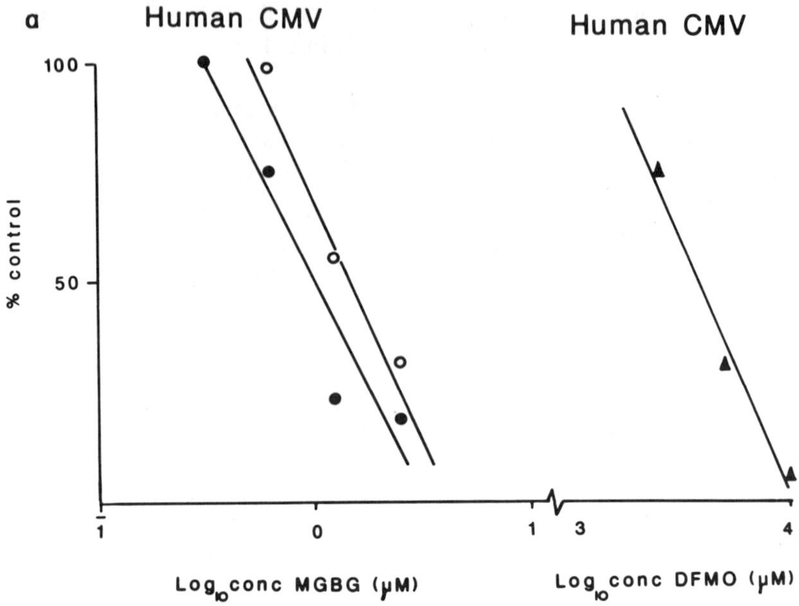
a
Human CMV
Human CMV
100
50
% control
1̄
0
1
3
4
Log10 conc MGBG (μM)
Log10 conc DFMO (μM)

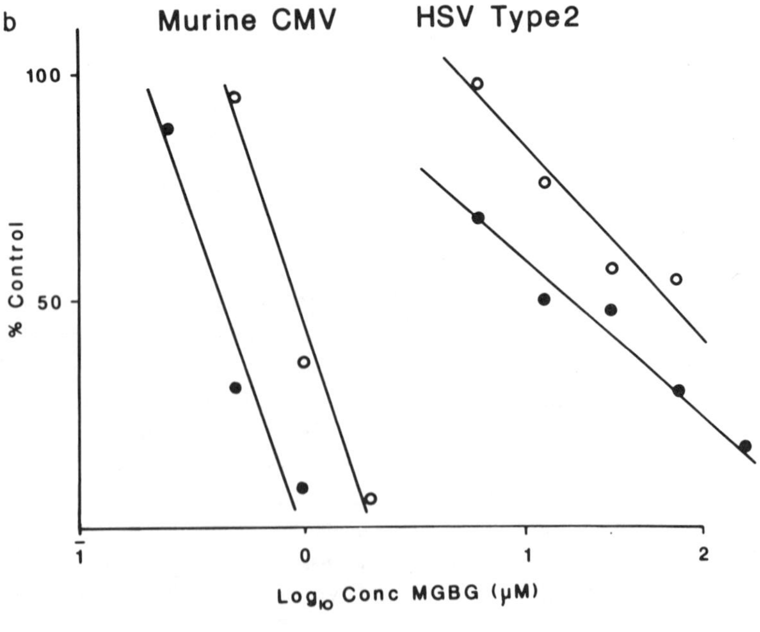
b
Murine CMV
HSV Type2
100
50
% Control
1̄
0
1
2
Log10 Conc MGBG (μM)

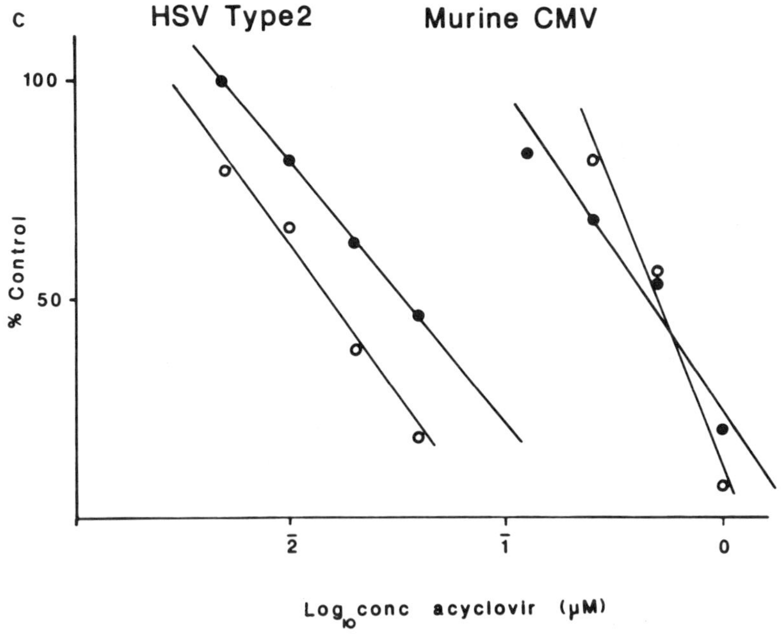

Fig. 4. Dose–response curves were obtained by plaque reduction assay using the method of Tyms *et al.* (1981) with different concentrations of MGBG or acyclovir in the presence or absence of DFMO. (a) Human embryo fibroblast cells were infected with a low dose (0.001 pfu/cell) of cytomegalovirus (strain AD-169) and treated with varying concentrations of MGBG with (●) or without (○) 2 m*M* DFMO. The dose–response for DFMO alone (▲) is shown in the right-hand section. (b) Whole embryo mouse (WEM) fibroblast cells were infected with a low dose of the Osborn strain of murine CMV or a genital strain of HSV type 2 and treated with varying concentrations of MGBG with (●) or without (○) 5 m*M* DFMO. (c) WEM cells were infected as in (b) but treated with varying concentrations of acyclovir, with (●) or without (○) 5 m*M* DFMO. The mean plaque number was determined from triplicate counts. In all experiments shown, DFMO treatment alone did not reduce the plaque count when judged against the untreated control cells. [The data are taken from Rawal and Tyms (1986)].

decreases in the ED_{50} for HSV type 2 were observed in these conditions with either 5 m*M* DFMO (Fig. 4c) or 2 m*M* DFMO (results not shown). The ED_{50} values recorded for acyclovir against these viruses were in the range reported previously (Tyms *et al.*, 1981).

Once again, the unpredictability of modulating polyamine levels by DFMO treatment in the cell systems used makes it difficult to interpret fully the results obtained in the combination studies described. The use of more potent ODC inhibitors will enable better depletion of spermidine and spermine in order to study the effects this has on virus-coded enzyme activities. Work is in progress concerning the investigation with 9-(1,2-dihydroxypropoxymethyl)guanine (DHPG) against infections with human CMV. This acyclic nucleoside has a mode of action similar to that of acyclovir (Cheng *et al.*, 1983), with proven virological and clinical efficacy against serious and life-threatening infections due to human CMV (Felsenstein *et al.*, 1985; Bach *et al.*, 1985; Shepp *et al.*, 1985). Side effects have been documented after DHPG treatment and protocols that enable lower doses to be used could greatly improve the performance of this drug.

VI. CONCLUDING REMARKS

It may be said that the use of polyamine inhibitors in virological studies has opened new avenues of research which may otherwise have remained undiscovered. This has enabled the intimate dependence of the replication of certain animal viruses on polyamine metabolism to be discerned. In terms of molecular virology, polyamine inhibitors will continue to be invaluable tools for the recognition of polyamine-dependent events in the virus replication cycle. This information, based on the expression of foreign genes in an intracellular milieu, may be usefully extrapolated to help our understanding of the physiological functions of polyamines.

In clinical terms, the polyamine story came of age with the reports that DFMO treatment helped cure disease in man due to parasitic infections (see Chapter 16). This drug fulfilled the main objective of antimicrobial chemotherapy, that is, the elimination of the causative agent in conditions of low or no toxicity. The mechanism of reproduction of protozoan parasites in the animal system is rather different from the obligatory association of the virus to its host cell. At this time there is no clear explanation for the selective effect of DFMO in protozoan disease. It seems most probable that the limited uptake of the compound by animal cells coupled to the greater demands for polyamines created by the highly proliferative nature of the parasite provides an acceptable therapeutic index. To our knowledge, no deliberate attempts have been made as yet to treat with polyamine inhibitors virus infections in man. Incidental observations on CMV infection were made by Golden and his colleagues (1984) during DFMO therapy of *Pneumocystis carinii* pneumonia in patients with acquired immune deficiency syndrome. There appeared to be no consistent antiviral effect in three out of six patients with concomitant CMV infection, although all patients responded to treatment for the protozoan disease. It was clear from studies of *T. b. brucei* infections in rats that maximum benefits from DFMO treatment are achieved only in the presence of an intact immune system (Bitonti *et al.*, 1986). It will be important in any assessment of the clinical potential of polyamine inhibitors as antiviral agents that the nature of the illness and immunological status of the patient be taken into account.

The observations that ODC inhibitors, if administered in the form of pro drugs, were beneficial to the treatment of trypanosomal infections in animals (Bitonti *et al.*, 1985) may be clinically relevant to the therapy of viral disease. In our hands the methyl ester of Δ-MFMO, which derives its intracellular activity as the parent compound, was much more effective than DFMO against CMV. This compound is being used for detailed studies of the mode of action of polyamine inhibitors against CMV but will also enable a closer investigation of the polyamine requirement of other viruses. For example, the investigation of Δ-MFMO methyl ester along with inhibitors of enzymes in the catabolic pathway targeted against amine oxidase activities should prove fruitful in deciding the nature of the polyamine requirement during herpes simplex replication. A main problem with herpesviruses is the ability to establish latent or persistent infections in man. The present generation of antivirals with proven clinical efficacy are targeted at DNA synthesis per se and are uneffective against latent infections, a major reservoir for these viruses in man. It has been proposed that early virus-directed events, including the activation of cellular genes, may be essential to maintain the latent state.

Inhibitors of polyamine biosynthesis alone or in combination may have a role in modulating this type of virus–cell interaction. In addition, the action of the most effective antiherpes drugs is dependent on enzymes coded by the virus genome. Mutations in these viral genes can result in the development of drug-resistance, so the future efficacy of such compounds may be affected, particularly if parallels with the widespread misuse of antibiotics are to be drawn. The requirement for alternative strategies of antiviral chemotherapy may be satisfied in part, at least, by future developments in the design of polyamine inhibitors.

ACKNOWLEDGMENTS

We gratefully acknowledge Action Research for the Crippled Child and an anonymous donor for their support of the studies in this laboratory which are reported here. We also thank Mrs. Helen Brown for the preparation of the manuscript.

REFERENCES

Alarcon, R.A., Foley, G.E., and Modest, E.J. (1961). *Arch. Biochem. Biophys.* **94,** 540–541.

Alhonen-Hongisto, L., Seppänen, P., and Jänne, J. (1980). *Biochem. J.* **192,** 941–945.

Allen, J.C., Smith, C.J., Curry, M.C., and Gaugas, J.M. (1977). *Nature (London)* **267,** 623–625.
Ames, B.N., and Dubin, D.T. (1960). *J. Biol. Chem.* **235,** 769–775.
Ames, B.N., Dubin, D.T., and Rosenthal, S.M. (1958). *Science* **127,** 814–816.
Archard, L.C., and Williamson, J.D. (1971). *J. Gen. Virol.* **12,** 249–258.
Bach, M.C., Bagwell, S.P., Knapp, N.P., Davis, K.M., and Hedstrom, P.S. (1985). *Ann. Intern. Med.* **103,** 381–382.
Bachrach, U., and Don, S. (1971). *J. Gen. Virol.* **11,** 1–9.
Bachrach, U., Don, S., and Weiner, H. (1974). *J. Gen. Virol.* **22,** 451–454.
Bitonti, A.J., Bacchi, C.J., McCann, P.P., and Sjoerdsma, A. (1985). *Biochem. Pharmacol.* **34,** 1773–1777.
Bitonti, A.J., McCann, P.P., and Sjoerdsma, A. (1986). *Biochem. Pharmacol.* **35,** 331–334.
Byrd, W.J., Jacobs, D.M., and Amoss, M.S. (1977). *Nature (London)* **267,** 621–623.
Carney, W.P., and Hirsch, M.S. (1981). *J. Infect. Dis.* **144,** 47–54.
Carney, W.P., Rubin, R.H., Hoffman, R.A., Hansen, W.P., Healey, K., and Hirsch, M.S. (1981). *J. Immunol.* **126,** 2114–2116.
Cheetham, B.F., and Bellett, A.J.D. (1982). *J. Cell. Physiol.* **110,** 114–122.
Cheetham, B.F., Shaw, D.C., and Bellett, A.J.D. (1982). *Mol. Cell. Biol.* **2,** 1295–1298.
Cheng, Y.-C., Goz, B., and Prusoff, W.H. (1975). *Biochim. Biophys. Acta* **390,** 253–270.
Cheng, Y.-C., Ostrander, M., Derse, D., and Chen, J.-Y. (1979). *In* "Nucleoside Analogues" (R.T. Walker, E. Clercq, and F. Eckstein, eds.), pp. 319–335. Plenum, New York.
Cheng, Y.-C., Huang, E.-S., Lin, J.-C., Mar, E.-C., Pagano, J.S., Dutschman, G.E., and Grill, S.P. (1983). *Proc. Natl. Acad. Sci. U.S.A.* **80,** 2767–2770.
Clarke, J.R., and Tyms, A.S. (1986a). *Med. Lab. Sci.* **43,** 258–261.
Clarke, J.R., and Tyms, A.S. (1986b). In preparation.
Clarkson, A.B., Bacchi, C.J., Mellow, G.H., Natham, H.C., McCann, P.P., and Sjoerdsma, A. (1983). *Proc. Natl. Acad. Sci. U.S.A.* **80,** 5729–5733.
Cooke, B.C., and Williamson, J.D. (1973). *J. Gen. Virol.* **21,** 339–348.
Don, S., and Bachrach, U. (1975). *Cancer Res.* **35,** 3618–3622.
Felsenstein, D., D'amico, D.J., Hirsch, M.S., Neumeyer, D.A., Cederberg, D.M., de Miranda, P., and Schooley, R.T. (1985). *Ann. Intern. Med.* **103,** 377–380.
Ferrari, W., Loddo, B., Gessa, G.L., Spaneda, A., and Brotzu, G. (1964). *Life Sci.* **3,** 755–758.
Fout, G.S., Medappa, K.C., Mapoles, J.E., and Rueckert, R.R. (1984). *J. Biol. Chem.* **259,** 3639–3643.
Gazdar, A.F., Stull, H.B., Kilton, L.J., and Bachrach, U. (1976). *Nature (London)* **262,** 696–698.
Gibson, W. (1983). *Virology* **128,** 391–406.
Gibson, W., and Irmiere, A. (1984). *Birth Defects Orig. Art. Ser.* **20,** 305–324.
Gibson, W., and Roizman, B. (1971). *Proc. Natl. Acad. Sci. U.S.A.* **68,** 2818–2821.
Gibson, W., van Breemen, R., Fields, A., LaFemina, R., and Irmiere, A. (1984). *J. Virol.* **50,** 145–154.
Golden, J.A., Sjoerdsma, A., and Santi, D.V. (1984). *West. J. Med.* **141,** 613–623.
Goldstein, D.A., Heby, O., and Marton, L.J. (1976). *Proc. Natl. Acad. Sci. U.S.A.* **73,** 4022–4026.
Gordon-Smith, R.H., Brooke-Taylor, S., Smith, L.L., and Cohen, G.M. (1983). *Biochem. Pharmacol.* **32,** 3701–3709.
Gordon-Smith, R.H., Smith, L.L., and Cohen, G.M. (1985). *Biochem. Pharmacol.* **34,** 1809–1816.

Hershey, A.D. (1957). *Virology* **4,** 237–267.
Hodgson, J. (1976). Ph.D. thesis, London University.
Hodgson, J., and Williamson, J.D. (1975). *Biochem. Biophys. Res. Commun.* **63,** *308–312.*
Hoffmann, P.J., and Cheng, Y.-C. (1979). *J. Virol.* **32,** 449–457.
Hölttä, E., Vartio, T., Jänne, J., Vaheri, A., and Hovi, T. (1981). *Biochim. Biophys. Acta* **677,** 1–6.
Isom, H.C. (1979). *J. Gen. Virol.* **42,** 265–278.
Isom, H.C., and Pegg, A.E. (1979). *Biochim. Biophys. Acta* **564,** 402–413.
James, K., and Hargreave, T.B. (1984). *Immunol. Today* **5,** 357–363.
Jänne, J., Alhonen-Hongisto, L., Seppänen P., and Siimes, M. (1981). *Med. Biol.* **59,** 448–457.
Kallio, A., and McCann, P.P. (1981). *Biochem. J.* **200,** 69–75.
Kallio, A., McCann, P.P., and Bey, P. (1982). *Biochem. J.* **204,** 771–775.
Kramer, D.L., Paul, B., and Porter, C.W. (1985). *Cancer Res.* **45,** 2512–2515.
Kuchler, C., Kuchler, W., and Schulze, W. (1968). *Acta Virol. Engl. Ed.)* **12,** 441–445.
Lanzer, W., and Holowczak, J.A. (1975). *J. Virol.* **16,** 1254–1264.
McCormick, F. (1978). *Virology* **91,** 496–503.
McCormick, F., and Newton, A. (1975). *J. Gen. Virol.* **27,** 25–33.
Macher, A.M., Reichert, C., Straus, S.E., Long, P.L., Parrillo, J., Lane, H., and Fauci, A.S. (1983). *New Engl. J. Med.* **309,** 1454.
Mamont, P.S., Danzin, C., Kolb, M., Gerhart, F., Bey, P., and Sjoerdsma, A. (1986). *Biochem. Pharmacol.* **35,** 159–165.
Mamont, P.S., Siat, M., Joder-Ohlenbusch, A.-M., Bernhardt, A., and Casala, P. (1984). *J. Biochem.* **142,** 457–463.
Meyers, J.D., Fluornoy, N., and Thomas, E.D. (1982). *Rev. Infect. Dis.* **4,** 1119–1132.
Morgan, D.M.L. (1985). *Biochem. Soc. Trans.* **13,** 322–326.
Moussatche, N. (1985). *Biochim. Biophys. Acta* **826,** 113–120.
Obert, G., Tripier, F., Nonnemacher, H., and Kirn, A. (1980). *Ann. Virol. (Inst. Pasteur)* **131,** 13–24.
Ostrander, M., and Cheng, Y.-C. (1977). *Pharmacologist* **19,** 166–170.
Ostrander, M., and Cheng, Y.-C. (1980). *Biochim. Biophys. Acta* **609,** 232–245.
Pathak, S.N., Porter, C.W., and Dave, C. (1977). *Cancer Res.* **37,** 2246–2250.
Pett, D.M., and Ginsberg, H.S. (1975). *J. Virol.* **15,** 1289–1292.
Prakash, N.J., Sunkara, P.S., and Sjoerdsma, A. (1985). *Biochem. Pharmacol.* **34,** 1887–1890.
Raina, A., Tuomi, K., and Mantyjarvi, R. (1981). *Med. Biol.* **59,** 428–432.
Rawal, B.K., and Tyms, A.S. (1986). In preparation.
Remacle-Bonnet, M., Culouscou, J.M., Pommier, G., Rance, R., and Depieds, R. (1985). *Am. J. Reprod. Immunol. Microbiol.* **8,** 55–61.
Seppänen, P., Alhonen-Hongisto, L., and Jänne, J. (1981). *Eur. J. Biochem.* **118,** 571–576.
Sharkis, S.J., Luk, G.D., Collector, M.I., McCann, P.P., Baylin, S.B., and Sensenbrenner, L.L. (1983). *Blood* **61,** 604–607.
Shepp, D.H., Dandliker, P.S., de Miranda, P., Burnette, T.C., Cederberg, D.M., Kirk, L.E., and Meyers, J.D. (1985). *Ann. Intern. Med.* **103,** 368–373.
Sheppard, S.L., Bruness, A.T.H., and Boyle, S.M. (1980). *J. Virol.* **34,** 266–267.
Shortridge, K.F., and Stevens, L. (1973). *Microbios* **7,** 61–68.
Siimes, M., Seppänen, P., Alhonen-Hongisto, L., and Jänne, J. (1981). *Int. J. Cancer* **28,** 567–570.
Sissons, J.G.P. (1986). *J. R. Coll. Phys. London* **20,** 40–44.
Sjoerdsma, A., and Schechter, P.J. (1984). *Clin. Pharmacol. Ther.* **5,** 287–300.

Smith, L.L. (1985). *Biochem. Soc. Trans.* **13,** 332–334.

Smith, C.J., Hussain, J.I., and Allen, J.C. (1985). *Biochem. Soc. Trans.* **34,** 266–267.

Stagno, S., Reynolds, D.W., Tsiantos, A., Fuccillo, D.A., Long, W., and Alford, C.A. (1975). *J. Infect. Dis.* **132,** 568–572.

Stagno, S., Pass, R.F., Dworsky, M.E., Britt, W.J., and Alford, C.A. (1984). *Birth Defects Orig. Art. Ser.* **20,** 65–85.

Sunkara, P.S., Prakash, N.J., Mayer, G.D., and Sjoerdsma, A. (1983). *Science* **219,** 851–853.

Tabor, C.W., and Tabor, H. (1984). *Annu. Rev. Biochem.* **53,** 749–790.

Tuomi, K. (1983). *Med. Biol.* **61,** 199–202.

Tuomi, K., Mantyjarvi, R., and Raina, A. (1980). *FEBS Lett.* **121,** 292–294.

Tuomi, K., Raina, A., and Mantyjarvi, R. (1982). *Biochem. J.* **206,** 113–119.

Tyms, A.S., and Davis, J.R. (1986). Submitted.

Tyms, A.S., and Williamson, J.D. (1980). *J. Gen. Virol.* **48,** 183–191.

Tyms, A.S., and Williamson, J.D. (1982). *Nature (London)* **297,** 690–691.

Tyms, A.S., Scamans, E.M., and Williamson, J.D. (1979). *Biochem. Biophys. Res. Commun.* **86,** 312–318.

Tyms, A.S., Scamans, E.M., and Naim, H.M. (1981). *J. Antimicrob. Chemother.* **8,** 312–318.

Tyms, A.S., Rawal, B.K., Naim, H.M., and Williamson, J.D. (1983). *Adv. Polyamine Res.* **4,** 507–517.

Tyms, A.S., Stevens, R.J., Mobberley, M.A., Ryder, T.A., and Jeffries, D.J. (1984). *J. Gen. Virol.* **65,** 2129–2139.

Tyms, A.S., Davis, J.R., Clarke, J.R., and Jeffries, D.J. (1987). *J. Gen. Virol.* **68,** in press.

Umezawa, H. (1983). *Adv. Polyamine Res.* **4,** 1–16.

Wallace, H.M., Baybutt, H.N., Pearson, C.K., and Keir, H.M. (1980). *J. Gen. Virol.* **49,** 397–400.

Williams-Ashman, H.G., and Schenone, A. (1972). *Biochem. Biophys. Res. Commun.* **46,** 288–295.

Williams-Ashman, H.G., Seidenfeld, J., and Galletti, P. (1982). *Biochem. Pharmacol.* **31,** 277–288.

Williamson, J.D. (1976). *Biochem. Biophys. Res. Commun.* **73,** 120–126.

Williamson, J.D. (1984). *Nature (London)* **310,** 103.

Williamson, J.D., and Cooke, B.C. (1973). *J. Gen. Virol.* **21,** 349–357.

Williamson, J.D., and Tyms, A.S. (1984). *In* "Medical Microbiology" (C.S.F. Easmon and J. Jeljaszewicz, eds.), Vol 4, pp. 239–285. Academic Press, Orlando, Florida.

14

Inhibition of Polyamine Biosynthesis in Plants and Plant Pathogenic Fungi

ROBERT D. SLOCUM* AND ARTHUR W. GALSTON†

* *Department of Biology*
Williams College
Williamstown, Massachusetts 01267

† *Department of Biology*
Yale University
New Haven, Connecticut 06511

I. INTRODUCTION

In the last few years, advances in our understanding of the regulation of polyamine (PA) metabolism and biological functions of PAs have been driven primarily by the development of specific inhibitors of PA biosynthesis. The selective modulation of PA biosynthesis with these inhibitors, and resultant effects on tissue PA titers, growth, and differentiation in plants, have been examined in numerous studies, some of which have been summarized in recent reviews (Slocum *et al.*, 1984; Smith, 1985). In the present discussion, we will identify recent developments and potential applications of this research.

II. INHIBITION OF POLYAMINE BIOSYNTHESIS AND CATABOLISM IN PLANTS

A. Arginine and Ornithine Decarboxylases

Unlike animals and fungi, plants and bacteria are able to synthesize putrescine (Put), the precursor to the PAs spermidine (Spd) and spermine (Spm), either directly from ornithine, through the activity of ornithine decarboxylase (ODC; EC 4.1.1.17), or indirectly, via arginine and two intermediates, through a pathway initiated by arginine decarboxylase (ADC; EC 4.1.1.19). *In vitro,* the plant decarboxylases are specifically inactivated by the enzyme-activated, irreversible inhibitors α-difluoromethylornithine (DFMO; Metcalf *et al.,* 1978) and α-difluoromethylarginine (DFMA; Kallio *et al.,* 1981), respectively, as has been demonstrated in other organisms.

DFMO and DFMA are also potent *in vivo* inhibitors of these enzymes in a variety of plant tissues and have been used extensively in studies of the regulation of Put biosynthesis. These inhibitors exhibit a high degree of specificity and do not appear to be metabolized in plants, except in tissues containing high arginase activities. In these tissues, DFMA appears to be metabolized, to a small extent, to urea and DFMO, resulting in the partial inactivation of ODC by DFMO (Slocum and Galston, 1985b). This limitation of specificity by DFMA in plants has not been encountered in bacteria, which do not exhibit arginase activity (cf. Tabor and Tabor, 1985). However, the recent development of potent agmatine-based ADC inhibitors (Bitonti *et al.,* 1987) should allow specific inhibition of ADC, even in plant cells containing high arginase activities, since agmatine does not appear to serve as a substrate for arginase (R.D. Slocum, unpublished). DFMO does not serve as a substrate for ornithine carbamoyltransferase; thus its possible metabolism in the urea cycle via citrulline seems unlikely (R.D. Slocum, unpublished).

B. Spermidine Synthase

In plants, PA biosynthesis from Put is carried out through the action of Spd (and presumably Spm) aminopropyltransferases, or "synthases." These enzymes utilize decarboxylated *S*-adenosylmethionine (AdoMet) as the aminopropyl group donor, as has been shown in other eucaryotes and bacteria. Spd synthase (EC 2.5.1.16) has recently been purified (Sindhu and Cohen, 1984) and is inhibited by cylohexylamine (CHA)[1] as well

[1] Commercial preparations of "dicyclohexylamine" ("DCHA," (Sigma), are actually cyclohexylamine (Batchelor *et al.,* 1986), rather than dicyclohexylamine, as has been reported in the past. The correct term, cyclohexylamine, will be used here.

as the transition-state analog, *S*-adenosyl-1,8-diamino-3-thiooctane (Ado-Dato) (Pegg *et al.*, 1983). *In vivo,* Greenberg and Cohen (1985) found that CHA inhibition of Spd synthase led to a rapid depletion of Spd in chinese cabbage protoplasts. This effect was accompanied by a 10-fold increase in decarboxylated AdoMet levels and increased Spm levels, suggesting that decarboxylated AdoMet is normally limiting for Spm synthesis. This may also explain the observation by Biondi *et al.* (1986) that uptake of CHA and inhibition of Spd synthesis in pine cotyledons had no effect on growth, since increased Spm synthesis may have compensated for the loss of Spd.

C. S-Adenosylmethionine Decarboxylase

Decarboxylation of AdoMet by *S*-adenosylmethionine decarboxylase (AdoMetDC; EC 4.1.1.50) leads to the formation of decarboxylated Ado-Met, the aminopropyl group donor in PA biosynthesis from Put. Ado-MetDC has been purified (Yamanoha and Cohen, 1985) and appears to be similar to the bacterial enzyme (Mg^{2+} stimulated, Put insensitive). This enzyme is weakly inhibited by methylglyoxal bis(guanylhydrazone), or MGBG, and this drug has been employed in the affinity purification of tobacco AdoMetDC (Hiatt *et al.*, 1986).

In vivo, MGBG-treated cells have been shown to overproduce Ado-MetDC to levels 20 times that of untreated cells, an effect which is due to an approximate doubling of the half-life of the enzyme (Hiatt *et al.*, 1986). This paradoxical enhancement of AdoMetDC activity by MGBG has long been recognized in animal tissues as well (Pegg *et al.*, 1973) and appears to be the result of stabilization of the enzyme by the bound inhibitor. MGBG-resistant mutants of tobacco were also shown to overproduce a 35-kDa protein (Malmberg and McIndoo, 1983) which was later identified as AdoMetDC.

The regulation of AdoMetDC activity in plant tissues is a topic of intense interest since this enzyme represents an important metabolic crossroad controlling the utilization of a common AdoMet precursor in both PA and ethylene biosynthesis. Ethylene is a senescence-inducing hormone in plants, while PAs exhibit strong antisenescence properties. Several recent studies have shown that inhibition of either pathway impacts the other (cf. Slocum *et al.*, 1984, for review). Concerted efforts are being made to estimate the flux of methionine through 5′-methylthioadenosine (MTA; Giovanelli *et al.*, 1985), a byproduct both of aminopropyl group transfer to Put and of conversion of AdoMet to 1-aminocyclopropane-1-carboxylic acid (ACC), the precursor to ethylene. MTA analogs being developed by Kushad *et al.* (1985) may find important application in these studies.

D. Amine Oxidases

Putrescine and PAs in plants are catabolized by diamine oxidases (DAOs; EC 1.4.3.6) and polyamine oxidases (PAOs; EC 1.5.3.3), although the latter group of enzymes generally exhibits broad substrate specificities and can also oxidize diamines.

DAOs have been purified and characterized principally from legumes (Federico *et al.*, 1985), and are sensitive to both Cu^{2+} chelators and carbonyl reagents such as aminoguanidine and semicarbazide (Matsuda and Suzuki, 1981). Federico and Angelini (1986) have recently demonstrated an apoplastic location for DAO in pea epicotyl tissues.

PAOs have been found only among the cereals, where this enzyme has also been localized in the cell wall (Kaur-Sawhney *et al.*, 1981). Cereal PAOs are flavoproteins which are insensitive to Cu^{2+} chelators but are inhibited by 2-hydroxyethylhydrazine; they catalyze the oxidation of PAs to 1,3-diaminopropane and Δ^1-pyrroline (Smith, 1983). In contrast, fungal and mammalian PAOs oxidize PAs or, more commonly, acetylated PAs, to 3-aminopropionaldehyde plus Put and Spd or their acetylated derivatives. Acetylated PAs appear not to have been tested as substrates for plant PAOs, therefore it is not clear whether a new group of "suicide" substrate analogs of acetylated PAs (Bolkenius *et al.*, 1985) will find application as inhibitors of this plant enzyme.

III. EFFECTS OF POLYAMINE BIOSYNTHESIS INHIBITORS ON PLANT GROWTH AND DEVELOPMENT

In plants, attempts to analyze PA metabolism and its relationship to growth and developmental processes, through the use of inhibitors of PA biosynthesis, have met with only limited success. This may be explained, in part, by the fact that we know virtually nothing about (1) the subcellular location of PAs or their biosynthetic enzymes and (2) the uptake and transport of exogenous or endogenous inhibitors or PAs. However, these parameters are poorly understood in other eucaryotes, as well.

A. Correlating Changes in Polyamine Biosynthesis and Titers with Growth Responses in Inhibitor-Treated Plants

A problem unique to plants is their tendency to accumulate PAs to levels far in excess of those required for normal cellular functions. Using even potent inhibitors, it is notoriously difficult to reduce PAs to levels which are limiting for growth or cell division. Even when tissue PA levels are severely depleted, inhibition of growth may not be observed, whereas similarly treated animal cells would quickly stop growing. In many plants, such as tobacco, PAs accumulate to very high levels as amide conjugates of cinnamic acids, and there is some evidence that Put conjugates may be reversibly hydrolyzed to maintain "free" Put levels in tissues treated with DFMO or other Put biosynthesis inhibitors (Slocum and Galston, 1985b). In these tissues, inhibitor effects on growth are minimal, suggesting that the small pool of free Put, which is maintained near control tissue levels, is the physiologically important target of the inhibitor. In studies of other plants not exhibiting PA conjugates, relatively small changes in PA levels in inhibitor-treated tissues are correlated with large-scale growth effects. These observations suggest that PA biosynthesis inhibitors may modulate growth through their effects on a small pool of "physiologically active" PAs, which may be maintained against a larger pool of PAs that are physically sequestered within organelles or chemically "sequestered" as nonmetabolizable conjugates, etc., in some plants. Obviously, it would be useful to have some information on changes in the subcellular distribution of these PAs in inhibitor-treated tissues.

Another problem common to most plant inhibitor studies is the difficulty in correlating changes in PA biosynthesis and tissue PA titers with specific growth or developmental responses. Part of this problem stems from the recent recognition that apparent "ADC" and "ODC" activities may not bear any stoichiometric relationship to Put formation *in vitro* (Birecka *et al.,* 1985a). Nonenzymatic decarboxylation of pyridoxal phosphate–amino acid complexes is especially prevelant in tissue extracts containing phenolics (Mazelis, 1971), and the contribution of this nonenzymatic component to apparent decarboxylase activities should be evaluated more closely. However, it is clear that even when accurate estimates of ADC and ODC activities are obtained, changes in their activities frequently bear no relationship to endogenous PA titers and growth responses *in vivo* (Birecka *et al.,* 1985a).

B. Evidence for Compensatory Mechanisms Regulating Putrescine Biosynthesis in Inhibitor-Treated Tissues

Plant inhibitor studies are also complicated by the fact that Put may be synthesized via either the ADC or ODC pathway. Constitutive ADC and ODC activities are found in all plant tissues examined to date, but it is clear that these enzymes are differentially induced and serve distinctly different roles in cell division vis-à-vis growth and stress responses. In actively dividing cells, ODC activity appears to be induced many fold above resting levels and DFMO treatment results in decreased cell division growth in numerous tissues. ADC activity usually predominates during vegetative growth and is also rapidly induced by a variety of stress conditions, while DFMA inhibits these responses. One unanswered question is whether compensatory mechanisms exist which would allow either of these enzymes to overcome the inhibition of Put biosynthesis by inactivation of the other. Such mechanisms are known in bacteria, in which simultaneous inhibition of both the ADC and ODC pathways is required to effectively inhibit Put synthesis (Bitonti *et al.*, 1982). Similar inhibitor treatments, or use of "broad spectrum" inhibitors of PA biosynthesis, such as 1-aminooxy-3-aminopropane (Khomutov *et al.*, 1985), may overcome such compensatory mechanisms, if they exist.

Berlin and Forche (1981) have provided some indirect evidence for such compensation in DFMO-treated tobacco suspension cultures, since endogenous PA titers were not lowered, even with nearly complete *in vivo* inhibition of ODC. Hiatt *et al.* (1986), however, have presented data which support specialized, indispensible functions for both ADC and ODC in similar cultured tobacco cells. They observed a transient increase in ODC activity in the first several days after cells were subcultured into fresh medium and reported that DFMO-treated cultures stopped growing and died, whereas DFMA-treated cultures grew at the same rate as the controls. This suggested that Put synthesized by the ADC pathway could not compensate for the loss of ODC-derived Put, perhaps due to differences in the subcellular compartmentation of this diamine. Conversely, when callus tissue adapts to suspension culture conditions, there is a transient increase in ADC activity, and DFMA treatment of these cultures is also lethal. Such inhibition of ADC activity was proposed to be lethal only under stress conditions, which universally lead to accumulation of Put following ADC induction. Other workers have suggested that the accumulation of Put itself may be toxic to the cell, and inhibition of this response with DFMA has been explored as one approach to the regeneration of viable cereal plants from osmotically stressed protoplasts (Tiburcio *et al.*, 1986).

A different compensatory mechanism has been noted in mammalian cells grown continuously in the presence of DFMO. In these cells, which can synthesize Put only via the ODC pathway, ODC overproduction appears to result from the amplification of one or more members of an ODC gene family (Alhonen-Hongisto *et al.*, 1985). This response has not been documented in inhibitor-treated plant cells, although it is possible that such gene amplification mechanisms do exist.

C. Transport of Polyamines and Inhibitors of Polyamine Biosynthesis

There have been few studies investigating the mechanisms of transport of PAs or inhibitors of PA biosynthesis in plants. Long-distance transport of PAs in plant tissues is not appreciable. Young and Galston (1983) reported that radiolabeled PAs injected into etiolated pea cotyledons were poorly transported to epicotyl tissues. However, significant radioactivity could be recovered in Put or cadaverine from epicotyl tissues when [^{14}C]arginine or [^{14}C]lysine was supplied to cotyledons. These findings suggested that PAs were synthesized in, or proximal to, their source tissues, being translocated only short distances, while their amino acid precursors were quite mobile. Studies by Cheng and Kao (1983) provided further evidence for this, since the antisenescence effects of exogenous PAs on chlorophyll loss in excised leaves were limited to conductive tissues and cut surfaces; PAs did not appear to penetrate nonvascularized areas of the leaf.

Cellular transport of PAs has been studied more recently. Bagni and Pistocchi (1985) studied the kinetics of ^{14}C-labeled PA uptake in *Saintpaulia* petals. In this system, uptake of Put and Spd was dependent upon the pH of the external medium. At low concentrations, Put was accumulated against a concentration gradient, with pH 5–5.5 optimum, while Put accumulation followed a concentration gradient at higher external concentrations, with pH 8 optimum. The uncouplers 2,4-dinitrophenol and CCCP did not inhibit uptake, but sodium thiocyanate inhibited Spd uptake by 68% versus controls (Pistocchi *et al.*, 1986), suggesting an active transport process. The results of competition experiments indicated that Put and Spd did not utilize the same transport system. Cell fractionation studies also demonstrated that the labeled PAs were not associated with the apoplastic compartment, but most likely accumulated in the vacuole. Similar results have been reported for the fungus *Aspergillus nidulans* by Spathas *et al.* (1982).

It is not clear whether PA uptake occurs via amino acid transport systems, or if separate PA transporting systems exist. In NB-15 mouse neuroblastoma cells, Rinehart and Chen (1984) observed that preloading of the cells with transport system A amino acids stimulated [^{14}C]Put uptake 8- to 10-fold, suggesting the participation of this system in PA uptake. Feige and Chambaz (1985), however, found that PA uptake was independent of type A,L, and N amino acid transport systems and that Put and Spd uptake occurred at the same transport site. In plants, different transport systems for PAs and amino acids seem to be supported by Joshi *et al.*, (1983), who observed that the uptake of [^{14}C]Spd by cowpea protoplasts was age dependent and required a well-defined lag time. Conversely, [^{35}S]methionine uptake required no lag time and was age independent.

It seems likely, however, that amino acid analog inhibitors of Put biosynthesis, such as DFMA and DFMO, would utilize amino acid transport systems in both plants and animals. In plants, numerous studies have suggested that there is a common carrier for all amino acids (Reinhold and Kaplan, 1984), although a recent study has provided evidence for a system II (basic) amino acid transporter (Johannes and Felle, 1985), which might function in the uptake of DFMO, DFMA, and related inhibitors. On the other hand, MGBG, a Spd analog, appears to compete for the Spd binding site on membranes (Pleshkewych *et al.*, 1983) and may utilize the Spd transport system for uptake.

D. Development of Model Systems for Inhibitor Studies in Plants

Ideally, plants in liquid culture systems would lend themselves most readily to investigations of the effects of inhibitors of PA biosynthesis, since the inhibitors and PA metabolites could be supplied to and removed from the treated cells efficiently.

Perhaps the best-characterized system for examining the effects of PA biosynthesis inhibitors on plant growth and differentiation is the wild carrot somatic embryogenesis model. In this system, cells removed from an exogenous auxin source undergo cell division and embryogenesis. This response is accompanied by increased ADC activity and PA titers in the differentiating cells and is almost completely blocked by DFMA, while exogenous Put or Spd partially reverses this inhibition (Feirer *et al.*, 1984). Using a nonembryogenic mutant line, Fienberg *et al.* (1984) presented further evidence that increased PA levels were specifically associated with embryogenesis, rather than "growth" per se. These, and other data, suggest that this system may be useful in exploring specific roles for PAs in regulating gene expression during development.

IV. CONTROL OF PHYTOPATHOGENIC FUNGI WITH DFMO

Many economically important diseases of crop plants are caused by fungal pathogens, and the results of several recent investigations suggest that the specific inhibition of PA biosynthesis in these fungi may represent a useful method for their control. These studies are based on the observations that (1) infections by most phytopathogenic fungi are spread by sexual or asexual spores, (2) fungal spore germination, but not initial hyphal growth, is dependent upon *de novo* polyamine biosynthesis, and (3) Put biosynthesis in these fungi results from the activity of ODC only, while the plant host can synthesize Put via either the ADC or ODC pathway. This last observation offered the possibility that specific inhibition of fungal ODC with DFMO might prevent spore germination and spread of the infection, without significantly affecting PA biosynthesis in the host plant tissues.

Like animals, the yeasts and most filamentous fungi that have been studied synthesize Put exclusively via the ODC pathway (Tabor and Tabor, 1985). Numerous workers have been unable to document the presence of ADC activity in these organisms (Birecka *et al.*, 1986; Stevens, 1975), with the possible exception of the basidiomycete *Panus*, from which a biodegradative-type ADC enzyme (pH 5.2 optimum) was partially purified (Uhlemann and Reinbothe, 1977). In fact, studies carried out with arginaseless mutants of *Neurospora crassa* have shown that these cells are incapable of utilizing arginine as a substrate in PA biosynthesis (Davis and Paulus, 1983), again ruling out the ADC pathway to Put synthesis.

De novo PA biosynthesis, or uptake of PAs from exogenous sources, appears to be prerequisite to *in vitro* spore germination in a number of fungi. Germination of *A. nidulans* spores is accompanied by a transient "burst" of ODC and AdoMetDC activities and an accumulation of Put and PAs (Stevens and Stevens, 1981). A mutant Put auxotrophic strain of this fungus requires exogenous Put or Spd for spore germination, but not for postgermination hyphal growth (Stevens, 1975). A similar situation has been reported for zoospore germination in the water mold *Achlya ambisexualis* (Wright *et al.*, 1982).

More recent investigations with plant pathogenic fungi grown on defined media have demonstrated that specific inhibition of ODC with DFMO inhibits not only the germination of various fungal propagules, but also mycelial growth and sporulation, which allow the pathogen to successfully infect its host. Birecka *et al.* (1986) reported that 0.5–2.0 m*M* DFMO inhibited *in vitro* mycelial growth and sporulation of the southern corn leaf blight pathogen *Helminthosporium maydis*. This inhibition was reversible by exogenous Put, suggesting that growth inhibition resulted from the inhibition of Put biosynthesis via ODC. Rajam and Galston (1985) had previously observed a similar DFMO inhibition of growth in the basidiomycetous soil pathogen, *Rhizoctonia solani*, and in species of the imperfect ascomycetes, *Botrytis* and *Monilinia*. These investigators also reported that DFMA was slightly more effective than DFMO in inhibiting fungal growth and suggested that this inhibition might have resulted not from the inhibition of ADC but, rather, from an arginase-mediated hydrolysis of DFMA to urea + DFMO (Slocum and Galston, 1985b), with the resultant inhibition of ODC. Indeed, this indirect arginase-mediated inhibition of ODC with DFMA has been confirmed in a recent study of the *Verticillium dahliae* wilt pathogen of tomato (Mussell *et al.*, 1986). Both DFMO and DFMA inhibited the growth of this fungus *in vitro,* although only ODC activity could be detected in extracts of this pathogen. Arginase activity was also demonstrated in these extracts, and urea production could be measured with either L-arginine or DFMA as substrate. Since the effects of either DFMA or DFMO could be reversed by Put or PAs, fungal growth inhibition appeared to result from the specific inhibition of PA biosynthesis, leading to growth-limiting PA titers in these organisms. Birecka *et al.* (1986) also noted a weak DFMA inhibition of growth in the corn leaf blight pathogen, but this was not reversible by exogenous Put, suggesting that the inhibition was due to nonspecific effects of the drug.

Clearly, the sensitivity of various broad host-range plant fungal pathogens to DFMO *in vitro* suggested the potential use of this drug in controlling infections of crop plants by these organisms. Rajam *et al.* (1985) demonstrated that a single spray application of 0.5 m*M* DFMO to newly emerging leaves of pinto bean plants protected them against infection by the bean rust fungus, *Uromyces phaseoli,* an obligate parasite. DFMO conferred protection against infection even in unsprayed parts of the plant, suggesting translocation of the drug or some other inhibitor within the plant. DFMA, even at a 5 m*M* concentration, had no protective effect.

Perhaps the most interesting finding in these studies, however, was the persistence of the DFMO inhibition effect. Leaves infected 1 week after DFMO application were still partially protected from infection. A time-course analysis of bean rust inhibition by DFMO (Rajam *et al.,* 1986) indicated that foliar applications of DFMO up to 3 days postinnoculation still completely inhibited fungal growth, while 4- to 5-day postinnoculation applications significantly reduced the severity of the infection. An examination of rust uredospores on infected leaf surfaces confirmed that DFMO completely prevented germination, while application of Put or Spd to DFMO-treated leaves partially reversed this effect. Foliar spraying with DFMO or DFMA (0.01–1.0 m*M*) produced no significant growth effects or changes in PA titers in the leaves of either infected or uninfected bean plants.

The extreme sensitivity of this fungus, and apparent insensitivity of the host plant, to DFMO underscores the efficacy of this treatment. Indeed, Rajam *et al.* (1985) have estimated that as little as 25 g DFMO/acre would be adequate to protect bean plants from this rust pathogen in the field! The toxicity of DFMO to other organisms at such low concentrations would be minimal. In addition, Mussell *et al.* (1986) have shown that foliar applications of DFMO are also able to protect tomatoes from the soil-borne *Verticillium* wilt fungus, again suggesting that the inhibitory effect is translocated through the plant to its roots. Studies in progress have shown that wheat rust and powdery mildew pathogens are also sensitive to DFMO (L. Weinstein, personal communication), and it may be possible to control many other types of plant fungal pathogens using this drug.

REFERENCES

Alhonen-Hongisto, L., Kallio, A., Sinervirta, R., Seppänen, P., Kontula, K.K., Jänne, O.A., and Jänne, J. (1985). *Biochem. Biophys. Res. Commun.* **126,** 734–740.

Bagni, N., and Pistocchi, R. (1985). *Plant Physiol.* **77,** 398–402.

Batchelor, K.W., Smith, R.A., and Watson, N.S. (1986). *Biochem. J.* **233,** *307–308.*

Berlin, J., and Forche, E. (1981). *Z. Pflanzenphysiol.* **101,** 277–282.

Biondi, S., Bagni, N., and Sansovini, A. (1986). *Physiol. Plant.* **66,** 41–45.

Birecka, H., Bitonti, A.J., and McCann, P.P. (1985a). *Plant Physiol.* **79,** 509–514.

Birecka, H., Bitonti, A.J., and McCann, P.P. (1985b). *Plant Physiol.* **79,** 515–519.

Birecka, H., Garraway, M.O., Baumann, R.J., and McCann, P.P. (1986). *Plant Physiol.* **80,** 798–800.

Bitonti, A.J., McCann, P.P., and Sjoerdsma, A. (1982). *Biochem. J.* **208,** 435–441.

Bitonti, A.J., Casara, P.J., McCann, P.P., and Bey, P. (1987). *Biochem. J.* **242,** 69–74.

Bolkenius, F.N., Bey, P., and Seiler, N. (1985). *Biochim. Biophys. Acta* **838,** 69–76.

Cheng, S.H., and Kao, C.H. (1983). *Plant. Cell Physiol.* **24,** 1463–1467.

Davis, R.H., and Paulus, T.J. (1983). *In* "Methods in Enzymology" (H. Tabor and C.W. Tabor, eds.), Vol. 94, pp. 112–116. Academic Press, New York.

Federico, R., and Angelini, R. (1986). *Planta* **167,** 300–303.

Federico, R., DiLisi, F., and Angelini, R. (1985). *Plant Sci. Lett.* **38,** 9–12.

Feige, J.J., and Chambaz, E.M. (1985). *Biochim. Biophys. Acta* **846,** 93–100.

Feirer, R.P., Mignon, G., and Litvay, J.D. (1984). *Science* **223,** 1433–1435.

Fienberg, A.A., Choi, J.H., Lubich, W.P., and Sung, Z.R. (1984). *Planta* **162,** 532–539.

Giovanelli, J., Mudd, S.H., and Datko, A.H. (1985). *Plant Physiol.* **78,** 555–560.

Goldberg, R., and Perdrizet, E. (1984). *Planta* **161,** 531–535.

Greenberg, M.L., and Cohen, S.S. (1985). *Plant Physiol.* **78,** 568–575.

Hiatt, A.C., McIndoo, J., and Malmberg, R.L. (1986). *J. Biol. Chem.* **261,** 1293–1298.

Johannes, E., and Felle, H. (1985). *Planta* **166,** 244–251.

Joshi, S., Pleij, C.W.A., Haenni, A.-L., and Bosch, L. (1983). *Plant Mol. Biol.* **2,** 89–94.

Kallio, A., McCann, P.P., and Bey, P. (1981). *Biochemistry* **20,** 3163–3166.

Kaur-Sawhney, R., Flores, H.E., and Galston, A.W. (1981). *Plant Physiol.* **68,** 494–498.

Khomutov, R.M., Hyvönen, T., Karvonen, E., Kauppinen, L., Paalanen, T., Paulin, L., Eloranta, T., Pajula, R.-L., Andersson, L.C., and Pösö, H. (1985). *Biochem. Biophys. Res. Commun.* **130,** 596–602.

Kushad, M.M., Richardson, D.G., and Ferro, A.J. (1985). *Plant Physiol.* **79,** 525–529.

Malmberg, R.L., and McIndoo, J. (1983). *Nature (London)* **305,** 623–625.

Matsuda, H., and Suzuki, Y. (1981). *Plant Cell Physiol.* **22,** 737–746.

Mazelis, M. (1971). *In* "Methods in Enzymology" (H. Tabor and C.W. Tabor, eds.), Vol. 17B, pp. 606–608. Academic Press, New York.

Metcalf, B.W., Bey, P., Danzin, C., Jung, M.J., Casara, P. and Vevert, J.P. (1978). *J. Am. Chem. Soc.* **100,** 2551–2553.

Mussell, H., Osmeloski, J., Wettlaufer, S., and Weinstein, L.H. (1987). *Plant Dis.* **70,** in press.

Pegg, A.E., Corti, A., and Williams-Ashman, H.G. (1973). *Biochem. Biophys. Res. Commun.* **52,** 696–701.

Pegg, A.E., Bitonti, A.J., McCann, P.P., and Coward, J.K. (1983). *FEBS Lett.* **155,** 192–196.

Pistocchi, R., Bagni, N., and Cruens, J.A. (1986). *Plant Physiol.* **80,** 556–560.

Pleshkewych, A., Maurer, T.C., and Porter, C.W. (1983). *Cancer Res.* **43,** 646–652.

Rajam, M.V., and Galston, A.W. (1985). *Plant Cell Physiol.* **26,** 683–692.

Rajam, M.V., Weinstein, L.H., and Galston, A.W. (1985). *Proc. Natl. Acad. Sci. U.S.A.* **82,** 6874–6878.

Rajam, M.V., Weinstein, L.H., and Galston, A.W. (1986). *Plant Physiol.* **82,** 485–487.

Reinhold, L., and Kaplan, A. (1984). *Annu. Rev. Plant Physiol.* **35,** 45–83.

Rinehart, C.A., Jr., and Chen, K.Y. (1984). *J. Biol. Chem.* **259,** 4750–4756.

Sindhu, R.K., and Cohen, S.S. (1984). *Plant Physiol.* **74,** 645–649.

Slocum, R.D., and Galston, A.W. (1985a). *Plant Physiol. Suppl.* **77,** 45.

Slocum, R.D., and Galston, A.W. (1985b). *Plant Cell Physiol.* **26,** 1519–1526.

Slocum, R.D., Kaur-Sawhney, R., and Galston, A.W. (1984). *Arch. Biochem. Biophys.* **235,** 283–303.

Smith, T.A. (1983). *In* "Methods in Enzymology" (H. Tabor and C.W. Tabor, eds.), Vol. 94, pp. 311–314. Academic Press, New York.

Smith, T.A. (1985). *Annu. Rev. Plant Physiol.* **36,** 117–143.

Spathas, D.H., Pateman, J.A. and Clutterbuck, A.J. (1982). *J. Gen. Microbiol.* **128,** 557–563.

Stevens, L. (1975). *FEBS Lett.* **59,** 80–82.

Stevens, L., and Stevens, E. (1981). *FEMS Microbiol. Lett.* **11,** 229–232.

Tabor, C.W., and Tabor, H. (1985). *Microbiol. Rev.* **49,** 81–99.

Tiburcio, A.F., Kaur-Sawhney, R., and Galston, A.W. (1986). *Plant Physiol.* **82,** 375–378.

Uhlemann, A., and Reinbothe, H. (1977). *Biochem. Physiol. Pflanzen* **171,** 85–92.

Wright, J.M., Gulliver, W.P., Michalski, C.J., and Boyle, S.M. (1982). *J. Gen. Microbiol.* **128,** 1509–1515.

Yamanoha, B., and Cohen, S.S. (1985). *Plant Physiol.* **78,** 784–790.

Young, N.D., and Galston, A.W. (1983). *Plant Physiol.* **73,** 912–914.

15

Parasitic Protozoa and Polyamines

CYRUS J. BACCHI* AND PETER P. McCANN†

** Haskins Laboratories and Department of Biology*
Pace University
New York, New York 10038

† Merrell Dow Research Institute
2110 East Galbraith Road
Cincinnati, Ohio 45215

I. INTRODUCTION

Protozoa are one of the most interesting groups of eukaryotic microorganisms. Their diversity in morphology, habitat, life cycles, and metabolism is matched only by the devastating diseases caused by some of them and the toll of human suffering and economic loss which accompany them.

The World Health Organization's Special Program for Research and Training in Tropical Diseases has given priority to six parasitic diseases, four of which are caused by protozoa: African sleeping sickness (*Trypanosoma brucei* spp.), Chagas' disease (South American trypanosomiasis: *Trypanosoma cruzi*), malaria (*Plasmodium* spp.), and New World and Old World leishmaniases (*Leishmania* spp.). (The other two diseases are schistosomiasis and filariasis: WHO, TDR 7th Program Report, 1985.) All of these diseases are singularly important because of the large number of people afflicted, the frequent lack of adequate (and nontoxic) chemotherapy, and the development of drug resistance.

There are a number of model systems to study parasitic protozoa encompassing *in vitro* cell culture techniques and *in vivo* studies using laboratory animals, both of which have allowed for the effects of ornithine decarboxylase inhibitors to be investigated in numerous parasitic protozoa. Notable in this respect are the coculture systems for growth of bloodstream forms of African trypanosomes (Hirumi *et al.*, 1977; Brun *et al.*, 1981), synchronous growth of human malaria in erythrocytes (Jensen, 1985), and production of intracellular forms of *Leishmania* spp. in macrophage cultures (Berens and Marr, 1979). Recently, culture techniques have been developed for African trypanosomes which support the parasitic bloodstream forms without a mammalian feeder layer (Baltz *et al.*, 1985; Duszenko *et al.*, 1985). These systems all allow reasonable standardization of culture conditions to study the consequences of inhibition of polyamine biosynthesis within the mammalian form of the parasite. This chapter will concentrate on the biochemical and pharmacologic implications of inhibition of polyamine metabolism in parasitic protozoa, with particular emphasis on African trypanosomes, *Leishmania,* and *Plasmodium*.

II. FLAGELLATES

A. African Trypanosomes

These organisms develop in the salivary glands of the vector (tsetse fly) and include a group of subspecies of *T. brucei*. *Trypanosoma b. brucei* itself causes nagana—a wasting disease of livestock; *T. b. gambiense* and *T. b. rhodesiense* are agents of the human disease African sleeping sickness and annually cause at least 20,000 new cases/year, a low estimate in the opinion of epidemiologists (WHO, TDR 7th Program Report, 1985; Knuttgen, 1985). Related organisms such as *Trypanosoma congolense, Trypanosoma evansi,* and *Trypanosoma vivax* cause disease in domestic and wild animals over a 10-million square kilometer area of sub-Saharan Africa (Wilson *et al.*, 1963), an area roughly equal to that of the United States.

Trypanosoma b. brucei and related organisms have a complex life cycle (Fig. 1) in which infectious metacyclic forms are injected by the fly into mammals. Slender forms develop which multiply in the blood and eventually invade the central nervous system. Intermediate and short stumpy forms develop from slender forms and are believed to be infective for the fly (Vickerman, 1985). The morphological switch from slender to stumpy forms is accompanied by a dramatic shift in metabolism from strict glycolysis to a tricarboxcylic acid cycle with cytochrome-mediated oxidative metabolism (Opperdoes, 1985).

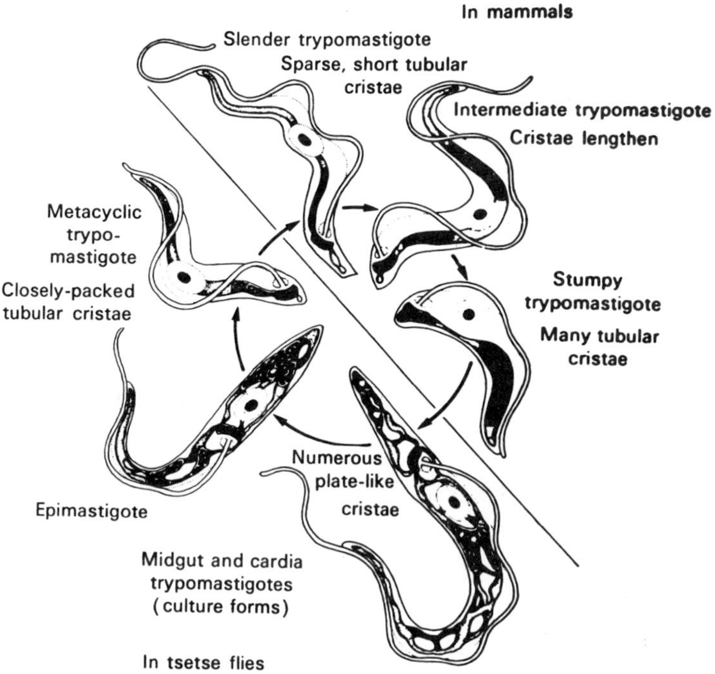

Fig. 1. Diagram showing changes in morphology of *Trypanosoma brucei* and changes in the mitochondrion during the life cycle. (From Vickerman, 1971.)

1. Polyamine Metabolism

Initial studies on polyamine metabolism in *T. b. brucei* bloodstream forms revealed that putrescine and spermidine but not spermine were present (Bacchi *et al.*, 1977) and that there was sufficient ornithine decarboxylase (ODC) activity to account for *de novo* synthesis of polyamines. Arginine decarboxylase was not detected although arginine was a precursor of putrescine in whole cells incubated *in vitro,* suggesting the presence of arginase (Bacchi *et al.*, 1979). All three biologically active polyamines were taken up by bloodstream forms but accumulation was 100-fold lower than in mammalian cells. *S*-Adenosyl-L-methionine decarboxylase (AdoMetDC) was present as indicated by blockade of [^{14}C]methionine incorporation into spermidine in intact cells and by inhibition of CO_2 production from [^{14}C]AdoMet in cell extracts by methylglyoxal bis(guanylhydrazone) (MGBG) (Bacchi *et al.*, 1979; Bitonti *et al.*, 1986a) and by MGBG inhibition of cell growth (Chang *et al.*, 1978). AdoMetDC in *T. b. brucei* was partially characterized and was found to have a K_m for AdoMet similar to that of the mammalian enzyme (30 μM); however, the trypanosome enzyme differed from its mammalian counterpart in other respects: putrescine and Mg^{2+} were not stimulatory and MGBG was a reversible inhibitor (Bitonti *et al.*, 1986a).

The ODC of *T. b. brucei* bloodstream trypomastigotes is similar in many respects to mammalian ODC: pyridoxal 5′-phosphate and dithiothreitol are required for optimal activity; the K_m for ornithine is 160 μM and the molecular weight is 107,000 as determined by gel filtration of crude enzyme preparations (Garofalo *et al.*, 1982). However, the enzyme is not especially sensitive to DL-α-difluoromethylornithine (DFMO) as reported earlier (Garofalo *et al.*, 1982): more recent work has demonstrated that the K_m for DFMO is 130 μM, higher than the K_i of 39 μM demonstrated for rat liver ODC (Bitonti *et al.*, 1985). Although ODC is rapidly inducible in procyclic forms *in vitro* (C.J. Bacchi and J. Garofalo, unpublished) and in the insect trypanosomatid *Leptomonas seymouri* (Hannan *et al.*, 1984), its half-life is prolonged (>2 hr) in bloodstream *T. b. brucei* incubated *in vitro* with cycloheximide (Garofalo *et al.*, 1982). Other basic features of polyamine metabolism in *T. b. brucei* include the presence of spermidine but not spermine synthase activity (Bitonti *et al.*, 1984), and the presence of methylthioadenosine (MTA) phosphorylase activity to regenerate adenine from MTA (Ghoda *et al.*, 1987).

African trypanosomes lack catalase and glutathione peroxidase and appear to depend on reduced glutathione (GSH) to scavenge free radicals produced from hydrogen peroxide and superoxide anion (Meshnick *et al.*, 1978). In mammalian cells, oxidized glutathione (GS–SG) produced in these reactions is recycled to GSH via glutathione reductase. Trypanosomes differ in that glutathione reductase activity is absent in dialyzed cell extracts. Instead, GS–SG and other disulfides are reduced in a two-step process which requires the presence of a low-molecular-weight dialyzable intermediate (Fairlamb and Cerami, 1985). This unusual compound, initially isolated from the nonpathogenic hemoflagellate *Crithidia fasciculata,* was N^1,N^8-bis(glutathionyl)spermidine (Fairlamb *et al.*, 1985). Since this spermidine-containing peptide was only found in *T. b. brucei, T. cruzi,* and *Leishmania mexicana,* and not in other nonhemoflagellate protozoa, it was given the trivial name trypanothione. Naturally isolated trypanothione and synthetic trypanothione functioned identically in the reduction of GS–SG by *T. b. brucei* enzyme preparations: first, trypanothione disulfide is reduced enzymatically in an NADPH-dependent reaction by "trypanothione reductase" to form dihydrotrypanothione; second, this compound then undergoes a nonenzymatic thiol–disulfide exchange with GS–SG, cystine, or other disulfides (Fairlamb and Cerami, 1985).

The recent purification of trypanothione reductase from *C. fasciculata* (Shames *et al.,* 1986) indicates that it is an FAD-containing dimer (M_r 53,800 for each subunit), similar in many respects to human glutathione reductase. However, trypanothione reductase is specific for trypanothione and does not use GS–SG as substrate. A 14-amino-acid sequence of the active-site peptide was found to be identical to that of glutathione reductase from human erythrocytes, suggesting that both enzymes possess a similar catalytic mechanism (Shames *et al.,* 1986).

Control of glutathione metabolism in African trypanosomes has been demonstrated to be a potential chemotherapeutic target. Blockade of glutathione synthesis through inhibition of glutamylcysteinyl synthase by buthionine sulfoxamine or melarsen oxide was lethal to *T. b. brucei* (Arrick *et al.,* 1981). Inhibition of polyamine synthesis seemingly would have a similar physiological effect since spermidine is required for trypanothione. Studies are now underway to determine whether the severe polyamine depletion caused by exposure to DFMO does in fact affect intracellular trypanothione levels in bloodstream trypomastigotes (Fairlamb *et al.,* 1987). The discovery of trypanothione is significant because it indicates one unique function of polyamines in flagellate protozoa and should thus help to encourage the search for logical drug targets in other such highly adapted eukaryotic parasites.

2. *DFMO Studies in Acute (Non-CNS) Laboratory Infections*

Initial *in vivo* studies using the noncompetitive enzyme-activated ODC inhibitor DL-α-difluoromethylornithine against an acute (non-CNS) *T. b. brucei* murine infection demonstrated that the agent was highly effective. Complete cures were obtained by administering 2% DFMO for 3 days in the drinking water of infected mice. In contrast, the competitive ODC inhibitor α-methylornithine was not effective against this infection (Bacchi *et al.,* 1980). DFMO also cured an acute murine *T. b. rhodesiense* infection in which animals were given 2% DFMO for 16 days (McCann *et al.,* 1981a) and it was similarly effective against *T. b. gambiense* in a *Microtus* model (McCann *et al.,* 1981b). DFMO given at 2 or 4% for 5 days cured infections of several strains of *T. congolense,* a major veterinary parasite (Karbe *et al.,* 1982; Schillinger and Gorton, 1984). The latter studies were carried out with a parasite resistant to diminazene aceturate (Berenil) and isometamidium (Samorin), veterinary trypanocides widely used in the field (Schillinger and Gorton, 1984). The action of DFMO is highly specific, since the physiological polyamines putrescine, spermidine, and spermine (administered ip for 3 days at 500, 100, and 50 mg/kg, respectively) block DFMO cures of *T. b. brucei* infection, while the "nonphysiological" diamines, diaminopropane and cadaverine (at 500 mg/kg), do not affect DFMO activity (Nathan *et al.,* 1981a).

The antitumor antibiotic bleomycin (Blenoxane) also cured acute *T. b. brucei* infections. Cures could also be blocked by coadministration of spermidine or spermine (Nathan *et al.,* 1981b). The rationale for polyamine antagonism of bleomycin appears to be related to the need for an amine side chain on the β-lactam ring of the molecule for biological activity (Umezawa, 1980). The logical extension of these experiments—use of DFMO and bleomycin in combination—was highly successful: the DFMO dose could be reduced 8-fold (to 0.25% for 3 days) and bleomycin levels 12-fold (to 0.25 mg/kg × 3 doses) while still achieving complete elimination of the parasites from blood (Bacchi *et al.,* 1982a,b). Cures resulting from DFMO/bleomycin combinations could also be blocked by coadministration of polyamines.

DFMO was highly synergistic with several standard trypanocides when examined in the acute *T. b. brucei* model. These agents include suramin (used in humans for early-stage trypanosomiasis); pentamidine (early-stage human disease); Berenil (a veterinary trypanocide); and melarsoprol, the only agent that has been available for late-stage (CNS) disease in humans. The degree of synergism varied but in general a fourfold reduction in the DFMO level was effective when used with one-tenth the curative dose of the trypanocides (McCann *et al.,* 1983).

The basis for this synergism is not clear, but with respect to combination of DFMO plus pentamidine or Berenil, the reason may be joint interference with polyamine synthesis and function in trypanosomes. Cures of the acute infection by standard doses of Berenil and pentamidine can be blocked by coadministration of spermidine or spermine (Bacchi *et al.,* 1983a; Bitonti *et al.,* 1986a), and these diamidines were found to be inhibitors of *T. b. brucei* AdoMetDC (Bitonti *et al.,* 1986a). Berenil in particular was a potent noncompetitive inhibitor of the enzyme (K_i 20 μM); it inhibited enzyme activity 70% and reduced spermidine content while increasing putrescine levels in treated bloodstream trypomastigotes (Bitonti *et al.,* 1986a). Berenil and DFMO are also particularly effective in curing a CNS model infection as well as an acute infection (H.C. Nathan and C.J. Bacchi, unpublished). It is difficult, however, to ascribe the trypanocidal activity of the diamidines to inhibition of a single enzyme, because of their many interactions with DNA and nucleotides as well as their rapid cytotoxicity (for review see Meshnick, 1984).

3. CNS Model Infections

Although the foregoing studies were significant in determining the susceptibility of African trypanosomes to DFMO, they employed acute bloodstream models which do not reflect the true clinical situation in which trypanosomes eventually invade the CNS and usually appear in cerebrospinal fluid. Jennings and associates (Jennings *et al.*, 1977; Murray and Jennings, 1983) have developed several reliable murine CNS models of *T. b. brucei* which should have value in predicting efficacy of agents for CNS disease. Typically in these models a CNS infection is established 10–15 days after infection and treatment is delayed until day 21 to ensure sufficient time for a CNS infection to develop. Evidence for CNS infection is demonstrated by failure of Berenil treatment to prevent recrudescence of bloodstream parasitemia. Recrudescence is the repopulation of blood by parasites arising from some cryptic site, usually the CNS. Berenil, a diamidine used to treat disease in cattle, will not penetrate the blood–brain barrier and does not eliminate parasites in the CNS.

Initial studies with a TREU 667 mouse CNS model demonstrated that neither DFMO, given up to 9 weeks at 2%, nor bleomycin, at 7 mg/kg (six daily doses, days 21–27), cured the infection (Clarkson *et al.*, 1983); however, combinations of 2% DFMO for 2 weeks (days 21–35 of infection) and a single daily bleomycin dose of 3.5 or 7 mg/kg for 3 days (ip, days 26–28) were 100% curative. Control experiments demonstrated that the infection was sensitive to Berenil, administered at day 4, and became refractory if treatment was delayed until day 21. Conversely, treatment with melarsoprol (3.5 mg/kg, iv, days 21–24) was 100% effective.

Despite the efficacy of the DFMO + bleomycin combination, the potential difficulties posed by introduction of two new agents for clinical use in Africa, plus the need to monitor patients receiving bleomycin for lung toxicity (Comis, 1978), appeared to make this combination less than ideal. As noted earlier, the synergism of DFMO and suramin in an acute model, and the reputation of suramin as the least toxic of trypanocides in current use, led to examination of this combination in the same CNS model. Suramin alone at 20 mg/kg does not cure the CNS infection, although it prolongs recrudescence of parasitemia for up to 55 days. However, 2% DFMO (days 21–35) plus a single 20-mg/kg (ip or iv) dose of suramin cured 100% of the animals (Clarkson *et al.*, 1984). It was later demonstrated that the suramin dose can be reduced to a single injection of 2 mg/kg without affecting cure rates. The low-dose regimen appears to be strain specific, since in a more virulent CNS model only the higher suramin dose was curative (Bacchi *et al.*, 1986).

DFMO dose regimens and blood levels have since been reexamined in the light of biochemical studies which clearly demonstrate that DFMO, when incubated *in vitro* with bloodstream forms, blocks putrescine formation from ornithine at 1–5 × 10^{-5} *M* (Bitonti *et al.*, 1985). Recent data indicate that up to 30% of the plasma serum levels of DFMO may be attainable in CSF (Levin *et al.*, 1983; Taelman *et al.*, 1984). Since at a clinical intravenous regimen of 15 g/day, plasma DFMO levels approach 1 m*M*, it thus seems reasonable that peak levels of 100–300 μM were obtained in CSF (Taelman *et al.*, 1984; Van Nieuwenhove *et al.*, 1985). DFMO at 50–100 μM completely inhibits the growth of bloodstream forms of *T. b. brucei in vitro* (Giffin *et al.*, 1986b). Continuous administration of high doses of DFMO also appears to be effective against the TREU 667 mouse model since 4% DFMO, given for 4 weeks, yielded a 90% cure rate (Bacchi *et al.*, 1986).

4. Clinical Trials

DFMO was initially studied as an antitumor agent, hence early phase I trials had already commenced in the United States just as antitrypanosomal activity was being demonstrated in animal models. A detailed analysis of clinical findings with DFMO for treatment of African trypanosomiasis is presented elsewhere in this volume (see Schechter, Barlow, and Sjoerdsma, Chapter 16). For purposes of continuity, those findings are summarized here.

Five studies discussing the use of DFMO in clinical trials for trypanosomiasis have been published: Sjoerdsma and Schechter (1984), Sjoerdsma *et al.* (1984), Van Nieuwenhove *et al.* (1985), Schechter and Sjoerdsma (1986), and McCann *et al.* (1986). These reports essentially state that (1) DFMO is effective, even against late-stage, melarsoprol-resistant trypanosomiasis when given alone at 200–500 mg/kg/day for up to 6–8 weeks; (2) DFMO appears much more effective if given intravenously for 2 weeks followed by 4–6 weeks of oral therapy; (3) the few relapses were only noted in cases treated solely by oral therapy; (4) DFMO was generally well tolerated in these trials; and (5) all cases reported were presumed, on epidemiological grounds, to have *T. b. gambiense*.

5. DFMO Treatment: Metabolic Effects on the Parasite

To assess the degree of metabolic disruption caused by DFMO, rats infected with an acute strain of the parasite were treated with 4% DFMO in the drinking water beginning 48 hr postinoculation. Parasites were isolated from host animals at 12 and 36 hr after treatment had commenced. Cells and extracts were prepared and results compared to trypanosomes harvested 72 hr postinfection from untreated animals (Bacchi *et al.*, 1983b). Rat serum DFMO levels have not been determined for this

TABLE I

Effects of DFMO on Metabolism and Morphology of T. b. brucei[a]

1. Rapid reduction in polyamine content: complete elimination of putrescine and a 60–75% decline in spermidine levels.
2. Total blockade of ODC activity after 12-hr treatment.
3. Increase in uptake of [^{3}H]-labeled polyamines after 12 hr-treatment: two- to fourfold increase.
4. Complete blockade of DNA synthesis after 32-hr treatment and 50–80% incubation of RNA synthesis.
5. Reduction (40–60%) in AdoMetDC activity and >1000-fold increase in intracellular decarboxylated AdoMet levels.
6. Increase in population of bizarre forms (multiple nuclei) and intermediate and stumpy forms. Latter expressly visible in coculture systems treated with DFMO. Development of mitochondrial cristae and NAD diaphorase.
7. Increase in cultivability of DFMO-treated slender forms from a monomorphic infection as procyclic forms *in vitro*.
8. Decrease in respiration (>50%) after treatment of bloodstream forms or culture of slender forms *in vitro* with DFMO. Reduction in O_2 consumption of mitochondria from DFMO-treated slender forms.
9. Increase in gene transcripts (cytochrome *b* and cytochrome oxidase subunits) arising from mitochondrial (kinetoplast) DNA during DFMO treatment of slender bloodstream forms.

[a] Results from Bacchi *et al.* (1983), deGee *et al.* (1984), Giffin *et al.* (1986a,b), and Feagin *et al.* (1986).

dose regimen, but mice, placed on a similar DFMO treatment regimen, had serum levels in excess of 500 μM as determined by high-pressure liquid chromatography analysis (A.J. Bitonti, personal communication).

This study and others relating to the metabolic effects of DFMO are summarized in Table I. Within 12 hr of treatment DFMO profoundly changed the metabolism of bloodstream trypomastigotes: ODC activity became undetectable and intracellular putrescine levels dropped to <10% of control values, while spermidine was 50–70% depleted. The concentration of spermidine fell to 10–25% of controls after 36 hr of exposure to DFMO. Again after 12 hr, AdoMetDC activity declined 30–70% while decarboxylated AdoMet levels rose >1000-fold, because of continued activity of AdoMetDC and blockade of aminopropyl group transfer due to complete depletion of intracellular putrescine (Bacchi *et al.*, 1983b). It has been suggested that decarboxylated AdoMet may serve as a growth inhibitor in DFMO-treated mammalian cells (Pegg, 1984).

After 36 hr of exposure to DFMO, DNA synthesis ceased as measured by incorporation of [^{3}H]thymidine. RNA synthesis ([^{3}H]uridine incorporation) decreased 50–80% while protein synthesis ([^{3}H]leucine incorporation) increased two- to fourfold (Bacchi *et al.*, 1983). The latter effect was unexpected and in contrast with findings in mammalian cells in which polyribosome content and quantity of [^{3}H]leucine incorporated decreased after exposure to DFMO (Rudkin *et al.*, 1984). However, it is possible that protein synthesis could occur in light of the striking morphological changes described below and the attendant revamping of energy metabolism normally accompanying such change (Opperdoes, 1985).

Accumulation of exogenously supplied polyamines increased two- to fourfold in 36-hr treated cells incubated *in vitro* with labeled polyamines (Bacchi *et al.*, 1983). This corresponds to similar but more dramatic changes in polyamine accumulation by DFMO-treated mammalian cells, (Alhonen-Hongisto *et al.*, 1980). An advantage of this alteration in polyamine uptake may be to enhance assimilation of structural analogs of polyamine (Porter *et al.*, 1984) which could be cytotoxic to trypanosomes when combined with DFMO.

DFMO also affects respiratory metabolism of slender forms: Giffin *et al.* (1986a) found that bloodstream-form trypomastigotes which were depleted of putrescine *in vivo* or *in vitro* (in coculture) have a 45 or 62% lower respiratory rate, respectively, than controls. The effect was specific for putrescine, since spermidine was not completely depleted in each instance. Moreover, only putrescine and not spermidine could restore parasite oxidative activity *in vitro* or *in vivo*. Furthermore, only the presence of 5 m*M* putrescine throughout the preparation of mitochondrial fractions from DFMO treated cells was able to increase a depressed O_2 consumption rate, while spermidine and $MgCl_2$ had no effect. Putrescine did not appear to act simply as a cofactor for mitochondrial O_2 uptake since it did not have a direct *in vitro* effect on respiration. The lowered respiratory rate noted after DFMO treatment may be the earliest indication of a DFMO-induced morphological shift (as detailed below), in which respiratory rates might be expected to drop because of decreased dependence on aerobic glycolysis and increased oxidative metabolism. Alternatively, putrescine could have a special role in the membrane integrity or substrate permeability of the mitochondrion.

One of the most interesting effects of DFMO on bloodstream *T. b. brucei* is the striking morphological changes occurring within 36 hr of treatment. In studies on a strain which normally only produces long, slender trypanosomes (Bacchi *et al.*, 1983), it was apparent that much of the population of slender trypomastigotes changed to intermediate and stumpy forms within 36 hr of initiation of DFMO treatment. These preliminary findings were substantiated in a strain of *T. b. rhodesiense* by electron microscopy and differential staining, revealing elongation of the mitochondrial tube, development of cristae, and NAD diaphorase activity as a rapidly occurring sequel to DFMO treatment (deGee *et al.*, 1984).

Giffin *et al.* (1986b) have used DFMO treatment of bloodstream infections and of slender forms grown *in vitro* in coculture with bovine fibroblasts to demonstrate progressive shifts in morphology from slender to intermediate to stumpy forms. DFMO treated organisms were transferred at various times after initiation of treatment into a semidefined medium supporting growth of procyclic (insect vector midgut) forms. This detailed study indicated that prolonged DFMO treatment (>6 days) promotes morphological changes to stumpy forms, which were then unable to transform to procyclic forms upon subculture in a semidefined medium. However, cells treated for shorter periods, and retaining the intermediate form, were able to transform to procyclics on subculture. Moreover, intermediate forms generated *in vitro* proved infective to animals (i.e., regenerating slender forms on inoculation). The DFMO-induced intermediate forms also showed evidence of an enlarged mitochondrion and NAD diaphorase activity indicative of their pivotal role between the bloodstream and vector forms. These relationships are summarized in Fig. 2.

In a recent study by Feagin *et al.* (1986), DFMO was used to induce slender to stumpy transformation *in vivo* and mitochondrial gene expression was studied during the transformation process. (The DNA transcribed here is the "maxicircle" portion of kinetoplast DNA. The kinetoplast is the ellipsoid organelle which contains the tightly coiled

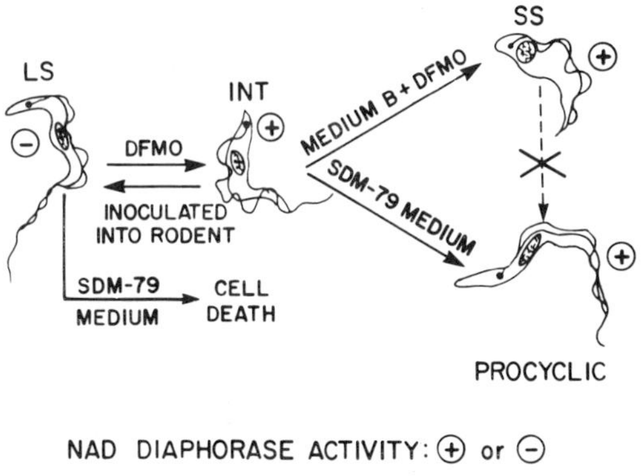

Fig. 2. Summary of DFMO-induced alterations in *Trypanosoma brucei brucei*. The exposure of long slender (LS) bloodstream-form trypanosomes to DFMO during *in vitro* culture or *in vivo* chemotherapy results in a rapid depletion of the intracellular putrescine pool. Cellular proliferation ceases, and the long slender form becomes an intermediate (INT) form. If the intermediate form is inoculated into a rodent, it can revert to a long slender form and produce a lethal parasitemia. Upon continuous exposure to DFMO the intermediate form becomes a short stumpy (SS) form, which does not transform further into procyclic trypomastigotes when introduced into SDM-79 medium. The intermediate form can successfully adapt to and transform in SDM-79 medium. (From Giffin *et al.*, 1986b.)

mitochondrial DNA in these organisms.) In this study, bloodstream populations consisting of 40% stumpy and intermediate forms were induced by administering 2% DFMO for 3 days to rats with a normally monomorphic infection. Transcripts of mitochondrial DNA studied included protein subunits of cytochrome oxidase and cytochrome *b,* proteins which do not appear in slender form. It was found that mitochondrial gene transcripts are more abundant in naturally occurring and drug-induced stumpy forms than in slender forms and that differential expression of transcripts takes place—i.e., not all mitochondrial genes are transcribed at the same rate, and some transcripts are more abundant in stumpy than in procyclic forms. Transcript patterns in organisms produced by DFMO treatment of slender forms were similar to those occurring in naturally occurring stumpy forms, except that quantitatively, some transcripts were more abundant in DFMO-induced stumpy forms (Feagin *et al.,* 1986).

The foregoing studies indicate that DFMO has a profound effect on the morphology of bloodstream forms and that, on the biochemical and molecular level, changes are induced which closely approximate those occurring in nature. Although additional work is needed to define more clearly the role of intermediate and stumpy forms and the degree of exposure to DFMO necessary to synchronize these changes *in vivo,* it appears that changes in polyamine content are part of the mechanism triggering life cycle changes.

6. Role of the Immune Response in DFMO Action

It is evident from the preceding discussion and other observations that DFMO is not cytotoxic to the parasite. In short-term infections treated with DFMO the parasite population is held stationary for 48–72 hr, and then rapidly disappears from the peripheral blood. In normal trypanosome infections, host antibody response is essential for blocking growth of a population having predominantly a single variant antigenic type of surface glycoprotein. The small percentage of survivors with a different antigenic type then repopulate the blood (Dempsey and Mansfield, 1983). As shown by deGee *et al.* (1983) and Bitonti *et al.* (1986b), a competent host immune response is required for elimination of the parasites after DFMO treatment. deGee *et al.* (1983) treated immunocompetent mice with the immunosuppressive agent cyclophosphamide prior to infection with a lethal strain of parasite and treatment with DFMO. Cyclophosphamide allows development of terminal parasitemia even if a low inoculum is used (Smith *et al.,* 1982). In these immunosuppressed mice, parasite growth was checked during DFMO treatment, but parasites increased in number after withdrawal of treatment and eventually killed the host. Resurgence of parasitemia was associated with absence of serum antibody titer to the infecting variant, as opposed to titers of up to 256 in DFMO treated mice not exposed to cyclophosphamide (deGee *et al.*, 1983). This study also demonstrated that DFMO efficacy was directly related to the natural genetic resistance of the host (i.e., ability to form antibody) when host mouse strains with low and high natural resistance to infection were compared in their response to DFMO. C3H mice, having a low genetic resistance to infection (having no antibody response before death under normal circumstances), survived with DFMO treatment: the antibody response occurred as a result of the increased survival time due to blockade of parasite division by DFMO (deGee *et al.*, 1983).

Bitonti *et al.* (1986b) demonstrated that 7 days of DFMO treatment cured rats of a lethal acute infection. However, in rats pretreated with another immunosuppressive agent, dexamethasone, DFMO initially cleared the blood of parasites, but, in the absence of antibody response, all eventually succumbed after relapsing. In the same study, athymic mice were also cured with DFMO treatment, indicating the immune response to trypanosomes is a T-independent B-cell response (Bitonti *et al.*, 1986b), in agreement with earlier suggestions (Campbell *et al.*, 1978). These studies are important since they clarify the role of the immune response in chemotherapy with DFMO and present a rationale for the success of long-term treatment in curing *T. b. gambiense* infections: the need for immune destruction of all parasites—including those in cryptic areas such as the CNS—to prevent clinical relapse.

7. Polyamine Biosynthesis Inhibitors as Trypanocides: Search for Specificity

As noted above, DFMO causes numerous perturbations in the metabolism of *T. b. brucei* which translate to cure of experimental infections and, ultimately, to clinical efficacy. A key question constantly arises throughout these studies: what is the basis for specificity of DFMO action? Although there is as yet no single clear answer, some possibilities have been eliminated. DFMO does not preferentially inhibit parasite rather than host ODC. In fact, the parasite enzyme seems to be less sensitive to DFMO than the mammalian enzyme (Bitonti *et al.*, 1985). DFMO is essentially not toxic in trypanosome culture systems where there is no immune response (Giffin *et al.* 1986b). Uptake of DFMO occurs by passive diffusion in mammalian cells (Erwin and Pegg, 1982), but in *T. b. brucei* uptake was inhibited by the specific trypanosome respiratory inhibitor salicylhydroxamic acid (Bacchi *et al.*, 1983). Later work, however, demonstrated that DFMO uptake was not saturable at up to 10 m*M*, could not be inhibited by excess concentrations of amino acids or polyamines, and was not inhibited by analogs of DFMO or the polyamines (Bitonti *et al.* 1986c). It appears, therefore, that the significant antitrypanosomal activity of DFMO cannot be explained on the basis of preferential cellular uptake.

An additional consideration for DFMO specificity is that, in mammalian cells, treatment leads to conversion of spermidine to spermine (Casero *et al.*, 1984), but in trypanosomes normally lacking spermine, only low levels (<1 nmol/mg protein) are accumulated even after 36 hr of exposure to DFMO (Bacchi *et al.*, 1983). The latter point may in part be a clue to parasite sensitivity to DFMO—in parasite cells having a more rapid division rate than mammalian counterparts, existing polyamines may be metabolized more rapidly or quickly diluted by cell division to a point below the minimum threshold level needed to sustain growth, as found in mammalian cells (Mamont *et al.*, 1976; Porter and Bergeron, 1983). In the absence of spermine in trypanosomes, metabolic changes due to complete polyamine deprivation should be more pronounced than in spermidine-containing mammalian cells (Casero *et al.*, 1984). It has been demonstrated, however, that spermine negates the effects of DFMO in trypanosomes *in vivo* and *in vitro* (Nathan *et al.*, 1981a; Giffin *et al.*, 1986b), hence it can be utilized as a physiologically active polyamine. Finally, although bloodstream forms of *T. b. brucei* will accumulate exogenously supplied polyamines, the rate of uptake is low, i.e., one-tenth the rate when compared to mammalian cells (Bacchi *et al.*, 1979), so that even if substantial levels of polyamines were present in the environment, uptake would appear to be minimal.

A key aspect of polyamine metabolism affecting chemotherapy is the high turnover rate of ODC in rapidly multiplying cells (Pegg, 1986). To counter this continuous production of enzyme, a constant level of ODC inhibitor must be present over long periods. The half-life of DFMO in mouse serum is <40 min. In man this is extended to ~180 min (Abeloff *et al.*, 1984), but clearly an agent having a longer serum half-life would increase pharmacologic efficacy.

Several analogs of ornithine and of putrescine have recently been synthesized and studied using an acute murine model of *T. b. brucei*. The most effective inhibitor of ODC, (*R*,*R*′)-methylacetylinic putrescine, had a K_i of 5 μM and also at 0.5 μM blocked synthesis of polyamines *in vitro* in bloodstream *T. b. brucei* (Bitonti *et al.*, 1985). Surprisingly, this compound was not curative or even partially effective when tested *in vivo* using the acute *T. b. brucei* murine model. However, an ornithine analog, monofluoromethyldehydroornithine methyl ester (Δ-MFMO·CH_3), was significantly more effective than DFMO *in vivo:* a 0.5% solution given for 3 days in drinking water cured >80% of the animals, whereas DFMO at 0.5% cured only 10%. Δ-MFMO·CH_3 apparently acts as a pro drug which must be hydrolyzed to the free acid before it becomes an inhibitor of ODC

(Mamont *et al.*, 1986). [*In vitro,* as the free acid, it is about nine times as effective as DFMO against *T. b. brucei* ODC (K_i 14 μM versus 130 μM for DFMO), although it is no more effective than DFMO in preventing polyamine synthesis in intact cells (Bitonti *et al.*, 1985).] One other aspect of the pharmacology of Δ-MFMO·CH_3 seems critical for its superior efficacy: its accumulation by the parasite at a rate approximately eight times that of DFMO (Bacchi *et al.*, 1987).

Δ-MFMO·CH_3 has now been tested in CNS model infections and the dose relationships noted previously with DFMO in acute model infections continue to apply: a minimum of 2% DFMO for 2 weeks + suramin (20 mg/kg) is necessary for >90% cures as compared to 0.5% Δ-MFMO·CH_3 + 20 mg/kg suramin. Dose levels of DFMO and Δ-MFMO·CH_3 given as percentage concentration in drinking water are not equivalent since mice drank 20% less of the Δ-MFMO·CH_3 analog, making the total dose approximately 1.1 g/kg/day Δ-MFMO·CH_3 versus 5.3 g/kg/day for DFMO (Bacchi *et al.*, 1987).

The observed efficacy and apparent low toxicity of these polyamine inhibitors is unprecedented in the treatment of the African trypanosomiases. Moreover, the apparent lack of immunosuppression as a side effect of DFMO treatment, as judged from clinical reports (Sjoerdsma *et al.*, 1984; Sjoerdsma and Schechter, 1984; Van Nieuwenhove *et al.*, 1985; Schechter and Sjoerdsma, 1986), and the high level of synergism observed between DFMO and conventional trypanocides or antitumor agents argue for the expanded use of polyamine inhibitors, alone and in combination chemotherapy.

B. Leishmania

Human leishmania infections are caused by a group of five organisms which are morphologically similar but differ in tissue specificity and cause diseases of varying degrees of severity. Visceral leishmaniasis is caused by *Leishmania donovani;* oriental sore is restricted to the skin (*Leishmania tropica, Leishmania major, and Leishmania mexicana*); mucocutaneous leishmaniasis is the often fatal and disfiguring disease of facial tissue caused by *Leishmania braziliensis* (Keithly, 1985).

Leishmania spp. have a simpler life cycle than African trypanosomes: elongated promastigotes are injected into the skin by bite of the sand fly (subfamily Phlebotominae); they are taken up by resident macrophages where they transform into small, ellipsoid (2–3 μm in diameter) bodies called amastigotes. These multiply and are dispersed to different organs, depending on strain tissue specificity. When amastigotes are ingested by the vector sand fly, they transform to and develop as promastigotes within the gut (Chang *et al.*, 1985).

In contrast to the African trypanosomes, the leishmanias are primarily intracellular in their mammalian hosts, living within phagocytic vacuoles of macrophages. The pathology of the diseases results from multiplication of the parasites and host reaction to their presence (Molyneux and Ashford, 1983).

Treatment of the almost 400,000 new leishmania infections/year relies heavily upon the antimony-containing compounds Na^+-antimony gluconate (Pentostam) and antimony-*N*-methylglutamine (Glucantime), the diamidine pentamidine, or amphotericin B. In general, response to antimonials varies according to the type of disease and various regimens must be given in different geographical areas (Marr, 1984). Toxicity and variable activity attend the use of antimony in visceral leishmaniasis, and many cases of mucocutaneous leishmaniasis remain refractory to any therapy (Marr, 1984; James and Gilles, 1985). There is clearly a need for novel approaches to chemotherapy of these organisms.

1. Polyamine Metabolism in Leishmania Species

Exploration of polyamine metabolism in *Leishmania* has lagged far behind that in trypanosomes. Much of the early work has been summarized (Bacchi, 1981). Putrescine, spermidine, and spermine have been identified and quantitated in *L. donovani* amastigotes (Morrow *et al.*, 1980) and in promastigotes of *L. donovani, L. tropica,* and *L. mexicana* (Bachrach *et al.*, 1979a). Both forms of the parasite were found to synthesize spermine with ornithine as its ultimate precursor. In amastigotes of *L. donovani* the distribution ratio of putrescine:spermidine:spermine approached 1.5:1:1 (Morrow *et al.*, 1980), while in promastigotes of *L. mexicana,* it was 4:1:0.1 (Coombs and Sanderson, 1985). These ratios differ from that of African trypanosomes, which have no spermine and have putrescine:spermidine ratios of 1:5 or more (Bacchi *et al.*, 1983).

Polyamine synthesis in various species of *Leishmania* promastigotes grown in culture was growth dependent, characterized by initial increases in putrescine levels (early log phase) followed by rapid accumulation of spermidine (Bachrach *et al.*, 1979a; Schnur *et al.*, 1979). In contrast, Coombs and Sanderson (1985) found elevated putrescine levels (120 nmol/mg protein) in late log phase of *L. mexicana* promastigotes. In this study, promastigotes in late log phase also excreted putrescine in large quantities (the final concentration in spent medium was 47 μM).

Early work on ODC activity in *Leishmania* spp. (Morrow *et al.*, 1980; Bachrach *et al.*, 1979a) was limited to indirect measurement of polyamine synthesis in intact cells. Later Bachrach *et al.* (1981) found rapid increases in ODC activity in crude cell-free extracts during growth of *L. tropica* promastigotes in culture. Increased ODC activity in infected mouse macrophages was found at 24 and 72 hr after inoculation with amastigotes. Polyamine content of leishmania-infected mouse skin and spleen was elevated 1 week after infection; however, it was not clear what proportion of activity was due to the amastigotes or the macrophages.

2. Drug Interaction with Polyamine Metabolism in Leishmania

There are several studies which appear to relate the effects of antileishmanial agents on polyamine metabolism. Bachrach *et al.* (1979b) found that ethidium bromide, MGBG, and pentamidine isethionate inhibited spermidine synthesis from [^{14}C]putrescine in intact promastigotes of *L. mexicana, L. tropica,* and *L. donovani* over a 5-hr incubation period. These data suggest a polyamine target site for ethidium and pentamidine; as noted previously, the diamidines Berenil and pentamidine inhibited *T. b. brucei* AdoMetDC activity (Bitonti *et al.*, 1986a). Bachrach *et al.* (1983) also demonstrated that ODC activity of host mouse skin increased dramatically after infection with *L. tropica major.* Treatment with paromomycin or pentamidine caused a reduction of putrescine, spermidine, and spermine levels in infected skin, and a parallel decrease in parasitemia. Paromomycin also completely blocked ODC activity in infected skin, concomitant with cessation of parasite division and elimination. Upon cessation of treatment, parasites resumed multiplication and ODC levels were restored. Although these studies suggest antagonism of polyamine metabolism as one probable target site, the numerous metabolic effects of these agents preclude pinpointing a specific target site until more detailed studies are available.

A more specific effect was found with the ability of DFMO to block amastigote to promastigote transformation in an *in vitro* system (Coombs *et al.*, 1983). DFMO, 1 m*M,* in the culture medium completely prevented transformation of *L. mexicana* amastigotes. At 0.5 and 0.1 m*M* there was substantial (>50%) inhibition of transformation. Promastigotes, however, were far less sensitive to DFMO since 50 m*M* failed to block division or affect parasite morphology after 7 days.

Recent studies by Kaur *et al.* (1986) examined susceptibility of *L. donovani* promastigote growth to DFMO. Remarkably, at DFMO concentrations as low as 50 μM, growth was completely blocked. Cells cultured with DFMO plus 10 or 100 μM putrescine grew normally. Cells incubated with DFMO for 70 hr were completely depleted of putrescine and had approximately 50% less spermidine than controls. Keithly *et al.* (1987), using promastigotes of *L. donovani* and *L. braziliensis guyanensis* grown in RE3 (defined) medium, obtained >80% inhibition of growth with 5 m*M*

DFMO. Inclusion of 1 m*M* putrescine completely reversed the action of DFMO. Differences observed in the inhibition of promastigote growth by DFMO in *L. mexicana* and *L. donovani* might be due to the strains or media employed. In particular, the high intracellular and extracellular putrescine levels observed in *L. mexicana* late log phase promastigotes (Coombs and Sanderson, 1985) may be responsible for the insensitivity of promastigotes of this species to DFMO (Coombs *et al.*, 1983).

Pertinent *in vivo* data have been obtained by Keithly *et al.* (1987) against experimental infections of *L. donovani* and *L. braziliensis guyanensis* in BALB/c ByJ mice. In this study, mice were dosed with DFMO, clinically used leishmanicides (Pentostam, pentamidine), or other agents (bleomycin, suramin) both singly and in combination. DFMO, given at 3% in the drinking water for 2 weeks beginning 24 hr preinfection, suppressed development of *L. donovani* amastigotes in spleen cells by 51%, as judged by direct cell counts of spleen impression smears and colony-forming units on blood agar. When treatment was initiated 1 week postinfection, DFMO was highly synergistic with suramin. Suramin alone (total dose = 20 mg/kg divided over 7 days) suppressed parasitemia by 50%. The combination of DFMO and suramin suppressed amastigote liver burdens by 96%.

Preliminary studies using DFMO in treatment of lesions of *L. braziliensis guyanensis* were also encouraging. Mice were infected 5 weeks prior to testing. DFMO, given at 3–5% in drinking water for 3 weeks 24 hr postinfection, was 50–74% suppressive to amastigote formation, as judged by reduction of tail lesion size and number of amastigotes in lesions. Treated animals retained reduced lesion size for at least 3 weeks after termination of treatment. DFMO was not synergistic with Pentostam or bleomycin in this model (J.S. Keithly *et al.*, unpublished). These initially promising results appear to be the first in more than 30 years in which a noncytotoxic agent was effective against *L. brazilensis*.

C. Other Flagellates

Polyamine metabolism in *Giardia lamblia* and *Trichomonas vaginalis* has been examined only in preliminary studies. A parasite of the small intestine, *G. lamblia* is cosmopolitan in distribution and produces severe diarrhea, particularly in children; *T. vaginalis* is a sexually transmitted parasite of the urethra and vagina of humans. *In vitro* growth of *G. lamblia* is susceptible to inhibition by DFMO; 10–20 m*M* increased the generation time twofold and sixfold, respectively. Normal growth in the presence of DFMO could be restored upon addition of spermidine (Gillin *et al.,* 1984). ODC was detected in the parasite, as were putrescine, spermidine, and spermine. The same authors reported that *in vitro* growth of *T. vaginalis* was not inhibited by up to 20 m*M* DFMO. ODC activity in this organism was$<$10% of that found in *G. lamblia,* but intracellular putrescine levels were sevenfold higher than in the latter organism (Gillin *et al.,* 1984). These findings are in agreement with work by White *et al.* (1983), who reported enormous intracellular putrescine levels (190 nmol/mg protein) for *T. vaginalis* and $>$200-fold increases in the putrescine content of spent culture medium. North *et al.* (1986) examined three species of trichomonads for polyamine biosynthesis, polyamine content, and the effect of DFMO on growth. Their results paralleled studies discussed above in that low levels of ODC activity were found, the intracellular putrescine concentration was elevated (50–100 nmol/mg protein), and putrescine levels were 3–5 times that of spermidine. DFMO at 4 m*M* blocked ODC activity and lowered putrescine levels in *T. vaginalis* 30–50% but did not inhibit growth. DFMO-treated parasites accumulated 5–7 times more [^{14}C] putrescine than untreated cells. These studies appear to demonstrate an unusual dependence on putrescine in the metabolism of trichomonads not characteristic of other flagellated protozoa.

Trypanosoma cruzi, the agent of Chagas' disease (American trypanosomiasis), is a flagellated protozoan with a life cycle and biochemical metabolism quite different than the African trypanosomes. It causes an intracellular infection in its mammalian host, particularly in heart and nervous tissues. The polyamine metabolism of *T. cruzi* has not been well studied except to report that it does contain putrescine and spermidine (Bacchi, 1981). Initial attempts failed to show activity of even high levels of DFMO *in vivo* against *T. cruzi* (Hanson *et al.,* 1982). Recently, an intriguing preliminary report has indicated some activity of DFMO against *in vitro* intracellular macrophage infection by blood forms of *T. cruzi* when the host cells were pretreated with DFMO. Even more interesting was the fact that DL-α-difluoromethylarginine, a specific irreversible inhibitor of arginine decarboxylase (an enzyme not found in mammalian cells), markedly inhibited the infection of mouse peritoneal macrophages by *T. cruzi* blood forms when the parasites were treated with this inhibitor. However, DFMO had no effect on the ability of these forms to infect macrophages (Kierszenbaum *et al.,* 1986). These data seem to imply that there may be a specific arginine decarboxylase involved in polyamine metabolism of *T. cruzi,* a finding which may ultimately yield a unique therapeutic target in this organism (Kierszenbaum *et al.,* 1987).

III. SPOROZOEA

A. Plasmodium (Malaria)

Protozoa of the genus *Plasmodium* are the agents of malaria, the most widely distributed and prevalent parasitic disease in the world today. *Plasmodium falciparum* is one of the species which causes acute disease in man and is widely distributed in South America, Africa, and Southeast Asia. Several stages of the malarial life cycle are potential targets for control by inhibition of polyamine biosynthesis via ODC. The mammalian phase of the life cycle begins with the bite of an infected mosquito and injection of *Plasmodium* sporozoites, which then invade liver parenchymal cells, transform into exoerythrocytic parasites, and undergo rapid multiplication (schizogony). Merozoites, released from exoerythrocytic schizonts in liver cells, invade erythrocytes and initiate repeated cycles of development. Some of these differentiate into sexual forms which, upon ingestion by a mosquito, undergo further development (sporogony) into sporozoites.

The initial report of the effects of DFMO on malaria was in a *P. falciparum in vitro* culture system in which, at concentrations of 5 m*M* or above, DFMO had a significant effect on the ability of parasites to undergo schizogony (McCann *et al.*, 1981a). This effect was related to a specific arrest of DNA synthesis, as found in mammalian cells and trypanosomes. Furthermore, blockade of schizogony and DNA synthesis could be reversed by concentrations of putrescine as low as 10 μM (McCann *et al.*, 1981a). Several other investigators confirmed the inhibition of schizogony by DFMO and its reversal by polyamines in *in vitro* systems (Heidrich *et al.*, 1983; Assaraf *et al.*, 1984; Whaun and Brown, 1985). Assaraf *et al.* (1984) also measured polyamine levels in the cultured parasites and later found that DFMO would actually synchronize >98% of cells in culture (Assaraf *et al.*, 1986). Whaun and Brown (1985), however, demonstrated that DFMO inhibited ODC activity and decreased putrescine and spermidine but not spermine levels of *P. falciparum* in infected erythrocytes.

Although all of these studies demonstrated >90% *in vitro* inhibition of malaria schizogony, no significant *in vivo* effects of DFMO could be detected in several animal models of erythrocytic infection measuring host survival, including *Plasmodium gallinaceum* (Hanson *et al.*, 1982) and *Plasmodium berghei* (Gillet *et al.*, 1982; Hollingdale *et al.*, 1985). As a result, Gillet *et al.* (1982) examined the effects of DFMO on exoerythrocytic schizogony (i.e., in liver cells) using a murine model, *P. berghei*. In two elegant studies (Gillet *et al.*, 1982, 1983) they found that DFMO could not only inhibit exoerythrocytic schizogony in mice but would also completely inhibit the sporogonous cycle in the mosquito vector. In the latter experiments, DFMO administered to mosquitoes was dissolved either in a sugar solution or was supplied through blood meals taken from treated mice. A recent study has also found that DFMO will induce significant protective immunity in mice initially inoculated with *P. berghei* sporozoites from mosquitoes and then rechallenged with parasites (Gillet *et al.*, 1986).

The exoerythrocytic effect of DFMO was confirmed by Hollingdale *et al.* (1985) and extended to an *in vitro* system as well wherein sporozoites were removed from salivary glands of mosquitoes and were then used to inoculate cultured human hepatoma cells. In both the *in vivo* and *in vitro* experiments, spermidine completely reversed the effects of DFMO on exoerythrocytic schizogony.

In testing several ornithine analogs Hollingdale *et al.* (1985) found that Δ-MFMO·CH_3 was approximately 10 times more potent than DFMO in inhibiting exoerythrocytic schizogony both *in vitro* and *in vivo* and that these effects were reversible by spermidine. The demonstration that Δ-MFMO·CH_3 was more effective than DFMO against malaria is consistent with findings in acute and CNS trypanosomiasis models (Bitonti *et al.*, 1985; Bacchi *et al.*, 1987). Malaria may well prove to be an appropriate disease target for chemotherapy by inhibition of polyamine biosynthesis, even in the erythrocytic cycle, as the earlier *in vitro* studies first suggested.

B. Eimeria (Coccidiosis)

Members of the genus *Eimeria* are parasites of most nonprimate mammals and aves, while the related genera such as *Isospora* and *Cryptosporidium* are occasional human pathogens. *Eimeria* spp. have a life cycle resembling that of *Plasmodium,* except that the entire life cycle (both asexual and sexual phases) takes place in cells of the intestinal mucosa or lumen. Transmission is by oocyst or sporocyst, both of which pass into the environment in feces. *Eimeria tenella* is the most common parasite of chickens, in which it destroys the intestinal mucosa and causes high mortality in young birds. It is of major economic concern in the poultry industry (Beaver *et al.*, 1984).

One of the most dramatic effects of DFMO in protozoan infections was reported by Hanson *et al.* (1982) wherein concentrations as low as 0.063% in the drinking water of chickens cured all animals exposed to from severe to lethal inocula of *E. tenella*. As seen with *T. b. brucei* and *P. berghei,* DFMO cures were completely reversible by coadministration of putrescine. Of work yet reported, the protozoan species *E. tenella* appears to be most sensitive to the growth inhibitory effects of DFMO. Furthermore, inoculated chickens in which coccidial lesions were suppressed by DFMO resisted subsequent challenge with *E. tenella*. Thus, the few parasites shown to be present in the site of infection of the chickens and whose growth was suppressed by DFMO allowed the host to develop an acquired immunity to the parasites, again consistent with a mode of action of polyamine depletion noted in numerous other types of cells (Pegg and McCann, 1982).

One of the most puzzling findings of the coccidiosis studies was the fact that while *E. tenella* was found to be most sensitive to DFMO, subsequent experiments showed that other species of chicken coccidia were much less or only partially sensitive. One species, *Eimeria brunetti,* was completely unaffected by DFMO (Hanson *et al.,* 1982). It remains to be determined whether these differences in sensitivity are biochemical in origin or are due to differences in location of the site of infection in the host.

C. Pneumocystis carinii

Pneumocystis carinii pneumonia has recently come to light as being the most common lethal infection in patients with acquired immune deficiency syndrome. Because of its morphology, *P. carinii* has been regarded as belonging to the class Sporozoea, subclass Coccidia, e.g., *Eimeria* spp. (Beaver *et al.,* 1984). Currently, there is a significant problem in treating AIDS patients with this infection, and as a result, the first use of DFMO against *P. carinii* was in a clinical setting. Initial results were remarkably encouraging (see Chapter 16, this volume; Golden *et al.,* 1984; Sjoerdsma *et al.,* 1984; McCann *et al.,* 1986). As clinical trials progressed, almost simultaneously DFMO studies with animal models were undertaken. In a model, DFMO at a dose of approximately 1.5% in drinking water (2.5 g/kg/day) had no effect on the development of dexamethasone-induced *P. carinii* lung infections (Hughes and Smith, 1984). Subsequent *in vitro* experiments with *P. carinii* in culture, however, showed that the growth of these organisms was clearly inhibited by DFMO and that this effect was reversible by putrescine (Cushion *et al.,* 1985). Very recent *in vivo* studies using the dexamethasone suppression model have demonstrated that if the dose of DFMO is increased by a factor of at least two, there is a clear suppression of the parasitosis (A.B. Clarkson, Jr., personal communication). Although laboratory findings have lagged behind the initial and ongoing positive clinical findings, more intense work with *P. carinii* laboratory models should generate further dose response and drug synergism data.

D. Other Sporozoa

Cryptosporidium spp., opportunistic parasites of the class Sporozoea, are considered to be coccidian species related to *Eimeria* and are often found in AIDS patients (Current and Haynes, 1984). Several cases of cryptosporidiosis in AIDS have been successfully treated with DFMO (see Chapter 16, this volume). DFMO was initially used simply because of the lack of availability of any other chemotherapy. However DFMO was not effective when used to treat a newborn pig experimentally infected with a human strain of *Cryptosporidium,* despite the fact that DFMO significantly reduced putrescine and spermidine levels in the cryptosporidia-infected pig intestinal mucosa (Moon *et al.,* 1982).

Toxoplasma gondii is yet another coccidian which has been causing toxoplasmic encephalitis in an unprecedented number of cases of patients with AIDS. DFMO was tested in an *in vivo* murine model and an *in vitro* macrophage culture and was ineffective in both, even at very high concentrations, e.g., 20 m*M* in culture media and 10% in host drinking water (Hofflin *et al.,* 1985).

IV. AMOEBAE

Pathogenic amoebae include those that invade the intestinal tract such as *Entamoeba histolytica,* agent of amoebic dysentery, and free-living, soil-dwelling amoebae such as *Naegleria* spp. and *Acanthamoeba* spp., which are facultative parasites causing primary amoebic meningoencephalatis. Although occurrence of the latter disease is rare—only 130 cases had been reported in the United States until 1981—it is resistant to treatment and usually fatal. Amoebic dysentery remains a common disease in temperate as well as tropical areas (Beaver *et al.,* 1984). Until recently, polyamine metabolism in these organisms had not been studied.

Gillin *et al.* (1984) demonstrated that the mucosal-dwelling human pathogen *E. histolytica* contained high (>90 nmol/mg protein) levels of putrescine, but very low quantities of spermidine (2.6 nmol/mg protein) and negligible amounts of spermine. ODC was present but growth of cultures was insensitive to DFMO at concentrations up to 20 m*M*. The levels of putrescine present in *E. histolytica,* as well as the insensitivity of growth to DFMO, are reminiscent of findings with *T. vaginalis* reported earlier (White *et al.,* 1983; North *et al.,* 1986), and may indicate alternative pathways of polyamine metabolism in this organism or presence of a DFMO-insensitive enzyme. Schuster and McCann (1982), however, found that pathogenic *Naegleria* spp. were sensitive to the effects of 5 to 10 m*M* DFMO but that species of the soil-dwelling pathogenic *Acantha-*

moeba (*Acanthamoeba culbertsoni* and *Acanthamoeba polyphaga*) were unaffected by DFMO concentrations up to 10 m*M*. To further complicate matters, Kim *et al.* (1987) have demonstrated that DFMO and Δ-MFMO·CH_3 will block multiplication of *Acanthamoeba castellanii* and, under different conditions, will also induce encystment of this organism, a differentiation event. The presence of polyamines reverses both effects of the ODC inhibitors. Further study of polyamine synthetic pathways and the activity of ODC inhibitors in various amoeba species is required to unravel the disparate effects cited above. It is clear, however, that polyamines play an important role in their cellular functions.

V. CONCLUSIONS

It has been less than 10 years since the first attempts at quantitation of polyamines in trypanosomes and less than 6 years since the initial cures of laboratory trypanosome infections with DFMO, yet in this brief time it has become apparent that use of inhibitors of polyamine metabolism against parasitic protozoa represents a unique opportunity for advances in chemotherapy as well as in metabolic regulation. Therefore, the likelihood is that continued study of polyamine antagonism may result in significant therapeutic progress in the treatment of major parasitic diseases. At the very least, a new field potentially important to the understanding of protozoan life cycles is being probed.

ACKNOWLEDGMENTS

Portions of the work presented here were done with the support of National Institutes of Health (NIH AI17340) and a grant from the U.N.D.P./World Bank/World Health Organization Special Programme for Training in Tropical Disease (WHO 850014).

The authors thank Drs. A.J. Bitonti, A.B. Clarkson, Jr., A. Fairlamb, S.H. Hutner, J. Keithly, and K. Stuart for supplying manuscripts, preliminary data, and editorial suggestions.

REFERENCES

Abeloff, M.D., Slavik, M., Luk, G.D., Griffin, C.A., Hermann, J., Blanc, O., Sjoerdsma, A., and Baylin, S.A. (1984). *J. Clin. Oncol.* **2,** 124–130.

Alhonen-Hongisto, L., Seppänen, P., and Jänne, J. (1980). *Biochem. J.* **192,** 941–945.

Arrick, B.A., Griffith, O.W., and Cerami, A. (1981). *J. Exp. Med.* **153,** 720–725.

Assaraf, Y.G., Golenser, J., Spira, D.T., and Bachrach, U. (1984). *Biochem. J.* **222,** 815–819.

Assaraf, Y.G., Golenser, J., Spira, D.T., and Bachrach, U. (1986). *Exp. Parasitol.* **61,** 229–235.
Bacchi, C.J. (1981). *J. Protozool.* **28,** 20–27.
Bacchi, C.J., Lipschik, G.Y., and Nathan, H.C. (1977). *J. Bacteriol.* **131,** 657–661.
Bacchi, C.J., Vergara, C., Garofalo, J., Lipschik, G.Y., and Hutner, S.H. (1979). *J. Protozool.* **26,** 484–488.
Bacchi, C.J., Nathan, H.C., Hutner, S.H., McCann, P.P., and Sjoerdsma, A. (1980). *Science* **210,** 332–334.
Bacchi, C.J., Nathan, H.C., Hutner, S.H., McCann, P.P., and Sjoerdsma, A. (1982a). *Biochem. Pharmacol.* **31,** 2833–2836.
Bacchi, C.J., McCann, P.P., Nathan, H.C., Hutner, S.H., and Sjoerdsma, A. (1982b). *Adv. Polyamine Res.* **4,** 221–31.
Bacchi, C.J., Garofalo, J., Mockenhaupt, D., McCann, P.P., Diekema, K.A., Pegg, A.E., Nathan, H.C., Mullaney, E.A., Chunosoff, L., Sjoerdsma, A., and Hutner, S.H. (1983). *Mol. Biochem. Parasitol.* **7,** 209–225.
Bacchi, C.J., Nathan, H.C., Clarkson, A.B., Jr., Bienen, E.J., Bitonti, A.J., McCann, P.P., and Sjoerdsma, A. (1987). *Am. J. Trop. Med. Hyg.* **36,** 48–54.
Bachrach, U., Brem, S., Wertman, S.B., Schnur, L.F., and Greenblatt, C.L. (1979a). *Exp. Parasitol.* **48,** 457–463.
Bachrach, U., Brem, S., Wertman, S.B., Schnur, L.F., and Greenblatt, C.L. (1979b). *Exp. Parasitol.* **48,** 464–470.
Bachrach, U., Abu-Elheiga, L., Talmi, M., Schnur, L.F., Elon, J., and Greenblatt, C.L. (1981). *Med. Biol.* (*Helsinki*) **59,** 441–447.
Bachrach, U., Abu-Elheiga, L., and Schnur, L.F. (1983). *Exp. Parasitol.* **55,** 280–288.
Baltz, T., Baltz, D., Giroud, C., and Crockett, J. (1985). *EMBO J.* **4,** 1273–1277.
Beaver, P.C., Jung, R.C., and Cupp, E.W. (1984). "*Clinical Parasitology*." Lea & Febiger, Philadelphia.
Berens, R.L., and Marr, J.J. (1979). *J. Protozool.* **26,** 453–456.
Bitonti, A.J., Kelly, S.E., and McCann, P.P. (1984). *Mol. Biochem. Parasitol.* **13,** 21–28.
Bitonti, A.J., Bacchi, C.J., McCann, P.P., and Sjoerdsma, A. (1985). *Biochem. Pharmacol.* **34,** 1773–1777.
Bitonti, A.J., Dumont, J.A., and McCann, P.P. (1986a). *Biochem. J.* **237,** 685–689.
Bitonti, A.J., McCann, P.P., and Sjoerdsma, A. (1986b). *Biochem. Pharmacol.* **35,** 331–334.
Bitonti, A.J., Bacchi, C.J., McCann, P.P., and Sjoerdsma, A. (1986c). *Biochem. Pharmacol.* **35,** 351–354.
Brun, R., Jenni, L., Schonenberger, M., and Shell, K.-F. (1981). *J. Protozool.* **28,** 470–479.
Campbell, G.H., Esser, K.M., and Philips, S.M. (1978). *Infect. Immun.* **20,** 714–720.
Casero, R.A., Jr., Bergeron, R.J., and Porter, C.W. (1984). *J. Cell. Physiol.* **121,** 476–482.
Chang, K.-P., Steiger, R.F., Dave, C., and Cheng, Y.-C. (1978). *J. Protozool.* **25,** 145–149.
Chang, K.-P., Fong, D., and Bray, R.S. (1985). *In* "Leishmaniasis" (K.-P. Chang and R.S. Bray, eds.), pp. 1–30. Elsevier, Amsterdam.
Clarkson, A.B., Jr., Bacchi, C.J., Mellow, G.H., Nathan, H.C., McCann, P.P., and Sjoerdsma, A. (1983). *Proc. Natl. Acad. Sci. U.S.A.* **80,** 5729–5733.
Clarkson, A.B., Jr., Bienen, E.J., Bacchi, C.J., McCann, P.P., Hutner, S.H., and Sjoerdsma, A. (1984). *Am. J. Trop. Med. Hyg.* **33,** 1073–1077.
Comis, R.L. (1978). *In* "Bleomycin: Current Status and New Developments" (S.K. Carter, S.T. Crooke, and H. Umezawa, eds.), pp. 279–291. Academic press, New York.
Coombs, G.H., and Sanderson, B.E. (1985). *Ann. Trop. Med. Parasitol.* **79,** 409–415.
Coombs, G.H., Hart, D.T., and Capaldo, J. (1983). *J. Antimicrob, Chemother.* **11,** 151–162.
Current, W.L., and Haynes, T.B. (1984). *Science* **224,** 603–605.

Cushion, M.T., Stanforth, D., Linke, M.J., and Walzer, P.D. (1985). *Antimicrob. Agents Chemother.* **28,** 796–801.
deGee, A.L.W., McCann, P.P., and Mansfield, J.M. (1983). *J. Parasitol.* **69,** 818–822.
deGee, A.L.W., Carstens, P.H.B., McCann, P.P., and Mansfield, J.M. (1984). *Tissue Cell* **16,** 731–738.
Dempsey, W.L., and Mansfield, J.M. (1983). *J. Immunol.* **130,** 405–411.
Duszenko, M., Ferguson, M.A.J., Lamont, G.S., Rifkin, M.R., and Cross, G.A.M. (1985). *J. Exp. Med.* **162,** 1256–1263.
Erwin, B.G., and Pegg, A.E. (1982). *Biochem. Pharmacol.* **31,** 2820–2823.
Fairlamb, A.H., and Cerami, A. (1985). *Mol. Biochem. Parasitol.* **14,** 187–198.
Fairlamb, A.H., Blackburn, P., Ulrich, P., Chait, B.T., and Cerami, A. (1985). *Science* **227,** 1485–1487.
Fairlamb, A.H., Henderson, G.B., Bacchi, C.J., and Cerami, A. (1987). *Mol. Biochem. Parasitol.* (in press).
Feagin, J.E., Jasmer, D.P., and Stuart, K. (1986). *Mol. Biochem. Parasitol.* **20,** 207–214.
Garofalo, J., Bacchi, C.J., McLaughlin, S.D., Mockenhaupt, D., Treuba, G., and Hutner, S.H. (1982). *J. Protozool.* **29,** 389–394.
Ghoda, L.Y., Savarese, T.M., Northup, C.H., Parks, R.E., Jr., Garofalo, J., Katz, L., Ellenbogen, B.B., and Bacchi, C.J. (1987). *Mol. Biochem. Parasitol.* (in press).
Giffin, B.F., McCann, P.P., and Bacchi, C.J. (1986a). *Mol. Biochem. Parasitol.* **20,** 165–171.
Giffin, B.F., McCann, P.P., Bitonti, A.J., and Bacchi, C.J. (1986b). *J. Protozool.* **33,** 238–243.
Gillet, J.M., Bone, G., and Herman, F. (1982). *Trans. R. Soc. Trop. Med. Hyg.* **76,** 776–777.
Gillet, J.M., Charlier, J., Bone, G., and Mulamba, P.L. (1983). *Exp. Parasitol.* **56,** 190–193.
Gillet, J., Bone, G., Lowa, P., Charlier, J., Rona, A.M., and Schechter, P.J. (1986). *Trans. R. Soc. Trop. Med. Hyg.* **80,** 236–239.
Gillin, F.D., Reiner, D.S., and McCann, P.P. (1984). *J. Protozool.* **31,** 161–163.
Golden, J.A., Sjoerdsma, A., and Santi, D.V. (1984). *West. J. Med.* **141,** 613–623.
Hannan, J.C., Bacchi, C.J., and McCann, P.P. (1984). *Mol. Biochem. Parasitol.* **12,** 117–124.
Hanson, W.L., Bradford, M.M., Chapman, W.L., Waits, V.B., McCann, P.P., and Sjoerdsma, A. (1982). *Am. J. Vet. Res.* **43,** 1651–1653.
Heidrich, J.E., Hunsaker, L.A., and Vander Jagt, D.L. (1983). *IRCS Med. Sci.* **11,** 929–930.
Hirumi, H., Doyle, J.J., and Hirumi, K. (1977). *Science* **196,** 992–994.
Hofflin, J.M., Guptill, D.R., Araujo, F.G., and Remington, J.S. (1985). *J. Infect. Dis.* **152,** 1101.
Hollingdale, M.R., McCann, P.P., and Sjoerdsma, A. (1985). *Exp. Parasitol.* **60,** 111–117.
Hughes, W.T., and Smith, B.L. (1984). *Antimicrob. Agents Chemother.* **26,** 436–440.
James, D.M., and Gilles, H.M. (1985). "Human Antiparasitic Drugs: Pharmacology and Usage." Wiley, New York.
Jennings, F.W., Whitelaw, D.D., and Urquhart, G.M. (1977). *Parasitology* **75,** 143–145.
Jensen, J.B. (1985). *In* "*In Vitro* Cultivation of Protozoan Parasites" (J.B. Jensen, ed.), pp. 155–192. CRC Press, Boca Raton, Florida.
Karbe, E., Bottger, M., McCann, P.P., Sjoerdsma, A., and Freitas, E.K. (1982). *Tropenmed. Parasitol.* **33,** 161–162.
Kaur, K., Emmett, K., McCann, P.P., Sjoerdsma, A., and Ullman, B. (1986). *J. Protozool.* **33,** 518–521.
Keithly, J.S. (1985). *Clin. Microbiol. Newslett.* **7,** 153–155.

Keithly, J.S., Bacchi, C.J., McCann, P.P., and Sjoerdsma, A. (1987). *Antimicrob. Ag. Chemother.* (submitted).

Kierszenbaum, F., Wirth, J.J., McCann, P.P., and Sjoerdsma, A. (1986). *Fed. Proc., Fed. Am. Soc. Exp. Biol.* **45,** 245.

Kierszenbaum, F., Wirth, J.J., McCann, P.P., and Sjoerdsma, A. (1987). *Proc. Natl. Acad. Sci. U.S.A.* (in press).

Kim, B.G., McCann, P.P., and Byers, T.J. (1987). *J. Protozool.* (in press).

Knuttgen, H.J. (1985). *Trop. Med. Parasitol.* **36** (Abstr.), 17.

Levin, V.A., Csejtey, J., and Byrd, D.J. (1983). *Cancer Chemother. Pharmacol.* **10,** 196–199.

Mamont, P.S., Bohlen, P., McCann, P.P., Bey, P., Schuber, F., and Tardif, C. (1976). *Proc. Natl. Acad. Sci. U.S.A.* **73,** 1626–1630.

Mamont, P.S., Danzin, C., Kolb, M., Gerhart, F., Bey, P., and Sjoerdsma, A. (1986). *Biochem. Pharmacol.* **35,** 159–165.

Marr, J.J. (1984). *In* "Parasitic Diseases" (J.M. Mansfield, ed.), Vol. 2, pp. 201–227. Dekker, New York.

McCann, P.P., Bacchi, C.J., Hanson, W.L., Cain, G.D., Nathan, H.C., Hutner, S.H., and Sjoerdsma, A. (1981a). *Adv. Polyamine Res.* **3,** 97–110.

McCann, P.P., Bacchi, C.J., Clarkson, A.B., Jr., Seed, J.R., Nathan, H.C., Amole, B.O., Hutner, S.H., and Sjoerdsma, A. (1981b). *Med. Biol. (Helsinki)* **59,** 434–440.

McCann, P.P., Bacchi, C.J., Nathan, H.C., and Sjoerdsma, A. (1983). *In* "Mechanisms of Drug Action" (T.P. Singer and R.N. Ondarza, eds.), pp. 159–73. Academic Press, New York.

McCann, P.P., Bacchi, C.J., Clarkson, A.B., Jr., Bey, P., Sjoerdsma, A., Schechter, P.J., Walzer, P.D., and Barlow, J.L.R. (1986). *Am. J. Trop. Med. Hyg.* **35,** 1153–1156.

Meshnick, S.R. (1984). *In* "Parasitic Diseases" (J.M. Mansfield, ed.), Vol. 2, pp. 165–199. Dekker, New York.

Meshnick, S.R., Blobstein, S.H., Grady, R.W., and Cerami, A. (1978). *J. Exp. Med.* **148,** 569–579.

Molyneux, D.H., and Ashford, R.W. (1983). "The Biology of *Trypanosoma* and *Leishmania,* Parasites of Man and Domestic Animals." Taylor & Francis, London.

Moon, H.W., Schwartz, A., Welch, M.J., McCann, P.P., and Runnels, P.L. (1982). *Vet. Pathol.* **19,** 700–707.

Morrow, C.D., Flory, B., and Krassner, S.M. (1980). *Comp. Biochem. Physiol.* **66B,** 307.

Murray, P.K., and Jennings, F.W. (1983). *In* "Animal Models for Experimental Bacterial and Parasitic Infection" (G. Keusch and T. Wadstrom, eds.), pp. 343–354. Elsevier, Amsterdam.

Nathan, H.C., Bacchi, C.J., Hutner, S.H., Rescigno, D., McCann, P.P., and Sjoerdsma, A. (1981a). *Biochem. Pharmacol.* **30,** 3010–3013.

Nathan, H.C., Bacchi, C.J., Sakai, T.T., Rescigno, D., Stumpf, D., and Hutner, S.H. (1981b). *Trans. R. Soc. Trop. Med. Hyg.* **75,** 394–398.

North, M.J., Lockwood, B.C., Bremner, A.F., and Coombs, G.H. (1986). *Mol. Biochem. Parasitol.* **19,** 241–249.

Opperdoes, F.R. (1985). *Br. Med. Bull.* **41,** 130–136.

Pegg, A.E. (1984). *Cell Biochem. Function* **2,** 11–15.

Pegg, A.E. (1986). *Biochem. J.* **234,** 249–262.

Pegg, A.E., and McCann, P.P. (1982). *Am. J. Physiol. Cell. Physiol* **243,** C212–C221.

Porter, C.W., and Bergeron, R.J. (1983). *Science* **219,** 1083–1085.

Porter, C.W., Miller, J., and Bergeron, R.J. (1984). *Cancer Res.* **44,** 126–128.

Rudkin, B.B., Mamont, P.S., and Seiler, N. (1984). *Biochem. J.* **217,** 731–741.
Schechter, P.J., and Sjoerdsma, A. (1986). *Parasitol. Today* **2,** 223–224.
Schillinger, D., and Gorton, E. (1984). *Drugs Exp. Clin. Res.* **10,** 677–679.
Schnur, L.F., Bachrach, U., Greenblatt, C.L., and Ben Joseph, M. (1979). *FEBS Lett.* **106,** 202–206.
Schuster, F.L., and McCann, P.P. (1982). *J. Protozool.* **29,** 291.
Shames, S.L., Fairlamb, A.H., Cerami, A., and Walsh, C.T. (1986). *Biochemistry* **25,** 3519–3526.
Sjoerdsma, A., and Schechter, P. (1984). *Clin. Pharmacol. Ther.* **35,** 287–300.
Sjoerdsma, A., Golden, J.A., Schechter, P.J., Barlow, J.L.R., and Santi, D.V. (1984). *Trans. Assoc. Am. Phys.* **97,** 70–79.
Smith, C.J., Levine, R.F., and Mansfield, J.M. (1982). *Am. Soc. Trop. Med. Hyg.* **31,** 1098–1102.
Taelman, H., Marcelis, L., Sonnet, J., Schechter, P.J., and Wery, M. (1984). *Abstr. Int. Congr. Trop. Med. Malaria, 11th, Calgary* p. 150.
Umezawa, H. (1980). *In* "Anticancer Agents Based on Natural Product Analogs" (J.M. Cassady and J.D. Douros, eds.), pp. 147–166. Academic Press, New York.
Van Nieuwenhove, S., Schechter, P.J., Declerg, J., Bone, G., Burke, J., and Sjoerdsma, A. (1985). *Trans. R. Soc. Trop. Med. Hyg.* **79,** 692–698.
Vickerman, K. (1971). *In* "Ecology and Physiology of Parasites" (A.M. Fallis, ed.), pp. 58–91. Univ. of Toronto Press, Toronto.
Vickerman, K. (1985). *Br. Med. Bull.* **41,** 105–114.
Whaun, J.M., and Brown, N.D. (1985). *J. Pharmacol. Exp. Ther.* **233,** 507–511.
White, E., Hart, D., and Sanderson, B.E. (1983). *Mol. Biochem. Parasitol.* **9,** 309–318.
Wilson, S.G., Morris, K.R.S., Lewis, I.J., and Krog, E. (1963). *Bull. WHO* **28,** 595–598.
World Health Organization (1985). Tropical Disease Research: 7th Program Report (1 Jan. 1983–31 Dec. 1984). WHO, Geneva.

16

Clinical Aspects of Inhibition of Ornithine Decarboxylase with Emphasis on Therapeutic Trials of Eflornithine (DFMO) in Cancer and Protozoan Diseases

PAUL J. SCHECHTER,* JOHN L. R. BARLOW,† AND ALBERT SJOERDSMA†

** Merrell Dow Research Institute*
16, rue d'Ankara
F 67084 Strasbourg Cedex, France

† Merrell Dow Research Institute
2110 East Galbraith Road
Cincinnati, Ohio 45215

I. INTRODUCTION

In all mammalian and many other eucaryotic cells, decarboxylation of ornithine by ornithine decarboxylase (ODC) is an obligatory step in the biosynthesis of polyamines such as putrescine (a diamine), spermidine, and spermine, which are ubiquitous in living cells and are thought to play important roles in cell division and differentiation (Pegg and McCann, 1982). Thus, inhibition of this enzyme with specific enzyme-activated irreversible inhibitors such as DL-α-difluoromethylornithine[1] (DFMO;

[1] Supplied as monohydrochloride·monohydrate.

eflornithine; Ornidyl) appeared to be an approach to control abnormalities of these cellular activities.

Our early therapeutic goal was the treatment of neoplasia, in part because of reports of increased urinary excretion of polyamines in cancer patients and increased polyamine concentrations in tumors; both increased excretion and concentrations decrease with antitumor chemotherapy (Russell, 1971; Heby and Russell, 1973). However, despite impressive effects on many tumors in culture and in laboratory animals (cf. Sunkara, Baylin, and Luk, Chapter 6; and Porter and Jänne, Chapter 10, this volume), results of treatment with eflornithine in human cancer have thus far been generally disappointing (Sjoerdsma and Schechter, 1984).

The discovery that polyamines in *Trypanosoma brucei brucei* are also synthesized from ornithine (Bacchi *et al.,* 1979) suggested testing eflornithine against African trypanosomiasis. In rodents infected with *T. b. brucei* (Bacchi *et al.,* 1980), *T. b. rhodesiense* (McCann *et al.,* 1981a), *T. b. gambiense* (McCann *et al.,* 1981b), and *T. congolense* (Schillinger and Gorton, 1984), it was found that eflornithine notably increased host survival and produced apparent cures. These effects were associated with a number of biochemical and morphological effects on the parasite as described in the preceding chapter (Bacchi and McCann, Chapter 15). As a result of this information, the clinical use of eflornithine against *T. b. gambiense* was begun in 1981 in the Sudan with a dramatically successful outcome (Sjoerdsma and Schechter, 1984; Van Nieuwenhove *et al.,* 1985).

Another parasitic protozoan has recently become a scourge in North America. *Pneumocystis carinii* pneumonia (PCP) is now the most common cause of death in patients with acquired immune deficiency syndrome (AIDS). Morphologically, the organism has been likened to a protozoan of the general class Sporozoea, being similar, for example, to Coccidia, some species of which have been shown to be extremely sensitive to the effects of eflornithine *in vivo* (Hanson *et al.,* 1982). The first use of eflornithine against *P. carinii* was in 1983 in the clinical setting of PCP AIDS, patients whose pneumocystosis could not be cleared with other drugs (pentamidine and/or cotrimoxazole, the latter being a trimethoprim/sulfamethoxazole combination) due to refractoriness or intolerance. Results were remarkably positive in the first six patients by both clinical and objective criteria (Sjoerdsma *et al.,* 1984; Golden *et al.,* 1984).

This chapter will summarize the results of the use of eflornithine in various neoplastic diseases and in both of the above protozoan diseases, as well as comment on future clinical applications of the biochemical approach of inhibiting ornithine decarboxylase.

II. MALIGNANCIES

The striking efficacy of eflornithine in retarding the growth of malignant cells studied *in vitro* and *in vivo* in animal models (cf. Pegg and McCann, 1982; and Sunkara, Baylin, and Luk, Chapter 6, this volume) prompted its use in patients with malignancies. Beginning in 1979, eflornithine was administered orally or intravenously to patients with far-advanced disease, who were not candidates for alternate therapeutic modalities, in order to evaluate tolerance to the drug. Untoward events occurring in some cases during oral dosage of ≥10 g/day (given in four divided doses, every 6 hr) included gastrointestinal effects (primarily diarrhea, less frequently nausea, emesis, abdominal pain, and anorexia), thrombocytopenia, anemia, and bilateral hearing loss (cf. Abeloff *et al.*, 1984). Fewer such effects were seen with intravenous eflornithine administration, and only at much higher doses (e.g., ≥20 g/day) (Maddox *et al.*, 1985; Griffin *et al.*, 1984, 1987). All adverse effects associated with eflornithine treatment were generally reversible upon decreasing the dose or discontinuation of treatment. They were not life threatening and, with the exception of the gastrointestinal disturbances, did not appear until several weeks after beginning treatment.

In approximately 500 patients with a variety of far-advanced malignancies, eflornithine therapy (usually oral) did not convincingly influence tumor growth or disease progression. In these studies, stabilization of disease progression occurred in occasional patients and there were actually a few complete responses, but no general pattern of response was evident. Despite the cytotoxic effect in cell lines of human small-cell carcinoma *in vitro* (Luk *et al.*, 1981, 1982) and *in vivo* (Luk *et al.*, 1983a, 1986), only one partial and unsustained response was seen (Abeloff *et al.*, 1984, 1985; M.D. Abeloff, unpublished data) in 27 patients with this condition treated with oral eflornithine. However, in 1986, Meyskens *et al.* reported a study of 21 evaluable patients with metastatic malignant melanoma who received eflornithine orally at a dose of 2–4 g/m^2 every 8 hr. Of these 21 patients, one had complete disappearance of evidence of disease after 2 months of treatment which was sustained for almost 8 months. In 7 patients, the disease was stabilized for a median duration of 8 weeks (range 4–12 weeks). These results indicated some activity in this condition.

Laboratory studies suggesting unique interaction between eflornithine and methylglyoxal bis(guanylhydrazone) (MGBG; methyl-GAG), a spermidine analog whose cytotoxic effects have been related in part to interference with polyamine metabolism, have provided an apparent rationale for combination therapy in man (cf. Pegg and McCann, 1982; Porter and Jänne, Chapter 10, this volume). Basically, depletion of intracellular polyamine concentrations by eflornithine in cultured cells or in tumor cells in animals resulted in greatly enhanced cellular uptake of MGBG (Alhonen-Hongisto *et al.*, 1980; Seppänen *et al.*, 1981). Furthermore, this effect occurred preferentially in tumor cells versus normal cells (see Porter and Jänne, Chapter 10, this volume).

The first clinical trial using eflornithine priming followed by MGBG treatment involved five children with advanced leukemia (Siimes *et al.*, 1981). The combination therapy produced an impressive clinical response with mild or absent toxicity. Unfortunately, attempts to reproduce this early promising finding in leukemia, as well as in other types of malignancy, have not been successful (except vide infra). A variety of different treatment sequences, doses, and schedules were tried for the two drugs, but most often the addition of eflornithine to MGBG therapy was found to enhance both the antitumor effects and the toxic effects of MGBG (Warrell *et al.*, 1983; Splinter and Romijn, 1986; Herr *et al.*, 1986).

Two investigators found differing intermittent schedules in which tumor response was observed and in which there was acceptable tolerance. Natale *et al.* (1985) reported an uncontrolled study of 25 evaluable patients to whom eflornithine was administered orally in daily divided dosage of 6.0 g/m^2 for 5 consecutive days, with MGBG given intravenously on the fifth day at a dose of 400 mg/m^2. They reported three complete responses (two cases of metastatic thyroid carcinoma and one of lymphoma) of 20, 15, and 12 months duration and three partial responses (one Hodgkins, one lymphoma, and one sarcoma) in these 25 patients. In this study, dose-limiting toxicities consisted predominantly of diarrhea. Leukopenia, thrombocytopenia, and anemia occurred infrequently.

Most interesting and possibly significant are the findings of Levin *et al.* (1987) with the combination of eflornithine and MGBG on brain tumors. Eflornithine was given 1.33 g/m^2 tid orally for courses of 14 days with 1-week intervals between courses. MGBG was given intravenously at a dose of 200 mg/m^2 on the fourteenth day of each course. The subjects were patients with recurrent primary brain tumors (e.g., astrocytomas). Sustained response or stable disease was obtained in 10 out of 16 patients treated with this regimen.

A synergism between eflornithine and α-interferon has been demonstrated in the suppression of a transplantable mouse melanoma (Sunkara *et al.*, 1983). This finding has stimulated a number of clinical trials with this combination. In the first of these, Talpaz *et al.* (1986) reported on a phase I study of 25 patients (12 of whom had malignant melanoma) in which dosage of eflornithine ranged from 1.5 to 2.25 g/m^2 every 6 hr orally and that of intramuscular human leukocyte interferon (IFN-α) ranged from 0.4 to 6.4 $\times$ 10^6 U/m^2. The maximum tolerated dose observed was eflornithine 1.5 g/m^2 every 6 hr and IFN-α 3.2 $\times$ 10^6 U/m^2 daily. Of the 12 patients with metastatic malignant melanoma, two experienced partial remissions and two had sustained minor responses. There were also minor responses in a patient with colon carcinoma and a patient with renal carcinoma. Other studies in other institutions are still ongoing.

Eflornithine was also combined with standard chemotherapy and standard radiotherapy in the treatment of patients with a wide selection of malignant tumors. In some studies, control groups treated with standard chemo- or radiotherapy only were run in parallel, while in other studies historical controls were used. No unequivocal added benefit of eflornithine could be ascertained in these trials. In contrast, eflornithine often did contribute its established side effects to the patients treated.

There seem to be some apparent discrepancies between the striking effects of eflornithine on tumors in animal models and in the clinical situation. A possible explanation would involve both the dosage schedules (in relation to pharmacokinetics) and the protocols of the studies in animal tumor models. Eflornithine must be administered frequently (on a practical level, usually by providing rodents with 1–3% eflornithine in the drinking water) due to a short eflornithine plasma half-life in rodents (i.e., approximately 80 min; Grove *et al.*, 1981) and because ornithine decarboxylase itself has the very rapid enzyme turnover of about 20–30 min (Pegg and McCann, 1982). In contrast, the plasma half-life of elfornithine in man is considerably longer, around 200 min (Haegele *et al.*, 1981). When plasma eflornithine concentrations in mice drinking 1% eflornithine in the drinking water were followed over a 24-hr period, concentrations varying between 0.29 and 0.97 m*M* (average = 0.5 m*M*, n = 12) were measured. The plasma levels achieved in man are comparable (Abeloff *et al.*, 1984). Thus, a pharmacokinetic explanation for a general lack of clinical antitumor efficacy of eflornithine does not seem reasonable.

The experimental protocols thus seem to be a key point in explaining the above relationship. In most animal studies published to date, eflornithine treatment is begun prior to, concomitant with, or shortly after tumor inoculation. These transplanted cells, as with cultured cells, then generally enter a logarithmic growth phase with rapid proliferation and differentiation, in which polyamine requirements are maximal. An inhibition of polyamine biosynthesis would be expected to be most effective at this time. In the clinical situation, on the other hand, treatment obviously is not considered until the diagnosis of malignancy is made, at which time the most rapid growth phase of most tumors is likely to have already occurred.

Thus, a general, more feasible use for eflornithine as monotherapy might be in a "prophylactic" situation wherein it would prevent tumor development over a period of time. Such possibilities have been discussed by Luk and Baylin (1984) for the treatment of familial polyposis where the development of dysplastic polyps, and ultimately colonic cancer, might be prevented by eflornithine. Similar ideas have been expressed by Loprinzi and Verma (1985) who have shown that very low doses of eflornithine can inhibit ODC activity induced by a tumor promoter in mouse skin. Earlier studies have also shown that long-term administration of eflornithine will almost completely block tumor formation in animal models where a carcinogen or tumor promoter has been given (see Verma and Boutwell, Chapter 11, this volume).

III. AFRICAN TRYPANOSOMIASIS

At the present time, despite more than 50 years of struggle against *T. b. gambiense* sleeping sickness, this disease is still endemic in 23 African countries with a global annual incidence of at least 10,000 new cases and an estimated population of 45 million at risk of infection (WHO, 1982). Among the African countries, Zaire reports the largest number of new cases (5000) per year (Kazyumba, 1979). A majority of patients with African trypanosomiasis can be treated effectively with the currently available trypanocidal agents. Suramin and pentamidine are usually curative in the early (hemolymphatic) stage (Stage I) of the disease, while melarsoprol (Mel B, Arsobal), a trivalent arsenical, is the drug of choice for the late (encephalomeningeal) stage (Stage II) of gambiense and rhodesiense sleeping sickness. All these drugs must be given parenterally and serious side effects are associated with their use. Toxicity is most pronounced with melarsoprol (Apted, 1980), the sole drug available for the late stage of the illness. Mortality due to reactive arsenical encephalopathy is estimated to occur in 5% or more of melarsoprol-treated patients (Bertrand *et al.*, 1973; Haller *et al.*, 1986). A further problem with regard to treatment with melarsoprol is the resistance of the organism in increasing numbers of cases (Ruppol and Burke, 1977; Ginoux *et al.*, 1984). Therefore, an effective, alternative drug with acceptable side effects would be most welcome, particularly for the late stage of the disease.

Cautious use of eflornithine in humans with trypanosomiasis was begun in July 1981 by Dr. S. Van Nieuwenhove at a field station in the Southern Sudan (Van Neiuwenhove *et al.*, 1985). The first patient (Case 1 in Sjoerdsma and Schechter, 1984; Patient 17 in Van Neiuwenhove *et al.*, 1985) was a 20-year-old male, weighing 58 kg, who complained of fatigue, daytime somnolence, and nighttime insomnia. On examination cervical lymph nodes were present, he was confused, and had a tremor. Trypanosomes were found in blood and in a lymph node aspirate. A lumbar puncture procedure produced cerebrospinal fluid (CSF) with >500 WBC/mm^3 (normal $\leq$ 5/mm^3), a protein concentration of 0.70 mg/ml (normal $\leq$ 0.30 mg/ml), and the presence of trypanosomes. There was no previous history of the disease. Treatment with oral eflornithine, 15 g/day (258 mg/kg/day) evenly divided in four 6-hourly doses, was begun. This dose was empirically chosen based on experience of acceptable tolerance in cancer patients. Within 2 days of beginning eflornithine treatment trypanosomes were absent from wet blood films and Giemsa-stained thick blood films; by 4 days trypanosomes were no longer seen in lymph node aspirates. There was a concomitant rapid clinical response. When the next lumbar puncture was performed, 1 month after commencing eflornithine, WBCs in CSF had decreased to 55/mm^3, protein to 0.55 mg/ml, and no parasites were found. All signs and symptoms had disappeared. Treatment was continued for 44 days. By this time there were 44 WBC/mm^3 and 0.4 mg/ml protein in CSF. Eflornithine treatment was well tolerated with no side effects noted. The patient left the clinic but returned 10 weeks later with apparent relapse. Although symptomless, CSF WBC count was 138/mm^3 and trypanosomes were present. Thus, from this initial case it appeared that in man oral eflornithine, at the dose given, was able to rapidly clear trypanosomes from body fluids, decrease the elevated CSF WBC count and protein content, and ameliorate clinical symptoms and signs. However, relapse could occur.

Subsequently, numerous additional cases of Stage II sleeping sickness have been treated with eflornithine orally in the Sudan. Based on experience with the initial case, the dose of eflornithine was increased to 400 mg/kg/day (in four divided doses). In each case, eflornithine treatment was associated with clearing of trypanosomes from CSF, decreases in CSF WBC count and protein content, and reversal of clinical signs and symptoms (Van Nieuwenhove *et al.,* 1985).

In early 1983 a patient (Case 9 in Sjoerdsma and Schechter, 1984) in Belgium, who had progressed to a comatose state, thereby dictating parenteral administration of drug, was given eflornithine by the intravenous route (Taelman *et al.,* 1987). This treatment was dramatically successful and well tolerated. This observation along with the obviously improved bioavailability (Haegele *et al.,* 1981) and better tolerance of the intravenous route influenced our decision to use eflornithine intravenously for at least the first 2 weeks of treatment.

As of May 1986, a total of 129 cases with gambiense sleeping sickness (one of these in the United States; Petru *et al.,* 1985) had been treated

TABLE I

Summary of 129 *T. b. gambiense* Sleeping Sickness Patients Treated with Eflornithine

Data	Number of cases
Total cases	129
Early-stage disease cases	5
Late-stage disease cases	124
Refractory to arsenical treatment	111
Treatment: Oral	62
Intravenous	16
Intravenous + Oral	51
Cases in which eflornithine treatment decreased or normalized CSF WBCs and protein, and/or eradicated trypanosomes	123
Documented unrelapsed cases (58) with varying follow-up times of at least 3 months	
From end of treatment: ≥3 mo	26
≥6 mo	16
≥12 mo	10
≥24 mo	6
Cases with documented relapse (or reinfection)	
After oral treatment	6
After iv or iv + oral treatment	0
Deaths	13
Causes of death were complex and included intercurrent infection, cachexia, severe CNS disease, severe anemia, and unknown factors	

with eflornithine (the etiological agent presumed on the basis of clinical history and course and geographical origins) (Table I). Five cases presented with Stage I (Hemolymphatic) disease while 124 patients, mostly with advanced clinical disease, had laboratory evidence of central nervous system disease (i.e., trypanosomes in CSF, elevated CSF WBC counts, and increased protein concentration). Of these latter patients, 111 had been previously treated with one to four full courses of organic arsenicals and, therefore, were considered arsenical refractory (Table I). Sixty-two patients received eflornithine by the oral route only, generally at daily doses of 400 mg/kg for 4–6 weeks; 16 intravenously only in daily doses of 400 mg/kg for 4–5 weeks; and 51 received both intravenous and oral eflornithine, generally 400 mg/kg/day iv for 2 weeks followed by 300 mg/kg/day po for 3–4 weeks.

Eflornithine treatment could be evaluated for efficacy in all 5 early

(Stage I) cases and in 118 of the 124 late-stage cases because posttreatment body fluids were available and clinical examination of the patient was possible. In *all* these "evaluable" cases, eflornithine treatment was associated with a decrease in CSF WBC counts (for late-stage cases), invariable, rapid (within a few days) disappearance of trypanosomes from body fluids (when parasites had been detected prior to treatment), and improvement in abnormal biological parameters. These changes were generally accompanied by rapid and dramatic amelioration of the clinical signs and symptoms of the disease.

Thirteen patients, all with late-stage disease, died during or shortly after eflornithine treatment. The causes of death were complex and difficult to establish precisely but included intercurrent infection, cachexia, and severe CNS disease. Treatment duration of these patients varied from only a few days to several weeks. In eight of these patients CSF was obtained prior to death and in each case trypanosomes were absent and CSF WBC count had decreased.

Individual patients have been followed for as long as 31 months after eflornithine treatment (Table I). Six patients have been followed for at least 24 months and 10 additional patients for at least 1 year (5 of these up to 22 months). Six of the patients treated exclusively with oral eflornithine have been documented as relapsing (or being reinfected) within 10 weeks to 10.5 months posttreatment. None of the patients treated with eflornithine intravenously is known to have relapsed.

During treatment with eflornithine a variety of adverse events were noted (Table II), most of which could be attributed to the treatment itself but some may have been caused by other factors (e.g., compromised physical, nutritional, and social state). Diarrhea could usually be controlled by symptomatic treatment or by temporarily discontinuing or decreasing the eflornithine dose. Nonspecific abdominal pain accompanied the diarrhea in about one-third of patients with that complaint. Reversible diarrhea and soft stools have previously been reported as a frequent side effect of eflornithine treatment (Sjoerdsma *et al.*, 1984; Abeloff *et al.*, 1984; Griffin *et al.*, 1987). Likewise, anorexia, emesis, and other gastrointestinal symptoms have also been described after both oral and intravenous eflornithine (Abeloff *et al.*, 1984; Griffin *et al.*, 1984, 1987; Maddox *et al.*, 1985). Thus, although diarrhea has been reported in association with African trypanosomiasis, it is likely that most of the gastrointestinal disturbances, all of which were reversible and seldom severe enough to discontinue therapy, seen in our patients are attributable to the drug treatment.

Seizures occurred in about 10% of the patients, all with late-stage disease. In at least one individual, seizures were reported to have occurred prior to treatment and in one case signs of increased intracranial pressure

TABLE II

Adverse, but Transitory, Events Associated with Eflornithine Treatment in 129 Patients with *T. b. gambiense* Sleeping Sickness

Event	Number of Cases
Diarrhea	77
Abdominal pain	26
Anorexia	8
Emesis	13
Vertigo (and/or dizziness)	8
Seizures	13
Alopecia	8
Hearing loss	2
Cutaneous eruption	3
Anemia	43
Thrombocytopenia	5
Leukopenia	9
Gastric ulcer	1
Opportunistic infections	17

were noted. Seizures have been reported in a similar percentage of untreated late-stage trypanosomiasis cases (Antoine, 1977; Ngandu-Kabeya, 1976; Giordano *et al.*, 1977). Alopecia (usually patchy) of the scalp occurred in 8 patients. This side effect has been seen sporadically when eflornithine was used to treat other conditions. Anemia, usually documented by a decrease of ≥2 g/dl in hemoglobin concentration with hematological testing, but occasionally suspected from clinical findings, occurred in 43 patients. Thrombocytopenia and leukopenia were less frequent, although platelet and WBC counts were not always routinely followed in all patients. Anemia and, less commonly, leukopenia and thrombocytopenia are found in patients with sleeping sickness (Wery *et al.*, 1982; Robins-Browne *et al.*, 1975; Davis, 1982; Greenwood and Whittle, 1980). On the other hand, eflornithine treatment is also associated with anemia and thrombocytopenia, but seldom leukopenia (Sjoerdsma *et al.*, 1984; Abeloff *et al.*, 1984; Griffin *et al.*, 1987). Thus, it is likely that in some cases of anemia, eflornithine may have been responsible; however, the influence of the disease process itself on hematological factors is a confounding factor. Various infectious processes were noted in 17 patients during eflornithine treatment. These included intestinal parasites (frequently), an amoebic abscess of the liver, generalized varicella, purulent pneumonia, rubeola, and cutaneous abscesses.

We strongly conclude, as we have earlier with fewer numbers of patients (Sjoerdsma and Schechter, 1984; Sjoerdsma *et al.*, 1984; Schechter and Sjoerdsma, 1986; McCann *et al.*, 1986), that eflornithine is a highly effective treatment of all stages of *T. b. gambiense* trypanosomiasis. Efficacy in arsenical refractory, late-stage cases is especially noteworthy. As yet, there are few results in cases of *T. b. rhodesiense* trypanosomiasis, and it is not yet known how effective eflornithine is in this form of sleeping sickness.

IV. PNEUMOCYSTIS CARINII PNEUMONIA (PCP)

At least in the United States, *Pneumocystis carinii* pneumonia, characterized by fever, dry cough, weight loss, progressive dyspnea, tachypnea, and ultimate respiratory failure, is the most frequent, life-threatening opportunistic infection in patients with acquired immune deficiency syndrome (leading article, *Lancet,* 1985; Kovacs and Masur, 1985; Fauci *et al.,* 1984). A mortality rate of 60% per episode has been quoted when standard treatment (Kovacs and Masur, 1985) with either cotrimoxazole or pentamidine is used. Moreover, both agents are associated with a high incidence of allergic and toxic reactions (leading article, *Lancet,* 1985; Kovacs and Masur, 1985; Fauci *et al.*, 1984; Haverkos, 1984; Jaffe *et al.*, 1983) in AIDS patients and therapeutic failure with either or both is associated with a worsened prognosis (Haverkos, 1984). Thus, current treatment is regarded as less than satisfactory. This section deals with the results of eflornithine therapy in patients with AIDS and PCP treated under an open emergency protocol which recruited patients refractory to standard therapy. The analysis is based on 151 patients in whom adequate data were available as of January 1986 to judge efficacy.

The rationale supporting the initiation of studies on the therapy of *P. carinii* pneumonia with eflornithine was based on the known activity of eflornithine against related protozoa, *Eimeria* spp. and *Plasmodia* spp. (Sjoerdsma and Schechter, 1984), its reported effect on increasing the helper/suppressor T lymphocyte ratio in normal mice (Sharkis *et al.*, 1983), and its *in vitro* activity against cytomegalovirus (Tyms and Williamson, 1984), at that time a putative causal agent for AIDS. More recently, but soon after the initiation of the clinical program, eflornithine was shown to be active against *P. carinii* in tissue culture (Cushion *et al.*, 1985; see Bacchi and McCann, Chapter 15, this volume). Following successful treatment of five patients with PCP and AIDS in a pilot study (Golden *et al.*, 1984), an emergency program by numerous investigators commenced in January 1984 in the United States.

The protocol allowed entry of AIDS patients with PCP confirmed by open lung or transbronchial biopsy or bronchoalveolar lavage. They must have failed treatment or had adverse reactions in the current or a previous episode sufficiently severe to have prompted discontinuation of therapy with pentamidine and cotrimoxazole. Signed informed consent by the patient had to be obtained and evidence of approval of the protocol by an independent review body had to be provided.

All patients were seriously ill, with a prognosis usually judged as poor, both on account of the failure of approved therapies and the presence of life-threatening concomitant infections such as disseminated *Mycobacterium avium-intracellulare,* pulmonary cytomegalovirus, or other bacterial or severe fungal infections. After microbiological confirmation of the diagnosis, baseline arterial blood gases, hematologic parameters, serum chemistry, and chest X ray were obtained. A posttreatment bronchoscopy was requested, with transbronchial biopsy, bronchial brushings or bronchoalveolar lavage, in order to attempt to obtain evidence of microbiological cure to reinforce the clinical observations. It evolved that most of those patients who returned to good health were unwilling to consent to this procedure.

Early in the program, a dose of 100 mg/kg every 6 hr given intravenously for 14 days followed by 4 to 6 weeks oral dosing at 75 mg/kg every 6 hr was universally recommended. This corresponds with the schedule previously demonstrated as successful in eradicating *T. b. gambiense* in African sleeping sickness (Sjoerdsma *et al.,* 1984). With the establishment of this dosage, it was determined that patients would be classed as unsuitable for evaluation of efficacy if the duration of therapy was less than 14 days.

Reduction in dosage in the event of adverse reactions was permitted, and in some instances patients entering with renal insufficiency or hematologic abnormalities (in particular, thrombocytopenia) started at an arbitrarily reduced dose. In special circumstances, patients have been given oral treatment as outpatients with the same dosage and the same schedule. Moreover, patients responding to intravenous dosing have been switched to oral therapy when fit to be discharged from the hospital. This has been in less than 14 days in some instances.

Of the 151 patients, 146 were either refractory or intolerant to cotrimoxazole and/or pentamidine for the treatment of the current episode (Table III) and 5 had histories of intolerance to these agents for the treatment of previous episodes of PCP. Without an effective therapeutic alternative these patients had little chance for improvement or survival. In addition, approximately 60% (90/151) of the patients were in frank respiratory failure requiring mechanical ventilation at the beginning of eflornithine treatment. Respiratory failure requiring assisted ventilation in the setting of AIDS complicated by PCP is almost uniformly fatal, and these patients,

TABLE III

Summary of Cases with Acquired Immune Deficiency Syndrome and *Pneumocystis carinii* Pneumonia Treated with Eflornithine on an Emergency Basis

Data	Number of Cases
Total cases	151
Refractory/intolerant to trimethoprim/sulfa and/or pentamidine	146
Requiring ventilator	
Yes	90
No	61
Total completed ≥14 days of eflornithine therapy	66
Started therapy requiring a ventilator	32
Started therapy not requiring a ventilator	34
Number of responders	49
Started therapy requiring a ventilator	16
Started therapy not requiring a ventilator	33
Number of patients with posttreatment biopsy, bronchoalveolar lavage, or bronchial brushing	16
Number negative for *Pneumocystis carinii*	16

after failing conventional therapy, would be expected to have an especially poor prognosis.

As shown in Table III, 66 of the 151 patients completed the treatment period of 14 days or longer required for evaluation of efficacy. A favorable clinical response was seen in 49 of these 66 patients (74%) as judged by defervescence, disappearance of symptoms, weaning from mechanical ventilation and/or return toward normal of arterial blood gases, improvement of chest X ray and/or gallium scan, and ultimate hospital discharge. Favorable clinical responses were seen in 33 of the 34 patients able to breathe spontaneously at study entry, and even more surprising were the favorable responses reported for 16 of the 32 patients in frank respiratory failure receiving mechanical ventilation at study entry.

In addition to the 49 responders who received treatment for 14 days or longer, two additional clinical responses were reported from the 85 patients treated for 13 days or less. The outcome for the remainder of these patients is currently undergoing evaluation but the majority of them died either of progressive disease or of overwhelming opportunistic infections. Sixteen patients underwent posttreatment biopsy, bronchoalveolar lavage, or bronchial brushing, and all were found to be clear of *P. carinii*.

TABLE IV

Adverse, but Transitory, Events Noted after Eflornithine Treatment (PCP)[a]

Event	Number of Cases
Thrombocytopenia	61
Leukopenia	12
Pancytopenia	1
Anemia	15
Alopecia	10
Bilateral hearing loss	8
Seizures	4
Cutaneous eruption	3
Alterations in hepatic function	2

[a] Diarrhea and nausea and/or vomiting were reported in a majority of patients on oral therapy, usually responding to dose adjustment or symptomatic medication.

During treatment with eflornithine a variety of essentially reversible adverse events occurred (Table IV), generally similar to the experience in trypanosomiasis. Thrombocytopenia, reported in 61 patients as a platelet count of less than 100,000/mm^3, was usually reversible following dosage reduction or interruption, as was leukopenia seen in 12 patients. In regard to the effects of low platelet counts, bleeding was rarely reported and in no instance has been life threatening. There were two instances of epistaxis, one of rectal bleeding with bright red blood, and one of slow oozing from thoracotomy wounds. A single instance of pancytopenia occurred late in treatment and resolved following discontinuation of therapy and transfusion. The relationship of anemia to eflornithine therapy was uncertain because of complicating nutritional factors, although there was precedence in the trypanosomiasis experience as well as in animal studies for such eflornithine-produced events (Luk *et al.*, 1983b).

Reversible patchy alopecia in 10 patients and high-frequency hearing loss in 8 patients were reported. Seizures were seen in 4 patients with suspected intracranial pathology. Cutaneous eruption occurred in 3 patients. Two patients were reported with alterations of hepatic function occurring during treatment, and one was rechallenged with oral eflornithine with no effect on liver function. In addition to the above listed effects, diarrhea and nausea and/or vomiting usually responding to dose reduction or symptomatic medication were seen in the majority of patients on oral medication.

The currently approved drug treatment regimens in the United States for *P. carinii* pneumonia are less than satisfactory. From the data reviewed here, it would appear that eflornithine is an acceptable therapy for PCP in AIDS and, surprisingly, for patients who have failed to respond to or are intolerant of standard therapy (Golden *et al.*, 1984; Sjoerdsma *et al.*, 1984; Gilman *et al.*, 1986; McCann *et al.*, 1986).

V. OTHER INDICATIONS AND FUTURE CLINICAL RESEARCH

There are a number of other indications for which eflornithine has been and most likely will be tried in the future. *Cryptosporidium*, which is considered to be a coccidian related to *Eimeria* (see Bacchi and McCann, Chapter 15, this volume), causes an opportunistic disease often found in AIDS patients and is manifested by severe gastrointestinal symptoms and diarrhea. There is, at present, no effective therapy for cryptosporidiosis. Fifty-five AIDS patients with cryptosporidiosis have been treated with eflornithine with inconclusive results as of May 1986. In light of several reported instances of anticryptosporidial activity in AIDS patients (Soave *et al.*, 1985; Rolston *et al.*, 1985; Dieterich *et al.*, 1986a,b), however, further clinical studies are warranted.

Another disease for which there has been a good deal of interest in the use of eflornithine has been psoriasis, a hyperproliferative cutaneous disorder characterized by increased polyamine production with associated elevated biosynthetic enzyme activities (Bohlen *et al.*, 1978). Various experimental models involving depletion of polyamines by the use of a combination of eflornithine and MGBG on mouse skin proved successful in inhibiting epidermal DNA synthesis (Kapyako *et al.*, 1982, 1983). A recent study with the same mouse skin model using eflornithine in a vehicle designed to obtain maximum percutaneous penetration showed even more promising results (McCullough *et al.*, 1985). (Topical delivery of eflornithine in a simple cream base had only marginal therapeutic activity in psoriasis; see Grosshans *et al.*, 1980.) Systemic therapy with eflornithine and other ODC inhibitors would also be of interest.

The contragestational effects of eflornithine have been extensively discussed by J.R. Fozard (Chapter 9, this volume), wherein he points out that inhibition of ornithine decarboxylase by eflornithine in laboratory animals completely blocks embryonic development at the stage of gastrulation and in essence acts as an exceptionally effective contraceptive agent (Fozard *et al.*, 1980a,b). Despite the clear evidence for the potential use of eflornithine as a contragestational agent, debatable ethical and political aspects centering on embryonic development in the human place clinical application in limbo for the time being.

Among other potential uses of eflornithine and other ornithine decarboxylase inhibitors, the results in treatment of experimental malaria with these agents appear increasingly persuasive. As indicated by C.J. Bacchi and P.P. McCann (Chapter 15, this volume) eflornithine is quite effective *in vivo* in blocking the exoerythrocytic reproductive cycle of *Plasmodium berghei*, a rodent malaria (Gillet *et al.*, 1982; Hollingdale *et al.*, 1985), and even in blocking the preinfective sporogonous cycle in the mosquito vector (Gillet *et al.*, 1983). Furthermore, eflornithine will induce significant protective immunity in mice initially inoculated with *P. berghei* sporozoites from mosquitoes and rechallenged with parasites (Gillet *et al.*, 1986). These findings, along with the observation of signficant effects *in vitro* of eflornithine on the erythrocytic cycle of the human pathogen *Plasmodium falciparum* (McCann *et al.*, 1981a; cf. Bacchi and McCann, Chapter 15, this volume), enhance the possibility that ornithine decarboxylase may be a potential target for the prophylactic use of inhibitors such as eflornithine in human malaria.

A recent finding has demonstrated that measurement of decarboxylated *S*-adenosylmethionine in the urine of patients treated with eflornithine represents an extremely sensitive and valid biochemical indicator of ODC inhibition in man (Haegele *et al.*, 1987). It was also determined that measurement of urinary polyamine and metabolite excretion was of no value in monitoring ODC inhibition. These results support the earlier finding that levels of decarboxylated *S*-adenosylmethionine are a true reflection of ODC inhibition in cells (Mamont *et al.*, 1982). This single measurement should have broad application in situations where a simple index of ODC inhibition is needed, e.g., determining optimal dosages of eflornithine for individual patients.

Such an index will also be extremely useful with the advent of even more potent inhibitors of ODC than eflornithine (cf. Bey, Danzin, and Jung, Chapter 1, this volume). Already it has been demonstrated that the α-monofluoromethyldehydroornithine methyl ester (Δ-MFMO$\cdot$$CH_3$) is significantly more potent than eflornithine in animal models against *T. b. brucei* (Bitonti *et al.*, 1985) and *P. berghei* (Hollingdale *et al.*, 1985). Furthermore, methyl acetylenic putrescine (MAP), a product analog irreversible inhibitor of ODC, has been shown to have notable antitumor effects *in vitro* and *in vivo* (Mamont *et al.*, 1984; see Sunkara, Baylin, and Luk, Chapter 6, this volume). MAP, in fact, has been used in phase I clinical trials (Cornbleet *et al.*, 1986). It is reasonable to suppose that if, as expected, eflornithine does turn out to have utility for treatment of neoplastic as well as protozoan diseases that other more potent inhibitors such as Δ-MFMO·CH_3 and MAP will also show similar activities.

VI. CONCLUSIONS

1. Ornithine decarboxylase inhibition, as indicated mainly by studies with eflornithine, is a promising approach to treatment of several clinical conditions characterized by rapid cell proliferation.

2. Although the antineoplastic effects of eflornithine alone have been disappointing, therapeutic value in combination with other drugs remains a strong possibility. The combinations of eflornithine with interferon (against malignant melanoma) and with MGBG (against recurrent brain tumors) appear most promising.

3. Eflornithine is highly effective treatment of both early and late stages of *T. b. gambiense* trypanosomiasis with dramatic efficacy noted even in arsenical refractory cases.

4. Eflornithine is an acceptable therapy for *P. carinii* pneumonia in AIDS, its use thus far being restricted to compassionate (emergency) use in patients who have failed to respond to or are intolerant of other drugs.

5. Other probable/potential uses of eflornithine discussed here include treatment of cryptosporidiosis, malaria, and hyperproliferative disorders such as psoriasis as well as the prevention of pregnancy.

6. In contrast to the results of measurements of polyamines and their metabolites in tissues and urine, the urinary excretion of decarboxylated *S*-adenosylmethionine has proved useful as a biochemical marker of ODC inhibition in man. Assays of decarboxylated *S*-adenosylmethionine should be helpful in future clinical studies with eflornithine and other ODC inhibitors.

REFERENCES

Abeloff, M.D., Slavik, M., Luk, G.D., Griffin, C.A., Hermann, J., Blanc, O., Sjoerdsma, A., and Baylin, S.B. (1984). *J. Clin. Oncol.* **2,** 124–130.

Abeloff, M.D., Rosen, S., Luk, G.D., Baylin, S.B., Zeltzman, M., and Sjoerdsma, A. (1985). *Proc. Am. Soc. Clin. Oncol.* **4,** 181.

Alhonen-Hongisto, L., Seppänen, P., and Jänne, J. (1980). *Biochem. J.* **192,** 941–945.

Antoine, P. (1977). *Ann. Soc. Belge Med. Trop.* **57,** 227–247.

Apted, F.I.C. (1980). *Pharmacol. Ther.* **11,** 391–413.

Bacchi, C.J., Vergara, C., Garofalo, J., Lipschik, G.Y., and Hutner, S.H. (1979). *J. Protozool.* **16,** 484–488.

Bacchi, C.J., Nathan, H.C., Hutner, S.H., McCann, P.P., and Sjoerdsma, A. (1980). *Science* **210,** 332–334.
Basson, W., Page, M.L., and Myburgh, D.P. (1977). *S. Afr. Med. J.* **51,** 453–457.
Bertrand, E., Rive, J., Serie, F., and Kone, I. (1973). *Med. Trop.* **33,** 385–390.
Bitonti, A.J., Bacchi, C.J., McCann, P.P., and Sjoerdsma, A. (1985). *Biochem. Pharmacol.* **34,** 1773–1777.
Bohlen, P., Grove, J., Beya, M.F., Koch-Weser, J., Henry, M.H., and Groshans, E., (1978). *Eur. J. Clin. Invest.* **8,** 215–218.
Cornbleet, M., Kingsnorth, A., Warrington, J., and Smyth, J. (1986). *Proc. Am. Assoc. Cancer Res.* **27,** 292.
Cushion, M.T., Stanforth, D., Linke, M.J., and Waltzer, P.D. (1985). *Antimicrob. Agents Chemother.* **28,** 769–801.
Davis, C.E. (1982). *Acta Trop.* **39,** 123–133.
Dieterich, D.T., Chachoua, A., and Faust, M. (1986a). *Gastroenterology* **90,** 1395.
Dieterich, D.T., Chachoua, A., Faust, M., and Moses, M. (1986b). *Am. J. Gastroenterol.* **81,** 871.
Fauci, A.S., Macher, A.M., Longo, D.L., Lane, C.H., Rook, A.H., Masur, H., and Gelman, E.P. (1984). *Ann. Intern. Med.* **100,** 92–106.
Fozard, J.R., Part, M.-L., Prakash, N.J., Grove, J., Schechter, P.J., Sjoerdsma, A., and Koch-Weser, J. (1980a). *Science* **208,** 505–508.
Fozard, J.R., Part, M.-L., Prakash, N.J., and Grove, J. (1980b). *Eur. J. Pharmacol.* **65,** 379–391.
Gillet, J.M., Bone, G., and Herman, F. (1982). *Trans. R. Soc. Trop. Med. Hyg.* **76,** 776–777.
Gillet, J.M., Charlier, J., Bone, G., and Mulamba, P.L. (1983). *Exp. Parasitol.* **56,** 190–193.
Gillet, J., Bone, G., Lowa, P., Charlier, J., Rona, A.M., and Schechter, P.J. (1986). *Trans. R. Soc. Trop. Med. Hyg.* **80,** 236–239.
Gilman, T.M., Paulson, Y.J., Boylen, C.T., Heseltine, P.N.R., and Sharma, O.P. (1986). *J. Am. Med. Assoc.* **256,** 2197–2198.
Ginoux, P.Y., Lancien, P., Frezil, J.L., and Bisadidi, N. (1984). *Med. Trop.* **44,** 149–154.
Giordano, C., Clere, M., Doutriaux, C., Doucet, J., Nozais, J.P., Bureau, J.P., and Piquemal, M. (1977). *Ann. Soc. Belge Med. Trop.* **57,** 213–225.
Golden, J.A., Sjoerdsma, A., and Santi, D.V. (1984). *West. J. Med.* **141,** 613–623.
Greenwood, B.M., and Whittle, H.C. (1980). *Trans. R. Soc. Trop. Med. Hyg.* **74,** 716–725.
Griffin, C., Abeloff, M., Slavik, M., Thompson, G., Luk, G., Hermann, J., Blanc, O., Sjoerdsma, A., and Baylin, S. (1984). *Proc. Am. Soc. Clin. Oncol.* **3,** 34.
Griffin, C.A., Slavik, M., Chien, S.C., Hermann, J., Thompson, G., Blanc, O., Luk, G.D., Baylin, S.B., and Abeloff, M.D. (1987). *Invest. New Drugs,* in press.
Grosshans, E., Henry, M., Tell, G., Schechter, P.J., Bohlen, P., Grove, J., and Koch-Weser, J. (1980). *Ann. Dermatol. Venereol.* **107,** 377–387.
Grove, J., Fozard, J.R., and Mamont, P.S. (1981). *J. Chromatogr. Biomed. Appl.* **223,** 409–416.
Haegele, K.D., Alken, R.G., Grove, J., Schechter, P.J., and Koch-Weser, J. (1981). *Clin. Pharmacol. Ther.* **30,** 210–217.
Haegele, K.D., Splinter, T.A.W., Romijn, J.C., Schechter, P.J., and Sjoerdsma, A. (1987). *Cancer Res.* **47,** 890–895.
Haller, L., Adams, H., Merouze, F., and Dago, A. (1986). *Am. J. Trop. Med. Hyg.* **35,** 94–99.
Hanson, W.L., Bradford, M.M., Chapman, W.L., Waits, V.B., McCann, P.P., and Sjoerdsma, A. (1982). *Am. J. Vet. Res.* **43,** 1651–1653.
Haverkos, H.W. (1984). *Am. J. Med.* **76,** 501–508.

Heby, O., and Russell, D.H. (1973). *In* "Polyamines in Normal and Neoplastic Growth" (D.H. Russell, ed.), pp. 221–237. Raven, New York.
Herr, H.W., Warrel, R.P., and Burchenal, J.H. (1986). *Urology* **28,** 508–511.
Hollingdale, M.R., McCann, P.P., and Sjoerdsma, A. (1985). *Exp. Parasitol.* **60,** 111–117.
Jaffe, H.S., Abrams, D.I., Amman, A.T., Lewis, B.T., and Golden, J.A. (1983). *Lancet* **2,** 1109–1111.
Käpyaho, K., Linnamaa, K., and Jänne, J. (1982). *J. Invest. Dermatol.* **78,** 391–394.
Käpyaho, K., Lauharanta, J., and Jänne, J. (1983). *J. Invest. Dermatol.* **81,** 102–106.
Kazyumba, G.L. (1979). *Med. Afr. Noire* **26,** 47–52.
Kingsnorth, A.N. (1986). *Ann. R. Soc. Surg. Engl.* **68,** 76–81.
Kovacs, J.A., and Masur, H. (1985). *In* "AIDS: A Basic Guide for Clinicians" (P. Ebbeson, R.J. Biggar, and M. Melbye, eds.), pp. 84–98. Saunders, Philadelphia.
Leading article (March 23, 1985). *Lancet* **1,** 676.
Levin, V.A., Chamberlain, M.C., Prados, M.D., Choucair, A.K., Berger, M.S., Silver, P., Seager, M., Gutin, P.H., Davis, R.L., Wilson, C.B. (1987). *Cancer Treat. Rep.,* in press.
Loprinzi, C.L., and Verma, A.K. (1985). *Cancer Lett.* **28,** 327–333.
Luk, G.D., and Baylin, S.B. (1984). *New Engl. J. Med.* **311,** 80–83.
Luk, G.D., Goodwin, G., Marton, L.J., and Baylin, S.B. (1981). *Proc. Natl. Acad. Sci. U.S.A.* **78,** 2355–2358.
Luk, G.D., Goodwin, G., Gordan, A.F., and Baylin, S.B. (1982). *Cancer Res.* **42,** 3070–3073.
Luk, G.D., Abeloff, M.D., Griffin, C.A., and Baylin, S.B. (1983a). *Cancer Res.* **43,** 4239–4243.
Luk, G.D., Sharkis, S.J., Abeloff, M.D., McCann, P.P., Sjoerdsma, A., and Baylin, S.B. (1983b). *Proc. Natl. Acad. Sci. U.S.A.* **80,** 5090–5093.
Luk, G.D., Abeloff, M.D., McCann, P.P., Sjoerdsma, A., and Baylin, S.B. (1986). *Cancer Res.* **46,** 1849–1853.
McCann, P.P., Bacchi, C.J., Hanson, W.L., Cain, G.D., Nathan, H.C., Hutner, S.H., and Sjoerdsma, A. (1981a). *Adv. Polyamine Res.* **3,** 97–110.
McCann, P.P., Bacchi, C.J., Clarkson, A.B., Jr., Seed, J.R., Nathan, H.C., Amole, B.O., Hutner, S.H., and Sjoerdsma, A. (1981b). *Med. Biol.* **59,** 434–440.
McCann, P.P., Bacchi, C.J., Clarkson, A.B., Jr., Bey, P., Sjoerdsma, A., Schechter, P.J., Walzer, P.D., and Barlow, J.L.R. (1986). *Am. J. Trop. Med. Hyg.* **35,** 1153–1156.
McCullough, J.L., Peckham, P., Klein, J., Weinstein, G.D., and Jenkins, J.J. (1985). *J. Invest. Dermatol.* **85,** 518–521.
Maddox, A.M., Keating, M.J., McCredie, K.E., Estey, E., and Freireich, E.J. (1985). *Invest. New Drugs* **3,** 287–292.
Mamont, P.S., Danzin, C., Wagner, J., Siat, M., Joder-Ohlenbusch, A.-M., and Claverie, N. (1982). *Eur. J. Biochem.* **123,** 499–504.
Mamont, P.S., Siat, M., Joder-Ohlenbusch, A.-M., Bernhardt, A., and Casara, P. (1984). *Eur. J. Biochem.* **142,** 457–463.
Meyskens, F.C., Keingsley, E.M., Glattke, T., Loescher, L., and Booth, A. (1986). *Invest. New Drugs* **4,** 257–262.
Natale, R.B., Meyer, K., and Forastiere, A. (1985). *Proc. Am. Soc. Clin. Oncol.* **4,** 137.
Ngandu-Kabeya, G. (1976). *Ann. Soc. Belge Med. Trop.* **56,** 85–93.
Pegg, A.E., and McCann, P.P. (1982). *Am. J. Physiol.* **243,** C212–C221.
Petru, A.M., Azimi, P.H., and Cummins, S.K. (1985). *Intersci. Conf. Antimicrob. Agents Chemother. Abstr.* p. 252.
Robins-Browne, R.M., Schneider, J., and Metz, J. (1975). *Am. J. Trop. Med. Hyg.* **24,** 226–231.

Rolston, K., Fainstein, V., Mansell, P., Sjoerdsma, A., and Bodey, G.P. (1985). *Int. Conf. Acquired Immunodeficiency Syndrome (AIDS), Atlanta, Abstr.* p. 77.

Ruppol, J.F., and Burke, J. (1977). *Ann. Soc. Belge Med. Trop.* **57,** 481–491.

Russell, D.H. (1971). *Nature* (*London*) **233,** 144–145.

Schechter, P.J., and Sjoerdsma, A. (1986). *Parasitol. Today* **2,** 223–224.

Schillinger, D., and Gorton, E. (1984). *Drugs Exp. Clin. Res.* **10,** 677–679.

Seppänen, P., Alhonen-Hongisto, L., and Jänne, J. (1981). *Biochim. Biophys. Acta* **674,** 169–177.

Sharkis, S.J., Luk, G.D., Collecter, M.I., McCann, P.P., Baylin, S.B., and Sensenbrenner, L.L. (1983). *Blood* **61,** 604–607.

Siimes, M., Seppänen, P., Alhonen-Hongisto, L., and Jänne, J. (1981). *Int. J. Cancer* **28,** 567–570.

Sjoerdsma, A., and Schechter, P.J. (1984). *Clin. Pharmacol. Ther.* **35,** 287–300.

Sjoerdsma, A., Golden, J.A., Schechter, P.J., Barlow, J.L.R., and Santi, D.V. (1984). *Trans. Assoc. Am. Phys.* **97,** 70–79.

Soave, R., Sjoerdsma, A., and Cawein, M.J. (1985). *Int. Conf. Acquired Immunodeficiency Syndrome (AIDS), Atlanta, Abstr.* p. 77.

Splinter, T.A.W., and Romijn, J.C. (1986). *Eur. J. Cancer Clin. Oncol.* **22,** *61–67.*

Sunkara, P.S., Prakash, N.J., Mayer, G.D., and Sjoerdsma, A. (1983). *Science* **219,** 851–853.

Taelman, H., Schechter, P.J., Marcelis, L., Sonnet, J., Kazyumba, G., Dasnoy, J., Haegele, K.D., Sjoerdsma, A., and Wery, A. (1987). *Am. J. Med.* **82,** 607–614.

Talpaz, M., Plager, K., Quesada, J., Benjamin, R., Kantarjian, H., and Gutterman, J. (1986). *Eur. J. Clin. Oncol.* **22,** 685–689.

Tyms, A.S., and Williamson, J.D. (1984). *Nature* (*London*) **297,** 690–691.

Van Nieuwenhove, S., Schechter, P.J., Declercq, J., Bone, G., Burke, J., and Sjoerdsma, A. (1985). *Trans. R. Soc. Trop. Med. Hyg.* **79,** 692–698.

Warrell, Jr., R.P., Coorley, C.J., and Burchenal, J.H. (1983). *Cancer Chemother. Pharmacol.* **11,** 134–136.

Wery, M., Mulumba, P.N., Lambert, P.H., and Kazyumba, L. (1982). *Semin. Hematol.* **19,** 83–92.

WHO Consultation, UNDP, World Bank (1982). *Bull WHO* **60,** 821–825.

Index

C

D

M

N

O

R

S

T

V